"This thoughtful and wide-ranging book is for those who wish to understand our predicament clearly, but especially for those looking for a glimmer of hope in our current darkness."

RAYMOND GEUSS, AUTHOR OF *PHILOSOPHY AND REAL POLITICS*

"Walking us through the flimsy defences of green capitalism, slicing through the nonsense with rapier analysis, Chaudhary explains why any workable climate future will need to be grounded in decolonization. The argument is careful, logical, and is destined to be a classic, a touchstone in global climate struggles to come. Chaudhary is a deeply gifted teacher, and *The Exhausted of the Earth* is a gift to a world burning to learn."

RAJ PATEL, *NEW YORK TIMES* BESTSELLING AUTHOR OF *A HISTORY OF THE WORLD IN SEVEN CHEAP THINGS*

"Ajay Singh Chaudhary has achieved the impossible: bringing together scientific, political, cultural and psychological thought to create the foundations of a new political 'climate realism,' blending global decolonisation and domestic emancipation. *The Exhausted of the Earth* opens new horizons for urgent and immediate climate action. A must-read for our times."

JULIA STEINBERGER, FACULTY OF GEOSCIENCES AND ENVIRONMENT, UNIVERSITY OF LAUSANNE

T0051530

"Have you felt exhausted lately? In this wonderfully rich inquiry into late climate politics, Ajay Singh Chaudhary zooms in on exhaustion as the predicament of a world too long subjected to the 'extractive circuit' of capital: the constant sapping of energies, returning to the planet as excess heat. The wretched of the Earth are today the exhausted — and if there is any way to fight back, it is, as Singh Chaudhary so convincingly argues, with southern resources, assembled by everyone from Frantz Fanon to Imam Mahdi. Bristling with ideas on every page, this book is the energy drink you need."

ANDREAS MALM, AUTHOR OF HOW TO BLOW UP A PIPELINE

"Written in a feisty and urgent style, *The Exhausted of the Earth* does the important work of not only showing that climate disruption and the Anthropocene are political, but also that they change what politics means. It shifts our attention in many, much needed ways."

MCKENZIE WARK, AUTHOR OF MOLECULAR RED: THEORY FOR THE ANTHROPOCENE

# THE EXHAUSTED
# OF THE EARTH

# THE EXHAUSTED OF THE EARTH

## POLITICS IN A BURNING WORLD

### AJAY SINGH CHAUDHARY

Published by Repeater Books

An imprint of Watkins Media Ltd

Unit 11 Shepperton House

89-93 Shepperton Road

London

N1 3DF

United Kingdom

www.repeaterbooks.com

A Repeater Books paperback original 2024

1

Distributed in the United States by Random House, Inc., New York.

ISBN: 9781915672117

Ebook ISBN: 9781915672124

Printed and bound in the United Kingdom by TJ Books Limited

*It's Hell, it's Heaven: the amount you earn*
*Determines if you play the harp or burn.*
*Gold in their mountains,*
*Oil on their coast;*
*Dreaming in celluloid*
*Profits them most.*

— Bertolt Brecht


*They poured the concrete and the columns stood,*
*laid bare the bedrock, set the cells of steel,*
*a dam for monument was what they hammered home.*
*Blasted, and stocks went up;*
*insured the base,*
*and limousines*
*wrote their own graphs upon*
*roadbed and lifeline.*

— Muriel Rukeyser

# Contents

# Prologue

The story goes, when I first started writing this book (I won't tell you exactly when but let's just say atmospheric carbon concentration hovered at a mere 400 parts per million), I was fairly confident I knew what I was going to argue. Some of which still holds: climate change is *inherently* political; climate change forces temporal pressures that are fundamentally at odds with, well, what most people think of as politics or political theory; *and* that it was vital for the *politics* of climate change to be far more informed, far more synthesized, with the genuinely unprecedented world that was being outlined in ever greater detail through the climate sciences. (And I wanted to write a book that hopefully could speak to the often-desired input from critical social and political analysis).

However, as I started to actually *do* that, really dig in, past other people's work, past headline claims, Intergovernmental Panel on Climate Change (IPCC) policymaker statements, news summaries, I realized I needed to drastically change course. That original story hinged on technologies I discovered were deeply oversold, on biophysical problems I hadn't yet encountered, and temporal pressures that make me embarrassed at the length of time writing this has taken. If this were – oh, let's say 1977 – perhaps some incrementalist nudge and investment program might have partially worked, at least for some of the most important aspects of climate mitigation and adaptation. (For the scientists reading this book, I am using "climate change" as a catch-all for related phenomena like biodiversity loss, ocean acidification, and so on.) A more aggressive, radical, but still systemically conservative plan might have done even more.

But it's 2023 and that hasn't been the case for some time now. A lot of what I am going to tell you about (descriptively and prescriptively) is significantly more brutal as there's no magical way to simply roll back the clock. Since we can't, let me start telling you the approximately 420 ppm story.

Climate change is not some nebulous impersonal abstraction, an unfortunate phenomenon we all come together and finally "solve." Climate change is political. Not the imaginary politics of universal consensus or the anti-politics of miraculous technological salvation — and, no, not the end of the world. The politics of climate change is a struggle between actually existing people over actually existing crises with actually existing differences, interests, and prospects. Climate change is about *power*. Not only the power generated in turbines, or the increased kinetic energy — one of the more elegant ways to describe global warming — coursing through the Earth's systems. It is the power of socioeconomic forces to shape our ecological niche, the power within those systems to change political systems, and the *potential political power* to alter the course of our burning world and knit together something unanticipated, something better instead.

So, here's where I come clean, I suppose. Maybe you were tipped off by the title of this book, paraphrasing one of the best-known anti-colonial thinkers of the twentieth century, the Afro-Caribbean psychologist and political theorist Frantz Fanon. Or by the fact that the critical education and research institute I founded in 2012 (Brooklyn Institute for Social Research (BISR)) is literally named after the Frankfurt School (someone's head just exploded at the Federalist Society). Yes, I am on the political left, a Marxist, albeit a heterodox one. So yes, you're going to hear about capitalism and how it is causally interrelated to climate change and climate politics. But I want to add that I came to these analyses honestly and share them with many who had quite different starting points. It would be a source of tremendous comfort if genuine climate mitigation and adaptation were possible within capitalism-as-we-know-it.

One of the things I discovered over the long course of this project is how climate change makes for unlikely bedfellows — analytically and sometimes politically — for those who start investigating climate from any angle with little more than the most mundane utilitarian assumptions: maximizing benefit for the largest number of people possible. I've seen some apolitical natural scientists become fire-breathing socialists in what seems like no time at all; business school professors, finance journalists, and the occasional *banker* quietly transform from Milton Friedman to Rosa Luxemburg. OK, not quite that far, but yes, even some dyed-in-the-wool capitalists become begrudgingly skeptical of the free market's ecology. And at the same time, I'm not actually making the case that, within the *very* pressing timeframe, we should be aiming for a classless society or the abolition of all private property. Climate politics is radical, unprecedented, but in some ways *lateral* to many conventional categories.

I've also noticed how much specific issues — time, again, for example — render certain hot-button policy debates largely moot. I critically assess a host of technologies in the third chapter but also some of the headline cases — right now, carbon dioxide removal (CDR) technologies are a hot topic — don't really matter. Time is so pressing for climate mitigation and adaptation that serious researchers and organizations who are confident in the technologies' potential, realistically don't see it playing any role until after a near-total energy transition — even in "hard-to-abate" sectors. And, as you'll read, you find people using every statistical or econometric trick there is, from shiny but obfuscating data visualizations to smuggling unfounded assumptions into economic models and measures.

I am not a "science" doctor (to borrow a joke from my colleague and BISR co-founder Abby Kluchin) but I take scientists' research seriously. This project has required learning much that I didn't know, and asking for collegial help in understanding what I could not. In the course of it I've done my best at digesting the key elements of a few thousand or so natural science papers. While I'm a political theorist, I've long had an interest in the natural sciences and knew I wanted to work on climate change in one way or another. (Last night, my mother, an actual science doctor, sent me a clipping I wrote on energy policy, not particularly good, in the *fifth grade*.) That said, I beg the forgiveness of anyone whose research I've misread. Another thing I discovered trying to write this comprehensively about this topic is that at some point you just have to give up adding new dimensions and updating to the latest data changing practically daily. Not every bit of every chapter is perfectly up to date. I have published short preview versions of a few of the chapters here before, but I've been augmenting and updating them for several years now, sometimes many times over.

I am though, above all else, a teacher. (I do a fair bit of administration too.) I've had the good fortune to teach material related to this book to waitresses and writers, coders and garbagemen, electricians and, yeah, some ecologists. And I've also done political work, from the most mundane volunteering to working on formal elections to advising organizations. I genuinely believe this has sharpened this book immensely, giving me windows onto a wide range of concerns and experiences.

About now, you're probably wondering where the requisite mindboggling fact-dump or standard summary introduction is. The answer is simple. I feel no need to frighten or bolster you. I'm neither a "doomer" nor a cheery "climate optimist." I'm a *radical climate realist*. What I mean by that, I'll address in the first chapter which, in reality, is an introduction of sorts. But I do want to "show the work" and give you a brief guide to the book.

Each chapter begins with, or contains, an illustrative narrative and proceeds to walk through the larger case in detail. In fact, there are almost two books in this book. The main text presents the full argument — the whos, whats, hows, and whens — of a climate politics with the most pertinent, direct details. There is also a fairly extensive notes section, filled with a lot more than citations. Early on, I considered some kind of avant-garde presentation where the longer notes would constitute a shorter parallel text full of details and commentaries which would appear right below. In this way, to use an example from the first chapter, while one is reading about economic facts like the extraordinary, relatively recent rise in foreign direct investment (FDI) or in the use of tax havens in terms of climate *politics*, one could glance down and read a parallel story about how such phenomena so seemingly distant from climate are actually, even within cautious, mainstream climate science research, connected to the accelerated rise of greenhouse gases or of biodiversity loss, respectively. However, in the end, and with the prudent guidance of my editors, these are now in easily accessible endnotes. If you are someone who is very interested in the longer explanations, references, data discussions, histories, or theories, you'll find *extensive* annotation in the endnotes. If you are interested in digging deeper in just a few areas, the notes clustered in each section provide direct citations, suggestions for further reading, and some extended commentaries. If you're not, though, the key arguments are all in the main text.

The first chapter, "We're Not in This Together," is the basic outline that sets up the *political* argument. The politics of climate change are not about denial or about extinction but essentially two different speeds that define two different, irreconcilable visions. In addition to introducing a number of lineaments, Chapter 1 sets up the concept of "right-wing climate realism" — a broad spectrum of positions that favor slower climate mitigation and adaptation, in rather ordinary terms, all the way to the fringes of climate barbarism and beyond to preserve or enhance existing wealth and power aboard a hierarchal, exclusive "armed lifeboat." Here we are talking about that wonderfully apt phrase, "business-as-usual," not in the sense of doing *nothing*, but of current trajectories as reflected in everything from the policies and practices of firms and governments to those in international political economy, treaties, and institutions. Far from the picture of climate denialism, or suicidal annihilation, we see why this is, horrifying as it might sound, a perfectly *rational* and *realistic* wager. In dramatic terms, think of this as the antagonist. (If you're an eco-fascist, or on this spectrum, you've purchased the wrong book. I'm sorry. All sales are final.)

The second chapter, "The Extractive Circuit," describes what is often euphemistically described in climate science literatures as "the dominant global socioeconomic system" (i.e., capitalism) in a full socioecological portrait. We trace lines on the circuit through specific people, production, and consumption processes, and through sites and zones of metabolism, extraction, and exhaustion. It's a contemporary and historical portrait. It is a picture of what capitalism really is. We begin to see the difficulty of "pricing-in" the "externalities" (in the simplest sense of mainstream economics, methods for moving currently unpriced, "free," natural sources and sinks back onto firm ledgers) and consider its multiple confounding characteristics in social and ecological terms. Moreover, a few dimensions are highlighted that I have found often absent in many literatures, in particular the qualities of increasing *speed* and *acceleration,* in a vicious cycle across socioecological metabolism. In dramatic terms, think of this as the setting.

The third chapter, "Climate Lysenkoism; Or, How I Learned to Stopped Worrying and Rescue Class Analysis," discusses the subordination of concrete material conditions to ahistorical, naïve, and often silly theoretical abstraction (i.e., Climate Lysenkoism), using two key examples from the Anglophone self-ascribed "left." In reality, there's nothing "left" at all about such cases, which have more in common with (and indeed, indirectly support) ecomodernist tech magnates like Jeff Bezos, Bill Gates, or Elon Musk. However, addressing a reactionary tendency sadly too present in the Global North gives us the opportunity to review a range of climate technologies and policies and political histories. In doing so, we also get to introduce theories of political affect and the "minor paradise of a sustainable niche," an outline of radical possibilities (not a utopian blueprint) of what a world turned away from the extractive circuit can be. In dramatic terms, think of this as Act II.

The fourth chapter, "The Exhausted of the Earth," introduces us to *the Exhausted* — the potential mass or "stretched" (adapting Fanon's language) class subject of left-wing climate realism. This chapter takes us from the negative ideal of individualized "resilience," through all the dimensions of "exhaustion," to discussions of actually existing climate movements, specific histories, and strategies the world over. This allows us finally to see not only where actual movements already are and the difficulties they face, but to theorize this mass subjectivity *primed* through the extractive circuit, possible modes of organization, and strategies and tactics, scaled to the daunting realities of climate mitigation and adaptation; the already-existing power of right-wing climate realism; and historical comparisons of how "radical transformations" on the scale of what's described in much climate literature have taken place and how they might take place now. In between meaningless reform and impossible revolution, we find mixed existing and historical models of formal state, civil, and guerilla strategies. Thinking with Fanon, we address layers of global colonization and the corrosive, tragic, and yet necessary brutality and affective charge of what a left-wing climate realism entails. In dramatic terms, the Exhausted is our protagonist and is — as you may already start to see — potentially you.

Finally, the fifth and concluding chapter, "The Long Now," is a meditation on how fundamentally climate change alters time (in ways that abrogate received traditions of political-time in conservative, liberal, Marxist, and other radical terms). In doing so, we not only review but expand key concepts already introduced, in particular the time-binds of Chapters 2 and 4, and the further aesthetic possibilities of a sustainable niche. In dramatic terms, this is our *denouement*.

And this is the story I am telling. It may not be the prettiest, but it does have the virtue, as close as I can judge, of being true. The great critical theorist Walter Benjamin once argued that "storytelling" (today a much abused term) is the way in which people can relate, record, and intermingle personal and social experience, passing the story along ever so slightly augmented. My hope is that you recognize yourself along the extractive circuit, among the Exhausted, that you already find yourself in a part of its potential international formation. And that you pass your story of exhaustion onto others who will share theirs; who will build, and struggle, and fight at your side for a flourishing Earth, and a better life, while it's still possible. Time is short, and politics immediate and urgent in a burning world.

# Chapter 1

# We're Not in This Together

*Is the Earth Fucked?*
— Brad Werner, presentation to the annual conference of the
American Geophysical Union, 2012

*Like a Good Neighbor, State Farm Is There*
— State Farm Insurance ad jingle

## Convenient Truths

In November of 2018, fires of "unprecedented speed and ferocity" broke out across Northern and Southern California. The "Camp Fire" in Northern California killed just under 90 people and destroyed approximately 19,000 structures. Even with modern safety protocols and building codes, it was the deadliest and most destructive fire in Californian history. The "Woolsey Fire" in Southern California, at the exact same time, burned nearly 100,000 acres. Fires are tricky things to understand. The fires that burn across most of central Africa, for example, on a yearly basis, are seasonal, mostly contained, and part of a decently well-maintained agricultural cycle. Californian wildfires, while certainly nothing new, are not. They may be sparked by simple heat or lightning strike, or by a recreational accident or a glitch in the utility grid, but their frequency, intensity, and duration have all unquestionably increased due to anthropogenic climate change.[1]

Within existing discussions of climate equity, it is often noted that the effects of climate change will impact countries and communities greatest who have contributed least to the overall phenomenon. Even though "national" accounts of, say, carbon emissions can be misleading, this is largely true. The United States is, however, a glaring exception: projected to receive, measured in economic impacts, the second-most devastating effects of climate change. This might seem cold comfort to the millions in the Maldives, Micronesia, and similar places whose entire geographies are at risk of a combination of submersion, erosion, and overall uninhabitability. But, in the words of journalist David Wallace-Wells, it seems "a case of eerie karmic balance."[2] But balance is not a characteristic of the Anthropocene, not historically and not now.

Rather, all that is apocalyptic melts into the politics of climate change. "God is dead, and so too is the Goddess," remarks McKenzie Wark.[3] Karma too. At first glance, the story of the 2018 California fires is one in which everyone stands together. Rich and poor, black and white, famous and unknown. Media reports observed celebrity actors and musicians like Miley Cyrus, Liam Hemsworth, and Gerard Butler all losing their homes (or parts of them) — just like everyone else. But among the various cultural detritus of the moment, TMZ (of all outlets) reported that Kanye West and Kim Kardashian had avoided this particular fate through the intervention of a private firefighting force. This was not some one-off procurement or out-of-the-ordinary solution. And, beyond their staggering wealth, it had little to do with West or Kardashian personally. Over the past decade and a half, major insurance companies like AIG and Chubb have begun to offer private emergency services "to elite policyholders." It is one of the more straightforward ways to think about the commodification of risk.[4] Incredibly expensive assets face catastrophic threat; the insurance company packages that threat and sells two incredibly attractive products. The first is insurance against climate damage; the second is access to services to prevent having to pay out when damage occurs. From the firm's point-of-view, they've sold two products at a tidy profit and avoided the exorbitant payout involved with actually covering insured losses. From the client's point-of-view, despite a steep cost, the financial and psychological burden of losing their home is avoided. For these two parties, it is, indeed, a win-win scenario.

However, there are "externalities," so to speak, in privatized social services. Protecting assets can be at odds with saving lives. Such private services skirt or run right over what tiny regulations exist for them. They physically impede and complicate public emergency services. And the same companies lobby for tax and infrastructure policies that necessitate their products, starving the public emergency services, which, in the case of California, deploys severely underpaid incarcerated people to enhance dwindling state capacity.[5] In recent years, insurers like State Farm, servicing lower tiers of customers, have announced restrictions on policies, rate hikes, and insurance moratoriums altogether. Meanwhile, elite insurance companies like AIG and the emergency service contractors themselves focus ever more exclusively on the high-end market, offering multifaceted concierge plans at prohibitive prices divorced from state-backed (and regulated) departments of insurance for ultra-high-net-worth clients.[6] This kind of tiered access to services is familiar to anyone who has encountered private or privatized social goods, like market-based health insurance. And yet it is also an extreme intensification where there is seemingly no bound to privatized governance. In this example, we can see one microcosm of what I call "right-wing climate realism."

One of the most common misconceptions concerning climate change is that it produces or even requires a united humanity. In that tale, the crisis in the abstract is a "common enemy," and a perfectly universal subject is finally possible in coming to "experience" ourselves "as a geological agent," through which a universal "we" is constituted in a "shared sense of catastrophe."[7]

The story I am telling you is different. In this story, there is no universal "we." Although there are so many stark political divides imaginable, the political divide of this moment is almost the quintessential, perfectly political divide. It is literally zero-sum. The divide is not, as many still argue, between those who accept and those who reject the overwhelming evidence of climate science.[8] It is about those who stand to gain — both in this moment *and* in the future as traditionally conceived — from fundamental system preservation or other modes of right-wing climate realism, and those whose current exhaustion is part of the *fuel* for that system as much as any petrochemical or industrial agricultural process. Climate change is *not* the apocalypse and it does not fall on all equally or, in at least a few senses, on everyone at all.

## Right-Wing Climate Realism

The idea of right-wing climate realism can strike many as odd in the first instance. Since, the assumption runs, climate change is universal, since we can turn to a "scientistic" politics, there is really only one "climate realism," and the primary task is to communicate, persuade, and teach the science from which the one-true-politick will emerge. All that stands in the way of this is the intransigence of non-believers. This is, perhaps ironically, a familiar story: it is a Christian story of gospel and evangelism. But the science does *not* have a single politics. Right-wing climate realism encompasses a *plausible* and thoroughly *realistic* — in terms of power and in terms of ecology — set of positions in which a business-as-usual scenario is absolutely worth it. Even given the staggering social and ecological costs, it might even be salutary, and not in some short-term or self-deluded sense. Right-wing climate realism then, in its simplest form, is a political-ecological scenario of the concentration, preservation, and enhancement of existing political and economic power.

Although the threat of ecologically articulated right-wing politics is quite real, has historical antecedents, and is visibly present in fringe and radical right movements in this moment, much of what might be better understood as right-wing climate realism need not be *articulated* as ecological at all.[9] It can simply build from what the right has already formidably established and wishes to pursue further. This is not something we see only in the Pinkertons and other private security agencies investing in climate-related protection for the wealthy,[10] but something we can observe in tax policies that go far beyond neoliberal catechism, maximizing accumulation while *disincentivizing* investment of any kind. It's not only a world in which what the International Monetary Fund (IMF) calls "phantom FDI" — foreign direct investment which goes towards no discernable actual investment but is rather just convenient avoidance of taxation and popular sovereign restriction — has shot up to 31% of all FDI.[11] It's a world in which even that cannot fully explain the 40% of "missing profits" that are simply unaccounted for.[12] It's not only the US military preparing for climate security scenarios and buildouts, it's the US's continued investment in the world's largest existing migrant detention, surveillance, and expulsion network. It's not only the development of what I call "detachable infrastructures" — luxury survival architecture built not only with internal power generation and potential for stockpiles, but to receive aerial deliveries and withstand floods or riots.[13] It's Amazon's infamous patented "airborne fulfillment center" which could connect far-flung supply chains with end-use consumption by drone delivery, furthering the development of ever-diverging narrow, luxury markets.

Taken alone, private security — particularly the Pinkertons — might seem *only* an extension of a longstanding parallel system of irregular violence for the protection of capital, alongside official policing. Even if today those services are beginning to outnumber the considerable mass of formal police forces in many parts of the world.[14] Migration policies might seem only extensions of existing ethno-nationalist principles and ideologies, even if today the right-wing ideal of ethnocracy has more purchase than it has had in decades. In isolation, privatized emergency services might just seem like one more neoliberal frontier. Phantom FDI and missing profits may just look like one further push into the freedom of capital already achieved in the neoliberal era. Over *half* the global profits of American-based transnational corporations are now held in opaque tax havens. When such havens began to be widely used in the aftermath of World War I, they held a scant half a percentage point of global wealth. Today, they are at an all-time high, holding over 10% of global GDP.[15] This could be as much as $21–32 *trillion* of assets — a torrent of profit escaping even the faintest glimmer of popular power. This can be the freedom to disappear, the freedom to "cash out" — although one does need somewhere to cash out to. If one is only looking at an economic portrait of infrastructure, this might look like the logic of finance taken to its most extreme: asset-stripping of even the most necessary goods for immediate return. The creation of luxury markets for the deep-pocketed high-end. Even the architecture, taken alone, might seem just a bit of prudence or caution or fear. This is not quite what people usually have in mind by "climate barbarism."

And for many, it is surely just these lanes or just these understandings; there is no necessity that such a politics be unified or ideologically coherent. But, understood within a larger climate portrait, this is building or at least preparing for a world that is an extension of what I call the "extractive circuit" — that is, capitalism in the twenty-first century in its full socioecological expression,[16] working for fewer and fewer. Or one of vigilant, restrictive, hyper-nationalisms. Or even structurally "neofeudal," a rather different "post-capitalist" vision for if or when the extractive circuit breaks. Or a world of all three. A world we can already see coming into its own in Fortress Europe or at the US border and in increasingly direct and punitive anti-popular governance in places like Puerto Rico or Greece, Flint or East Palestine. We can observe interstitial border states, like Turkey and Mexico, adopting the control of migrants as a geopolitical lever, but for a permanent "tier" below, providing a key service for the preservation of power in a world already experiencing migration on a scale not seen since World War II and projected to be in the hundreds of millions in the near term.

There is tremendous evidence that people deeply invested in fundamental system preservation do consciously proceed with "business-as-usual," knowing with a reasonable degree of certainty the likely climate outcomes. I often jokingly call this the "Rex Position," named for Rex Tillerson, CEO of ExxonMobil from 2006–2016, and Secretary of State under Donald Trump for parts of 2017–2018. Tillerson famously proclaimed, "my philosophy is to make money. If I can drill and make money then that's what I want to do." Tillerson is not a climate denier *per se*. He just doesn't share the urgency of his many critics. Tillerson is more than willing to talk about a shift to renewables, but there's no rush. People will adapt to climate change, there are "engineering" solutions. It's only those trapped within a different temporality who think otherwise. There are many clear and *bright* futures for business-as-usual in the meantime. There's a certain clarity and precision in even Tillerson's public arguments: "Our view reflects the reality that abundant energy enables modern life."[17] The Rex Position is not one single positive political ideology but the convergence point of the politics of right-now for a panoply of right-wing climate realisms.

The Rex Position is not shortsighted or irrational once one internalizes that climate change does not "produce" a universal human subject. One does not have to construe Tillerson as "evil." Tillerson and those he works with are not in some kind of shadowy conspiracy. The Rex Tillersons of the world have taken a look at the same data, the same trends, the same underlying social and political conditions and noticed that in the probable world in which little changes for them — business-as-usual — they end up on the "winning" side of a sharp global and local dividing line. Every structural incentive serves to reinforce such thinking. The *best* outcome in such a position is to push on with business-as-usual; the costs of climate change will largely be borne by those who already bear the cost today. Indeed, as I will argue, those people *bearing* those costs help keep the system going as long as possible and advance the Rex Position of maximal extraction for maximal maintenance or cash out that much better. Even modestly successful climate mitigation and adaptation for the vast majority of people *require* socioeconomic and political changes that would pose a steep loss to the Rex Position.

There might, among the generously minded, be an inclination to see this position as deeply impoverished. Surely the Tillersons of the world will be profoundly affected by such staggering human loss and damage, let alone the diminishment that is a part of the sixth mass extinction event in the Earth's history, in which some 30–50% of species may be facing extinction by the mid-twenty-first century.[18] In some meta-ethical way, such claims might register. But the systems we have inherited, the world capitalism has built, is not a world which inculcates, rewards, or incentivizes such a mode of thought or its practical applications. If Rex himself felt that way, there are a thousand Rexes equally well-placed, able, and at-the-ready to replace him. The Rex Position represents a kind of logic of self-interested self-preservation taken to an extreme.

## Really Real? Realism, Climate, and Conflict

It may be helpful to pause for a moment to think about what "realism" means when we are thinking politically. Realism can at first glance appear an odd concept in political discourse, particularly in Anglophone countries where political writing and rhetoric is dominated by ahistorical moralism and formal idealism. John Rawls' *A Theory of Justice* — probably the most influential work of relatively recent American political philosophy — is an exemplar of such moralism and idealism. As the political philosopher Raymond Geuss critically summarizes, [this view] assumes that one can complete the work of ethics first, attaining an ideal theory of how we should act, and then in a second step, one can apply that ideal theory to the action of political agents. As an observer of politics one can morally judge the actors by reference to what this theory dictates they ought to have done. Proponents […] go on to make a final claim that a 'good' political actor should guide his or her behavior by applying the ideal theory. Empirical details of the given historical situation enter into consideration only at this point. 'Pure' ethics as an ideal theory comes first, then applied ethics, and politics is a kind of applied ethics."[19]

In this way of thinking, there is no qualitative difference between ethics and politics, or morality and politics; politics is simply morality at scale. Although incredibly common (and not only in liberal thinking), there could be no more fundamental error.

Starting from first principles, an entire ideal system is designed based on ethical or moral scaffolding. Actual politics are then judged by their discrepancy with or distance from that system. It is important to underline that "ideal" as I'm using it here is meant in the philosophical way — idealism as opposed to materialism — and not as opposed to "principles" themselves or even something "idealistic" in the colloquial sense, like "utopianism." Realism can also denote *Realpolitik*, commonly associated with a figure like Otto von Bismarck.[20] And although this Bismarckian usage brings in *some* of the questions of power, interests, and conflict desperately needed to understand climate politics, it also expresses another form of realism, one readers are more likely to encounter in everyday political conversation: "politics is the art of the next best." This is political realism as in explicit *contradistinction* to political radicalism — i.e., politics as the art of compromise and other conventional clichés. The Italian political theorist Antonio Gramsci once wrote of such attitudes as "too much […] political realism," a version of the natural fallacy that confuses not only "is" and "ought" but also what, concretely, might be.[21] When discussing the above examples of right-wing climate realism, we see nothing of this limitation. In contrast, this so-called "political" faux-realism takes the status quo to define the parameters of the "realistic" and naïvely (or cynically) dismisses politics outside those parameters — ignoring actual necessity, actual life, actual power, and actually shifting material conditions.

Any climate politics must start from "the way things are" — those actual conditions; must work through historically situated conflicts therein; analytically, practically, and theoretically, the mutual exclusivity of contending positions results in modes of domination and conflict. Climate change presses the question. Radical realism in the context of climate doesn't mean all questions of compromise, of the imperfect, are discarded, but it begins from today's concrete, fundamentally interrelated ecological and social conditions; its baseline similarly builds from actually inseparable ecological and social necessities. In this broad way, realism within Western political thought can be found in various forms, perhaps most famously in the works of the Florentine political strategist Niccolò Machiavelli. But it is hardly reducible to a supposedly "Machiavellian" *praise* of power or an argument that the "ends justify the means." Realism in this sense asks how ends and means relate in the world as it stands, how existing power works, and how these might change. Machiavelli, himself a classical republican, and his political thought were no apology for existing power or dismissal of popular power. Gramsci came to read Machiavelli as providing a kind of "manifesto" for "those who are not in the know" — nothing less than the basic building blocks of a radical political theory.[22] But Machiavelli almost paradigmatically stands for the introduction of thinking about politics through the lens of power, of contending forces, of, put simply, *real* elements.

Political realism doesn't *have* to be divorced from moral considerations; when the North African Church Father Augustine of Hippo wrote, in his magisterial *City of God*, that there are virtues of statecraft (of the "City of Man") that are distinct from those of Christian faith (in the "City of

God"), he was certainly not dismissing the latter, but rather noting how we can distinguish between political and moral activities and realms. When Martin Luther King wrote of the "white moderate" who prefers the "negative peace which is the absence of tension to a positive peace which is the presence of justice," he was obviously not dismissing justice, but rather making a *politically realist* assessment of the balance of forces at that moment in the Civil Rights movement — who stands where and for what. And at the same time, we see that political realism can — sometimes *must* — be radical. Lenin's pithy "who whom?" (Or as Geuss elaborates: "Who <does> what to whom for whose benefit?") is not only a consummate expression of an aspect of political realism, it is also radical and even quasi-utopian. When Audre Lorde says "within the interdependence of mutual (nondominant) differences lies that security which enables us to descend into the chaos of knowledge and return with true visions of our future, along with the concomitant power to effect those changes which can bring that future into being," she expresses an ethical principle, a utopian ideal, and a realist assessment all at once. And political realism doesn't have just one political valence. Abraham Lincoln was a consummate political realist, and as such he served as an ideal case for the Nazi jurist and political philosopher Carl Schmitt. Whether explicitly stated or implicitly assumed, each of these examples contain versions of Schmitt's understanding of "the political" as a distinction between "friend and enemy."[23] This aspect of Schmitt's thought is itself though an appropriation of left-wing politics — from Lincoln to Lenin — recast and redefined for conservative revolution.

Not every moment poses the most extreme questions of political realism, but climate change — particularly in its full socioecological expression — demands it. Incumbent in moments which press against the material and sensed baseline of life, in moments of a "legitimation crisis" or when hegemony is fracturing, in which social, economic, and political conflicts become zero-sum, lies the stark antagonism of radical realism.

Or in Karl Marx's words: "Men make their own history, but they do not make it just as they please; they do not make it under circumstances chosen by themselves, but under circumstances directly found, given, and transmitted from the past."[24] As Frantz Fanon argued, there is an "urgency" and "intensity" to material decolonization which cannot wait; this is unquestionably the case with climate *for some*. Fanon's succinct phrasing of anti-colonial radical realism is particularly apt for understanding how such climate politics "cannot be accomplished by the wave of a magic wand, a natural cataclysm, or a gentleman's agreement."[25] Officially sanctioned "consensus politics," the recourse to technological "magic," or even climate "cataclysm" itself, however powerful, however, as we'll see, *palpable*, is insufficient politically. "The political" is already *inside* climate change, inextricably, not through oft-supposed qualities of "human nature," but through the historical, socioecological processes and relations that are bound up in its actual unfolding.

All too often stories of climate politics are divorced from this reality, whether they assume that, say, the particular form of "liberal democracy" associated with the postwar United States and similar countries is a kind of eternal reality or Platonic ideal of politics, or they try to graft without modification ecological questions onto pre-existing political theories whose coordinates climate fundamentally scrambles. In "climate realism" both terms are operative. It is a *climate* politics because it takes seriously how imbricated our ecological niche is with a political-economic reality and also how much this reality challenges, augments, and transforms operative principles in society, economy, and politics. It is a climate *realism* because it understands that those politics will proceed from such conditions, the multiple possibilities for conflict and antagonism in this moment, the structural importance of social position and the "passions," affects, and feelings shaped by these realities, and that such politics must proceed through holding, building, and/or mobilizing *power*.[26]

# Really Unreal? Denial, Deception, Climate, and Capitalism

Some of what might be construed as right-wing climate realism can seem, well, not particularly grounded in reality. Doomsday prepping may be a scam but the wealthy quietly buying real estate in safer geographies — whether inland or abroad — is not. Luxury buildings in flood zones which can generate their own power and utilize private floodgates; high-end skyscrapers with private drainage; these too are not.[27] As Sarah Miller describes in her beautifully wrought "Heaven or High Water" — an essay about Miami, climate change, and denial — it can be hard to neatly distinguish sincerity, ideological conviction, and con job. And, in the last case, who is conning who? It is a near certainty that Miami is going to be one of the coastal areas hardest hit by climate change in the United States. And yet construction goes on as if a 30-year mortgage has meaning in Miami. Climate discourse among the people Miller talks to hovers in between denialism and forms of techno-optimism (what I call "techno-mysticism").[28] When a real estate agent tells Miller, "there were just too many millionaires and billionaires here for a disaster on a great scale to be allowed to take place," it's difficult to tell whether this is a conviction based on a kind of right-wing climate realism, a sales pitch inflected with right-wing climate realism, or a kind of trickle-down prosperity gospel. But, as Miller realizes, she herself, purchasing a home in a Northern Californian town almost certainly about to go up in flames is, in her own way, a functional climate denier.

Companies boasting billionaire bunkers are more clever marketing and scam than anything else. In many ways, climate denialism and skepticism should be treated this way as well. Only it's not a scam to take advantage of the skittish wealthy, but rather one propagated on adjacent allies and unwitting enemies. It is now well-known that firms — particularly but not only in the fossil fuel sector — actively knew and accepted the basic dynamics of anthropogenic climate change but chose to fund organizations like the Global Climate Coalition and others to flood denialism into media, government, and cultural organizations.[29] From the point of view of propaganda, it's an extraordinary achievement. Although one of my goals in this book is to take seriously, to fundamentally integrate, how novel natural scientific knowledge about climate — even where the socioecological nexus requires critical intervention and scrutiny (indeed, where such scrutiny is often *invited*) — fundamentally alters the basic logics of even the most radical political theories, possibly the most effective outcome within climate *denialism* plays on methodological concerns within the natural sciences themselves. Through amplifying the vaguest sense that a consensus position has not been reached, many scientists rush in to rectify and reestablish consensus.[30] The popular discourse of climate science shifts into an absurd back-and-forth, refuting ever more outlandish climate claims.

Moving outside of strictly scientific discourse, what makes decades of climate denial promotion such a spectacularly brilliant *political* maneuver is that it reinforces a narrow *ideological* understanding of what Gramsci called "effective reality" such that acknowledgment of the fact of climate change remains the primary fulcrum for "good" or "bad" climate politics in much official discussion. This, ironically, has no relationship to what politics or policy might be suggested through engaging the actual substance of research, nor with actual public opinion on the subject. Think of the dreadful American film *Don't Look Up*. Bracketing aesthetic questions, its story *ideologically* characterizes a genuinely universal phenomenon — a world-destroying asteroid strike — with the radically divided phenomenon of climate change; it equates an *impending* crisis with one that is already happening; it frames the debate as one which turns on "belief." Such an ideological frame is pervasive in official, or normative, American political discourse — think of the questions posed in televised political debates or in Congress — even though *actual* reality and even the warped world of public opinion polling has long moved on.

Climate denial in its simplest form — what the sociologist Kari Marie Norgaard calls "literal climate denial" — is, even in the United States, its chief global home, largely now a non-issue. In 2011, as she notes, it was already the case that a majority of Americans "believed" in anthropogenic climate change. By 2018 one Pew survey reported that 50% of Americans, including 29% of Republicans, "believed" in anthropogenic climate change. A Gallup survey from the same period records 68% of Americans and 40% of Republicans expressing such a "belief." And still another survey from the Annenberg Center for Public Policy finds as much as 71% of all Americans and 61% of Republicans. What accounts for these discrepancies? Methodological research that analyzes such results points to relatively banal — true, if partial and incomplete — issues in the collection or wording of survey data. But such analyses miss the burning forest for the already dead trees. Denialism has moved on to greener pastures. Even acknowledging the considerable shortcomings of this kind of data, as of 2022, the Yale Program on Climate Change and Communication comprehensive survey found 72–76% of Americans endorsing the existence of climate change and, more remarkably, a consistently flat 12% engaging in outright denial, while Pew's most recent surveys have simply stopped asking the question altogether. Survey data of this kind is particularly useful not for what it confirms but for what it contradicts. In this way, such polling reflects the palpable experience of climate change itself as against official narratives — even if not fully integrated or connected with related socioeconomic phenomena.[31] The Yale and Pew surveys report strange contradictions: a majority of *Americans*, for example, endorsing out-and-out political emergency measures, the suspension of all institutional political limitation, to address climate change — one of the most extreme responses possible. And yet simultaneously climate is regarded as less important than predominantly social and economic "issues" — a product likely explained not only by actual continued disaggregation between fundamentally connected phenomena but by the

23

survey questions themselves which, like so much political thought, pose and project these "issues" as easily and simply disconnected.

Far more interesting than the simple denial that is largely long past, climate denialism takes on at least two further configurations. Norgaard's second category is "interpretative denial," in which facts are acknowledged but "euphemisms" and "technical jargon" are employed to obscure any meaning to be found in them. She cites the example of military officials calling civilian deaths in war "collateral damage." In more contemporary climate terms, one might think of the phrase "net zero." As Kate Aronoff pithily puts it, "everyone, it seems, wants to get to 'net-zero.' What exactly that means for companies that have only ever revolved around producing oil and gas is anyone's guess. For now, it's one-part creative accounting and many parts a P.R. strategy of waving around shiny objects like biofuels, hydrogen, and carbon capture and storage." Net zero, in other words, is often *not* zero; it is continued emissions and "magic." Or think of "carbon offsets," which don't actually offset carbon. (Recall Fanon's "magic wand.") Norgaard's third category is "implicatory denial," in which "information" about "climate change" is not rejected "*per se*" but *implications* in social terms — from the systemic to the subjective — are simply set aside, ignored, or not integrated into any kind of proportionate social action. Implicatory climate denial might be thought of here as a kind of (dis)functional climate denialism — acknowledging the reality of climate change but remaining unwilling to consider related socioeconomic and political questions. In Norgaard's terms, there is a "double reality": in one reality, a relatively non-obfuscated understanding of climate change is acknowledged, while at the same time, in another reality, the structures of everyday life are assumed immutable, integrating the two is "unrealistic."[32]

This is realism as *ideology*. Some readers may most frequently encounter ideology in its "positive," i.e., explicit, expressed form, for example, as a set of political and programmatic goals.

But what ideology means here is a much broader sense of the underlying assumptions, parameters, and languages, often unconscious (although not necessarily so), that structure, limit, or inform a sense of what is socially "real." Mark Fisher's phrase "capitalist realism" is a wonderfully succinct way of expressing this sense of ideology and realism. "Capitalist realism," Fisher writes, is the "widespread sense that not only is capitalism the only viable political and economic system but also that it is now impossible even to *imagine* a coherent alternative to it."[33] Realism here, although related to existing power, is not about expressing the reality of that power. Rather, in this form, historically contingent social structures, institutions, and systems are "naturalized." But whereas the laws of physics are immutable, social systems are not. First generation Critical Theorists like Walter Benjamin, Theodor Adorno, or Max Horkheimer argued in different ways how an empirical portrait of society should not be confused with a scientific portrait of nature, or how dynamic social qualities at any given time in no way imply static "natural" facts. Without denying the validity of scientific inquiry even into social realms, we must ask when, how, and why a given inquiry is conducted; where criteria for evaluation come from; and what are its bases and roles in economic production and socioeconomic reproduction. Why do some questions get asked and others not? Where do our standards of "better, useful, appropriate, productive, and valuable" come from? And — who, whom? — what power do they serve?[34] Without this critique, such portraits themselves become "theodicies" — apologetics for the social world as is, because the way it is, the broadest structures of "social totality," is the only way possible.

Realism here is not *about* reality. Rather, this kind of realism sets limits about what is deemed officially *realistic*, without regard to reality. It lies beneath what is broadcast as politically "possible." This form of realism infects everything from culture to understandings of nature itself. Why, critics ask, are only certain kinds of stories told? How are different kinds of people categorized, governed, or represented, and why? Some will note overarching systems like capitalism, structural racism, or patriarchy in explaining such phenomena. Others will — and this can occur across the political spectrum — chalk this up to simply random opinion, taste, or, more grandiosely — and tautologically — in the realm of production and consumption, the claim that prevalence reflects unmediated popular desires and preferences.[35] Supply and demand. What is interesting in this moment is that ideology is less "totalizing," is increasingly less effective, less hegemonic in the sense of successfully generating broad public consensus. In the distance between officially sanctioned "consensus" and even the most cautious interpretation of survey statistics above, although these hardly alone, it is clear, in Gramsci's famous phrasing around waning hegemony, "that the great masses have become detached from their traditional ideologies, and no longer believe what they used to believe previously, etc. The crisis consists in the fact that the old is dying and the new cannot be born; in this interregnum, a great variety of morbid symptoms appear."[36] Such symptoms are not limited to novel political formations — Gramsci has in mind, for example, fascism — but can be found in everything from increasingly coercive rule to fracturing political legitimacy. However much consensus political discourse likes to paint itself as realistic, it is literally at odds with reality *and* its popular perception in the most prosaic ecological terms.

Take, for example, the Paris Agreement. With the agreement first put in place in 2015, assuming universal compliance to the non-binding treaty, the stated targets — to keep global temperatures "well below 2 °C above pre-industrial levels" and ideally limit increase to $\Delta 1.5$ °C over the next 80 years — were technically impossible within the treaty's *framework* assumptions *three years* before the treaty itself was signed.[37] [The mathematical symbol $\Delta$ — delta — will refer to change above pre-industrial baseline throughout the rest of the book.] In the simplest terms, in 2012, assuming a theoretical reduction in the *rate* of carbon emissions, emissions in toto were on track to increase by about 3% per year. To hold to the goals of the climate treaty, *actual* emissions would have to decrease in the range of 2–3% per year by 2020. (Today this reduction must be *considerably* higher.) Not only has this not occurred, but even the *rate* of increase is accelerating as of 2023.[38] Thus, in most likely Paris framework scenarios, the Δ1.5 °C threshold will possibly be passed as early as the 2030s.[39] One of the many problems with Paris is that it encourages the view of Δ1.5 °C or Δ2 °C as thresholds as opposed to highly cautious, useful rubrics for what, at the most rudimentary level, one might expect in such a world. For just one example, a Δ2 °C increase world is one in which a vast number of people, particularly in regions like Central Asia, are regularly exposed to quite simply "deadly heat," not counting likely human migration effects, social and political instability, and other intuitively obvious effects of such a transformation. The IPCC special report on holding to a Δ1.5 °C world (a world in which the already-existing approximately Δ1.1–1.2 °C changes less than half a degree) paints a possibly bleak although *not* eschatologically apocalyptic picture. In social terms alone, destabilization of food systems, impossible prospects for many coastal — where the vast majority of people live and work — and dry-inland communities, disease, poverty, resource conflict, death.[40] Not, of course, equally distributed.

Paris, we are told, is the prudent, mature, *realistic* kind of climate policy. Bold and "progressive." Shortfalls or lack of ambition are not, though, primarily why the Paris Agreement, alongside so many other articles of existing "climate" governance, remain so trapped between ideological "realism" and reality. It is rather that it is predicated on fitting genuinely admirable and aggressive goals in tune with broad climate scientific understanding into a *framework* that assumes only modest — at best — deviations from "dominant" socioeconomic conditions.

For example, the Paris Agreement does not actually discuss decarbonization or even fossil fuels. Rather it suggests a "balance" to be achieved between emissions and sinks by 2100. (This timescale is part of a set of "inconsistencies," as one team put it, in the Paris framework. Many others, including the intrinsically conservative IPCC argue that this timeline, as another study observes, "contradictorily suggests that the best way of keeping warming to a specific level would be achieved by temporarily exceeding the set maximum level before 2100.") Article 6 of the Paris Agreement notoriously sets out a plan for carbon markets and other financial instruments as key to climate mitigation. (Including the aforementioned non-carbon-offsetting carbon offsets.) Evaluative questions are left to the other rubrics, like the United Nations Framework Convention on Climate Change (UNFCCC), which exclude entire sectors like aviation, shipping, and consumption-based emissions reflected in international trade (which even the World Trade Organization (WTO) estimates as 20–30% of emissions, almost certainly an undercount). Even the recent Glasgow Climate Pact revisions do not alter much: a fossil fuel "phaseout" is encouraged, particularly in "*unabated*" coal, i.e., coal with highly dubious carbon capture technology[41] remains in the energy mix for some time; while whole sectors, consumption-based, and "scope 3" (value chain) emissions are still absent (except in the final case, where they are included for construction). The Paris Agreement not only enshrines economic growth and markets, but even with revision the UNFCCC still, almost unbelievably, directly states: "the Parties should cooperate to promote a supportive and open international economic system [...] measures taken to combat climate change, including unilateral ones, should not constitute a means of arbitrary or unjustifiable discrimination or a disguised restriction on international trade." When a recent review by a massive interdisciplinary team led by ecologist Isak Stoddard chalks up the failures of Paris and other treaties to "the central role of power, manifest in many forms, from a dogmatic political-economic hegemony and influential vested interests to narrow techno-economic mindsets and ideologies of control" — that is, to theories of hegemony — they are actually *understating* the case.[42] The contradictions and inconsistencies are plain in the texts.

What we start to see is that the Paris Climate Accords

embody a generic problem in that there is no way to square climate mitigation and adaptation goals with the premise of fundamental system preservation, of a world constituted by capitalism-as-we-know-it. This is not simply the view of eco-Marxists and others already predisposed to such a position. The plain reading of climate science findings — coupled with a roughly utilitarian understanding of maximizing sustainability for the largest number of people possible — increasingly produces its own critique of capitalism. In their 2018 "Hothouse Earth" paper for the *Proceedings of the National Academy of Sciences* (actual title: "Trajectories of the Earth System in the Anthropocene," one of the most cited papers in the climate literature), the late atmospheric chemist Will Steffen, alongside 18 others, wrote, in the generally cautious mode of the natural sciences: "The present dominant socioeconomic system [...] is based on high-carbon economic growth and exploitative resource use. Attempts to modify this system have met with some success locally but little success globally in reducing greenhouse gas emissions or building more effective stewardship of the biosphere."[43] This is quiet, climate science speak for capitalism, or at least capitalism-as-we-know-it.

They are hardly isolated. "Climate change and large-scale contamination cast doubts on the sustainability of the current mode of global socioeconomic operation," argues another paper led by an atmospheric physicist. Or another team: climate change forces "the necessity to consider underlying social drivers such as capitalist competition and unequal power relations." "Tackling proximate impacts and causes through modifying thresholds and feedbacks will rarely have lasting or meaningful effects unless the underlying drivers (for example, of capitalism, power asymmetry, and exploitative and extractive lock-in) are also addressed," an interdisciplinary team of ecologists and geographers argue. They even explicitly contrast what is ecologically necessary with the political faux-realism of "the art of the possible." One of the IPCC's current lead authors simply states, in an article on how to hold to the Δ1.5 path: "achieving pronounced emission reductions requires a transformation of the global economy." "The roots of the current crisis," that is "capitalism" and "colonialism," a team led by a *biologist* argue, "rest in our societal paradigm. A proper understanding of its mechanisms and key actors is outside the comfort zone of academics studying natural sciences and ecology." "Under business-as-usual future scenarios — meaning that drivers of change do not deviate from the current socioeconomic and governance trajectory — nature in terrestrial, freshwater, and marine realms and most of its contributions to people will continue to decline sharply," writes a team led by the Argentinian ecologist — also an IPCC author — Sandra Díaz.[44]

This is truly just a tiny sample; even under the hood of official documents — often past the executive summaries which get turned into convenient headlines and talking points, in between claims larded in by BlackRock, Exxon, and the United States — are carefully worded calls which amount to nothing less than fundamental, structural challenges to capitalism-as-we-know-it. Anything less is simply *unrealistic*. There is a degree of vagueness and uncertainty — quite openly acknowledged — in such claims which some on the left attack as misleading, or even some kind of greenwashing of radicalism. What it really is, though, is an invitation: for the vast majority of people who are not already anti-capitalists, who *are* concerned or *feeling* generally uneasy about accelerating climate and/or interrelated social crisis, to consider just how radical reality is. And an invitation for social, political, and economic intervention, sharpening, and critique.

The market frameworks of the Paris Agreement and its previous antecedents don't call for *nothing*. Rather — in a newer language of delay — they take their time. Nearly every climate scientific analysis — regardless of positions on technologies, growth, and other topics of deep debate — emphasizes the need for immediate and extreme measures: phase*outs* starting now, not phase*downs* starting down the road, "keep it in the ground," rapid decarbonization and energy efficiency, demand reduction, and more. These measures are extraordinarily expensive and threatening to existing power and wealth; slower, more modest ones are not only safe but extraordinarily profitable in short and long terms. Capital is having a bonanza within the carbon trading and climate financializing spaces unleashed in these frameworks: heatstroke insurance (SoftBank), livestock insurance (ITK), weather futures (CME). Firms from JPMorgan to H&M are already trading "carbon capture credits" — "bets" as the *Financial Times* words it — not on any actual potential of the technologies but on the explosion of investment in them.[45]

It is a truth increasingly acknowledged that an economist in possession of an ideal model must be in want of a "market failure." And such financial instruments supposedly add the missing information into markets whose intrinsic efficiency will then produce the "optimal" outcome. The prevailing logic in treaties and official "realism" is that of "market failure." Nicholas Stern — the famous *liberal* economist on climate questions — famously called climate change the largest market failure ever. "It was a geophysicist, Brad Werner, who in 2012 argued precisely the opposite case — that we are in this mess not because the market system is not working well enough but because it is working too well," as Australian *business school* professors Daniel Nyberg and Christopher Wright remark, "the problem is not Stern's market failure but market success."[46] To think that this is "the politics" of "the science" is, though, still fundamentally constricted, limited, and in need of elaboration. All plausible portraits for mitigation and adaptation in the interests of the vast majority of people on Earth involve some turn — whether one considers it Marxist or not, ecosocialism or managed capitalism — to planning and decommodifying vast sectors of economic life. There really is a red interior to that lush green exterior.

There are multiple reasons — technical, ideological, political — why a political realism in this moment not only can but *must* be radical. "Incremental linear changes to the present socioeconomic system are not enough to stabilize the Earth System. Widespread, rapid, and fundamental transformations will likely be required to reduce the risk of crossing the threshold and locking in the Hothouse Earth pathway," Steffen et al. go on to argue. Again, some translation required: the nudges of market mechanisms, modest efficiency gains, simply are not credible *if* one assumes a trajectory for the vast majority. What can be hard to accept is that denialism in the form of the "acceptance" or "rejection" of climate science is largely irrelevant to the political divide of climate change. Rather the line is increasingly drawn by feelings, anxieties, wishes, and dreams that hardly flow in the same direction for all. The Rex Position, right-wing climate realisms, at this moment, characterize the far more developed, far more crystalized side of that line.

## Whose Apocalypse When?

Actually Existing Capitalism runs on stress and stressors, social and ecological. Ecological sustainability, a socioecological flourishing for the vast majority, for the many, requires addressing such stressors — such *exhaustion* — across ecological, economic, social, and political systems, otherwise the overall project is a constant, unstable, and, in terms of maintaining a niche capable of such flourishing, unsustainable shell-game. Right-wing climate realism embraces the shell-game by simply choosing the shells. To help, no matter what the cost, maintain capitalism in the twenty-first century.

Or worse. A gradual lateral exit from contemporary capitalism into forms of what would be more precisely termed neofeudalism. Simultaneously extending the life of our current global socioeconomic and political systems as long as possible for maximum real accumulation while "cashing out," toward more directly coercive forms of privatized rule. In this sense, against both a triumphalist liberal democratic narrative or some forms of Marxian analysis, capitalism would have been a massively costly *aberration*. A world-historical blip that expanded the wealth, power, and size of the ruling class and locked in a niche incompatible with the flourishing of some seven billion human beings. Through an ecological lens, what Thomas Piketty describes as "patrimonial capitalism" becomes a horrifyingly distorted mirror-image of the steady-state or circular economies often posited as goals in ecological literature. Wealth is held, maintained, and recirculated back to itself in close kinship networks. This is, of course, already a characteristic of the world as it is.

Neofeudalism is *also* a steady-state economy — one in which growth is essentially nil or close to it — but characterized not by some harmonious socially sustaining socioeconomic life. Rather, it would be one in which the rate of return on "capital"[47] lies at its historic norms of 4–5% for a rather much larger and much more comfortable ruling class than existed in pre-modern periods across the world, coupled with the conditions for a handful of distinct — if almost certainly overlapping — surplus populations: first, a massive number of socioeconomically expendable people (no longer needed for the basic stable reproduction of the sated and well-off) who face direct, permanent ecological adversity (think about the UN's estimate of 1–2 billion who will "no longer have adequate water"[48]). And second, an over-large body — also facing severe economic and ecological constriction — of what one might term "neoserfs": people who work in basic production and extraction, maintenance, and non-essential service functions. In between these groups and a ruling class would be a third mass: loyal retainers, if you will. Those who perform high-level services, especially governance and security.[49]

I want to be careful with my language as such a scenario might seem simply dystopian speculation. Or an exercise in the increasingly popular genre of climate doom-mongering. Climate change is *not* the apocalypse. Rather, coming to know right-wing climate realism and its worst possibilities is coming to know an enemy. To have a "clear-sighted" view of its concrete realities, its politics, and its passions, allows for the possibility that we might cultivate and *externalize* our own. This possible reactionary climate realism scenario is hardly certain. It could take differing forms across differing geographies. In some places we might see neofeudal company towns as described here, but in others virulently and exclusive nationalist states; in some places massive "sacrifice zones," as already exist around key points for the global extraction of rare earth metals or fracking; and in others, a rump state working in concert with private powers.

We can have Narendra Modi's neo-fascist Hindutva "spiritual environmentalism" — which simultaneously dismantles India's regulatory regime (99.82% of recent industrial projects have been swiftly permitted), boasts of its Paris commitments, pledges photovoltaic buildout, and preaches "special attention to yoga amidst discussions about climate change," while pinning social and ecological crises on Muslims, Dalits, and "secularists" and pursuing adaptation and mitigation in ways that sometimes (literally) pave over them.[50] Or we can have Jair Bolsonaro calling for extractive genocide in the Amazon without much reference to ecology at all. We can have the burgeoning Green-Black eco-fascist political formations coming back into European vogue that cite climate change as the *reason* for reactionary policies, from borders to eugenics. In January 2020, the right-wing People's Party of Austria (ÖVP) entered into a new governing alliance with Austria's Green Party. Goals of renewable energy by 2040 and investment in efficient, clean, and modernized public transportation were married to "preventative detention" of migrants and expansions of internal restrictions on Muslims in order to "protect both the climate and the borders." While the coalition still holds, the ultra-right Freedom Party (FPÖ) — with its own roots in classical Nazi environmentalism — leads upcoming polls. From the French National Rally, with its own explicit fascist roots, now blending romantic environmentalism and Heideggerian "rootedness" with more contemporary sounding arguments that "borders are the environment's greatest ally; it is through them that we will save the planet" to the Swiss People's Party (currently the largest in the Swiss Parliament) arguing "if you want to effectively protect the environment in Switzerland, you must fight mass immigration" — Green really is the new Black.[51] It can be difficult to tell the difference between an ecofascist manifesto from a mass shooter, lamenting that the only politically "realistic" response to climate change is nationalist population control, and a "liberal" *New York Times* columnist commenting that "we have a real immigration crisis […] the solution is a *high wall with a big gate — but a smart gate*," to allow in "desirable" people fleeing from economic, social, and ecological crises elsewhere.[52] But how different is this than the EU placing its notorious Frontex border control agency — which reiterates the new threats of climate migration — under the auspices of a commissioner to "protect our European Way of Life"? Or Dianne Feinstein saying it all just costs too much? All of these are variations on a theme of right-wing climate realism — a cornucopia of Rex Positions generating similar visions of the preservation of power and security for a select few whose primary difference is intensity and audience. Superstorm Sandy strikes New York and the lights remain on at Goldman Sachs, which, like the new luxury developments, has its own internal power generation system.

The point is that, in some form, right-wing climate realism is not only plausible and possible, but probable. Such a world need not be characterized by some cataclysmic break with this one, nor would it be unrecognizable. Indeed, part of what makes this such a plausible outcome is that it is another intensification of the existing world. Approximately 25% of the American workforce is *already* employed protecting wealth and surveilling other workers.[53] This is a trend, also seen in other countries, that tracks inequality. The business-as-usual world is a radically unequal one. But such an intensification of phenomena like this would be so significant as to be a qualitative shift. Such a world is not speculative fiction; it is an existing trend.

Things like the Rex Position are usually chalked up to corruption, greed, short-sightedness, or a host of cognitive biases. But the world projected in the vision of tiers of private protection is not shortsighted at all. If anything, it is the assumption that climate change is universally apocalyptic that is the exemplar of cognitive confusion and blurred vision. "'The easiest thing to do,' enthused the director of carbon and energy finance at one Australian financial services company, 'is to go carbon trading. There's a way to make money!'"[54] If I can drill and make money, then that's what I'll do.

The argument is that whatever protections wealth and power afford will evaporate at some degree of a $\Delta X$ scenario. Of course, there is some X at which universal effects to such an extent really do begin. But that is not what we are talking about. Some business-as-usual scenarios put the Earth on track for a $\Delta 3$–4 °C change by the end of the twenty-first century, $\Delta 5$ accounting for the possibility that feedbacks might trigger a faster rate of warming or lock in a particular pathway. Somewhere around $\Delta 4$ seems likely with current business-as-usual.[55]

And when one wants to add the already complex ecological picture to social systems — war, resource distribution, land-use — it is even more difficult. A $\Delta 4$ or $\Delta 5$ scenario by 2100 is almost certainly characterized by mass levels of direct and indirect deaths, rampant disease in some regions, billions in food and water insecurity, vast numbers of climate refugees, resource conflict, and so on. But this is not "existential" in the mundane sense of the word; it is not extinction. As a recent study focusing on worst-case scenarios put it, "existential risk [...] is not a solid foundation for a scientific inquiry." The authors instead focus on 10–25% mortality and fundamental institutional collapse — a world perfectly in line with the more neofeudal modes of right-wing climate realism. This is absolutely *catastrophic* — in no uncertain terms, that's in fact the point of the research: "exploring severe risks and higher-temperature scenarios could cement a recommitment to the 1.5 °C to 2 °C guardrail." But it's not the perfect universal eschatology that informs so much thinking around climate change. Or, as Kate Marvel remarks, "it's worth pointing out there is no scientific support for inevitable doom [...] there is a real continuum of futures, a continuum of possibilities."[56] Politically speaking, those possibilities have different weights, different logics, different existing or potential powers.

The more aggressive forms of what I have been calling right-wing climate realism are what Christian Parenti terms the "politics of the armed lifeboat." However, Parenti deems such politics as "bound to fail" — not on grounds of insufficient adaptation and mitigation or catastrophic outcomes, but rather because "if climate change is allowed to destroy whole economies and nations, no amount of walls, guns, barbed wire, armed aerial drones, or permanently deployed mercenaries will be able to save one half of the planet from the other."[57] But this is highly uncertain, not particularly likely, and certainly not automatic.

One 2012 report engaged the difficult task of synthesizing direct ecological and social systemic impacts and predicted, in terms of deaths alone, 100 million climate deaths between the report's publication and 2030. By then, six million people may die every year from climate change. This is a shocking number. But the same models note that the world *currently* experiences some 4.5 million climate deaths per year. A more recent — if less capacious — 2023 study remarkably also finds the same 100 million by 2030.[58] Even granting the haziness of many climate projections — not the science *per se*, although climate scientists will often, rightly, urge caution that models are not perfect predictions — particularly when we move into trying to predict the complex interrelation of society and ecology into the future, something quite different is apparent when looking at the often ignored socioeconomic and climatic impacts of today and a potential tomorrow.

Thinking about current realities of economic inequality, warfare, and grinding poverty, let alone historical parallels, suggests that some version of *this* world — that is, this global system — is far more capable of absorbing truly catastrophic climate impacts discussed in likely scenarios than many anticipate. A brief survey of colonial history shows that with even more staggering odds, and sometimes far greater access to wealth and resources than in the "bound-to-fail" scenario, there was no automatic reversal or chaotic karma for "the wretched of the Earth."

One of the most prominent examples of an early ecological-economic-political matrix is the Bengal famine of 1770. To think with that historical example, only a few hundred clerks, owners, and managers, and several thousand soldiers, subdued an army of 50,000 at the Battle of Plassey to establish the rule of the British East India Company in Bengal in 1757. At that time the size of Bengal was approximately 30 million people. By the time the East India Company captured Delhi at the beginning of the nineteenth century, its private security force had grown to some 260,000 men (twice the size of the British military at the time) and would come to rule an entire subcontinent of over 200 million.[59] At the height of company rule, the force was approximately 350,000.

This was not, of course, a benign period. And it is even one in which we can already see a set of social-ecological relations. As Mike Davis noted, "there is little evidence that rural India had ever experienced subsistence crises on the scale of the Bengal catastrophe of 1770" before company rule. Mughal India was "generally free of famine until the 1770s."[60] During the Bengal famine some ten million people — or, as one British report at the time estimated, a third of the entire population — died in one of the first great "natural" disasters that were anything but. Unlike contemporary political ecology in which human impacts can be causally associated with truly niche-wide ecological forces, these earlier episodes were more straightforward: new forms of capitalist imperialism transformed cyclical events like El Niño into "natural" disasters. The famine was not the result of ecological cycles, but rather of the metabolic relation between a host of company policies, especially grain export, and the local ecological system.

In the Bengal famine of the late eighteenth century, such policies included both the continued promotion of cultivating land for export crops but also company-imposed taxation, exacted with extraordinary violence. Both continued even in the face of four years of ecological adversity. The next century or so of East India Company and then direct British Raj rule was nothing short of "free-market economics as a mask for colonial genocide," similar to that of the Great Hunger in Ireland, and even in at least one case, directed by the same personnel. Millions would continue to die of famine all across the subcontinent throughout the period of British rule. Even areas where rainfall was adequate were inundated with the social spillover effects of capitalist colonization. Early "experiments" were conducted "that eerily prefigured later Nazi research on minimal human sustenance diets in concentration camps," in finding just the optimal level at which capital could simultaneously extract ecological value and still generate profit through productive activity. Just as capital will inexorably chase the incredible profitability that fossil energy provides, regardless of human cost, "grain merchants, in fact, preferred to export a record 6.4 million cwt of wheat to Europe in 1877–1878," one of the periods of intense famine, "rather than relieve starvation in India."[61]

Up to the end, such practices continued in British-ruled India. Winston Churchill reached back to the empirically untrue Malthusian argument that famine was the result of the rampant population growth of the poor, to explain what was happening in India. Indians were to blame for "breeding like rabbits," an argument you can hear echoed today in different forms from Davos to Washington, D.C. At least three million Indians died in the Bengal Famine of 1943 as Churchill and his government insisted on policies of maximal grain extraction and stockpiling for British use. This time, though, there was no particular "natural" disaster to pin the blame on. It was a "manmade" famine, as Amartya Sen wrote. A recent study of soil samples over the course of Indian history found that, for this particular famine, there was no corresponding weather event. As the study concludes, the deaths were attributable only to "complete policy failure."[62] Or, in more ordinary terms, genocide.

British rule in India lasted two centuries. At their high point, colonization forces never exceeded approximately 500,000 across the subcontinent, ruling over 300 million people. It is not that Indians did not revolt, strike, and fight back; they did, throughout the period of the Company and the Raj. But rather that one-fifth of 1% of the population were perfectly capable of withstanding these actions.

India is no isolated case either. Integration into the world market went hand-in-hand with new economic-ecological disasters. In the late nineteenth century, these could be witnessed not only in India but in Algeria, Egypt, Brazil, Angola, Queensland, Fiji, and Samoa. The hand of Empire and the Invisible Hand replaced staple crops with cash crops, "natural" famine followed. Anti-colonial and, on the other side of the world, slave revolts, rebellions, and revolutions did occur, sometimes even achieving legal and formal decolonization. The success of the Haitian Revolution at the beginning of the nineteenth century is one of the paradigmatic cases. Cedric Robinson gives half the story pertinent here: "in Haiti, between 1791 and 1804, slave armies managed to defeat the French, Spanish, and English militaries — the most sophisticated armies of the day," allowing Haiti to become "the first slave society to achieve the permanent destruction of a slave system" and "to achieve formal independence." But C.L.R. James provides the other: "if the Haitians thought that imperialism was finished with them, they were mistaken."[63] Haiti would eventually be, in the colder language of the twentieth century, "contained" by the global market, subordinated in that market and by its imperialist homes, frequently besieged by blockade or direct military intervention to this day.

But there was nothing automatic about any of these processes, which lasted in many cases a century or even centuries. And even after all that time and under such conditions, few would claim that the world is "decolonized" today. If anything, it is less that the world has decolonized than that colonial relations have become more omnipresent. Anti-colonial struggles are not automatically, inevitably, or historically bound to fail. There have been many real victories achieved in the struggle for decolonization. And the *politics* of a *left-wing climate realism* is, in many ways, a broadened mode of anti-colonial struggle —

Fanon's "stretched" Marxism, stretched and transposed further, as much in the metropole as in the periphery. It is, in its simplest form, the very real, very possible mitigation and adaptation scenario for a quasi-utopian flourishing for the vast majority of people on Earth. Put differently, left-wing climate realism is the politics of a world relieved from social, economic, and ecological despair and exhaustion.

This is absolutely possible, *pace* the doom-mongers. But, in the same way, right-wing climate realism is hardly "bound to fail." It is perfectly imaginable because the world as we know it already absorbs the scaling horrors of the Anthropocene.[64] It is perfectly imaginable because the colonial relation has, necessarily, adapted and spread over time. It is perfectly imaginable that some, even a modest, "amount of walls, guns, barbed wire, armed aerial drones, or permanently deployed mercenaries will be able to save one half of the planet from the other." It's perfectly imaginable because the world we actually know and the one we can observe through the historical record makes the "politics of the armed lifeboat" far from a bad gamble for those whose stake promises a payout.

## Disunion in Stereo

As is a constant refrain in intergovernmental reports and climate science literatures, it is overwhelmingly individuals in the Global South, particularly the global poor and working class, who will bear the brunt of climate change. "Over 90% of mortality assessed […] occurs in developing countries only," argues the same synthetic 2012 report. That is, of course, in terms of mortality only. As already mentioned, the United States has a surprisingly large exposure to the difficulties of climate change. However, this is far more likely to be accounted for by economic and social deprivation than by sheer death. In the United States, the ensuing conflicts are less likely to be out-and-out resource wars and complete social disintegration than drastic socioeconomic degradation to those outside the top-income deciles. Exacerbating existing inequalities and inequities from public health to political governance itself: an intrastate version of the globalized caste system. There are certainly cases where climate change will have extreme impact on wealthy settler-colonial states — think of the intensified 2019–2020 Australian bushfires — but even in such cases, it is the poor and the already dispossessed, for example the Aboriginal populations of New South Wales, which is also where fire concentrations were the highest, who face the greatest risks.[65] It is not simply that even in extreme cases there is no "karmic balance" or that there is an ironic, ecological cruelty. Existing inequalities and inequities increase exposure to climate impacts, even while, through exploitation, extraction, and enclosure, they are simultaneously *drivers* of further inequality and ecological stress. It may be that the wealthy in a place like Australia will only have the recourse to escape. But, at the very least, as climate change continues to intensify, the geographical maps of the global caste system will continue to be redrawn.

Between the statistical image of the most impacted — both internationally and within countries like the United States or Australia — as well as scenarios like the tiered protections of the Californian wildfires, climate-proof luxury construction, or more dramatic considerations of resource wars and instability, it begins to become clear that, although we all inhabit the same ecological niche, we don't really share *one* climate.

It is here that we can really start to see the political implications of probable outcomes between different degree changes over time. In looking at the reality of current and projected climate impacts in the near term, and how climate change truly is not a universal condition, we can finally begin to piece together a more accurate, stereoscopic view of the moment as it currently stands.

The stereoscope, invented in the nineteenth century, allows a viewer to see two images at once, one in each eye, such that the brain can perceive a three-dimensional image. Walter Benjamin introduced the idea of a stereoscopic view in his work on history, perception, and politics. One of the reasons that Benjamin is so powerful a theorist to think with ecologically — and why one finds his work in so many social scientific and humanistic texts on climate — is that he conceived human history and natural history as a continuum. Time was geological for Benjamin: the past does not go away but accretes, layer upon layer, into the present. Historical time was "biological" even for Benjamin, not to be understood as some long chain of events but as a mere split-second in the "history of all organic life."[66] At any given moment one could grasp a more accurate, more meaningful view of history by replacing a static, single image with a stereoscopic one, bringing an analytically or politically relevant "past" or layer or "constellation" into view that enhances our initial glance, that creates a three-dimensional portrait of the present.

If we learn how to see stereoscopically, "to educate the image-making medium within us, raising it to a more stereoscopic view,"[67] we find not the flat tableaux of the triumphal procession that characterizes official "universal history"; instead, a true three-dimensional image of the present emerges. It is just such a view that we need to understand the politics of climate change.

In one image is a vast number of people, many already feeling and experiencing the catastrophe that even a $\Delta 1.5$–2 °C world promises. Such experiences are in no way equal, stretching from out-and-out deaths of millions, millions more as refugees, others facing chronic food and other resource shortages and attendant conflicts, and still others locked into tiered castes of misery. In the other image is a world that has not only profited from 30 years of "denial" and delays but stands to continue to reap dividends and increase the already substantial *permanence* of such social structures, or at least their payout. Most climate research focuses, for good reasons, on how to hold to that $\Delta 1.5$–2 world, ideally the $\Delta 1.5$ goal. The *immediacy* and radicality of all that unprecedented socioeconomic transformation is predicated on the timescale that appertains only to the first image. Worst-case scenarios — difficult to predict as they are —

are actually *not* what is in question here. It may be difficult to project the long-term outcomes of worst-case scenarios, but there is a far greater certainty concerning who those hundreds of millions will be within *likely* trajectories, within business-as-usual or the leisurely ambitions outlined in existing Paris-based Nationally Determined Contributions (NDCs).

In that second image, though, there is no urgency between $\Delta 1.5$–2 and $\Delta 3$–4; the projected approximately $\Delta 2.7$ *promised* through current policies is more than adequate. It's not that "adaptation is not possible" in such scenarios; it is just that it is highly *restricted*. And, in the meantime, there is money to be made and power to be secured in heaven's high water. The better climate plan for some is not to just build lifeboats but protect them further still. Miami might be lost, but as one researcher quietly observed, off the record, that's not the case for Manhattan, which would likely be protected by seawalls or surge barriers. Only that form of adaptation would almost certainly come at the cost of the outer-lying areas of Brooklyn and southern Queens. Perhaps this might seem outlandish to some, but it is the basic logic of "the discount rate" — the mainstream macroeconomic approach to climate change — just relieved of some of its ideological distortions.

The economist William Nordhaus was widely celebrated for his "discount rate" model. To his credit, Nordhaus wanted to reattach the standard macroeconomic model to the actual, physical world, to plug it back in and account for the sources of energy, to "price in" the "externalities."[68] Stopping carbon emissions and bringing other environmental factors within planetary boundaries pose steep costs both in the present and even to future generations, the argument goes. Thus, "we" must be prudent in how much "we" weigh the needs of climate mitigation policy to balance present needs and present costs with future needs and future costs. The discount rate model postulates that there must be some optimal inflection point — usually to be "priced in" through a carbon tax or some form of cap-and-trade — at which future needs outweigh present benefits. A high discount rate indicates that present costs and growth needs outweigh immediate climate action in the favor of future generations.

Based on this discount rate model, Nordhaus suggested a relatively high social cost of carbon that grows at 3% per year from approximately now till 2050. In that time, this would suggest a carbon tax starting at approximately $19 per metric ton of $CO_2$ and topping out around $53. This is only a minute difference from his most well-known critics in the Stern Review whose proposals differ by as little as a few dollars per metric ton of carbon to about a $150 difference. The IPCC special report on staying within a $\Delta 1.5$ world, in contrast, suggests costs as high as $14,300 by the end of this decade. The point isn't the comical difference between these numbers. Nor how carbon taxing and cap-and-trade systems will never work or are ludicrously inadequate measures. Or even how the IPCC models incorporate "Negishi weights" and their consequences which, as the economist Elizabeth Stanton notes, "freeze the current distribution of income between world regions" and without which "IAMs [integrated assessment models] that maximize global welfare would recommend an equalization of income across all regions as part of their policy advice." But rather how the very idea of "the discount rate" is so perfectly a "victor's" story, how wonderfully it obfuscates reality.[69] There is no universal "we" whose present benefits are being maximized. Profits, rather, are maximized for the few at extraordinary socioeconomic and ecological costs to the vast majority. Climate change does not negatively impact only prospective "future" generations but is already exacting costs from *most* people currently alive while benefiting a rather smaller number immensely.

Nordhaus advocated, as increasingly "mainstream" and even "liberal" voices are normalizing today, a $\Delta 3$ world or so, right in the business-as-usual range.[70] This is well in line with many of the right-wing climate realism scenarios we've explored. This *is* the Rex Position. Although so seemingly technical or distant from the kind of theoretical ideas Benjamin was exploring in the late 1930s, the discount rate turns out to be an exquisite illustration of Benjamin's arguments about time, history, and power. Just as "we" are all the beneficiaries of growth, as the dominant story goes, "we" are all better off with a high discount rate. The idea that some universal "we" is better off not only in a Δ3 world but along that trajectory belies literally everything we've learned about *current* ecological realities, let alone projected ones. The idea that profit maximization actually benefits "everyone" belies a world of stark and growing inequality. Whether it promises doom, salvation, or even just modest improvements, the universal story truly does serve only the powerful.

Here we can finally see diverging "climates," diverging interests, and even diverging *times*. While it is not the case that everyone will die in the next ten to 20 years or even longer (indeed, as we've seen, the threat of human extinction is not really relevant to a political divide characterized by the structural *depth* and *speed* of climate policy[71]), it *is* the case for *billions* of people and even majorities in Global North countries that they have strong interests in immediate and radical transformation. This is expressed; this is *felt*, although not necessarily yet, as a climate politics. At the same time there are those that have no rush at all. This is not about people's moral character *per se*, but actually just a straightforward comparison of structural interests. Nordhaus is, in a weird way, *right*; it's not worth the trillions of dollars of losses or the trillions of dollars of costs to mitigate so quickly. It's that he's only right for one of those images, for one small set of people with inordinately large amounts of power. When we look stereoscopically, we can see the full three-dimensional image not of single pathways but of two diametrically opposed worlds, where even ecology does not unite all people. Simply put, we're not all in this together.

# Chapter 2

# The Extractive Circuit

*Isn't the shivering earth beneath your feet lonelier than you?*
*The prophets, to our century, carried their message of destruction*
*these continuous explosions*
*these poisonous clouds*
*are these the echoes of sacred verses?*
*Oh friend, oh comrade, oh kin,*
*when you reach the moon*
*write the history of flower massacres.*

— Forugh Farrokhzad, *Window*

The machinery — the actual form and function — of twenty-first-century capitalism is an extractive circuit which quite literally crisscrosses the world. Its global value chains stretch through physical infrastructure and "frictionless" financial flows at the speed allowed by fossil fuels, telecommunications, and geophysical, technological, psychosocial, and bodily limits and "optimizations." It connects economically and ecologically dispossessed agricultural communities in the Global South with regimes of hyperwork in the Global North, rare earth "sacrifice zones" with refugees, migrant labor with social reproduction, ocean acidification and atmospheric carbon with profitable opportunity. It has transformed states; it has ripped through biomes and through flesh. Capital often appears and is treated as a historical abstraction; this is doubly true of globalized, financialized capital. The extractive circuit is the leaden reality of a global human ecological niche organized for maximal profitability — no matter how difficult or costly to maintain. Its realities underscore the *generalization of a colonial social relation* in *socio*ecological terms, even as older modes of imperialism and neo-colonialism are hardly swept aside. Its speed, frenzy, coercion, and brutality reach into the very heart of the imperial metropole, far beyond where such relations were already present. Feelings of exhaustion — depression, desperation, fatigue, exasperation — course through its wirings, neurons, biochemicals, and sinews.

At every "node" along such a circuit, "inputs" — ecological, political, social, individual — are extracted and "exhausted." The circuit, like capital, crosses boundaries without entirely obliterating them, and, similarly, connects a vast potential political subject across disparate lines — Global North and South, gender, class, race, nationality, religion, and sexuality. The extractive circuit is the socioecological portrait of capitalism historically and its transformations to maintain profitability in the face of immanent headwinds.

Just as Marx once invited us to look behind the factory door — above which was inscribed "No admittance except on business" — to understand the way in which a nascent industrial capitalism was creating value, we need to "unbox" the extractive circuit, catalog its parts, and pry past a few bezels if we want to see Actually Existing Capitalism today.

## The Dismal Science

Consider the Philippines. Over the past several decades, the Filipino economy has become increasingly dependent on the export of low-cost labor, largely along gendered lines, in the form of care-workers to North America and Europe (mostly women) and extremely low-cost manual laborers to the Gulf states (mostly men). Remittances now make up 10% (or more) of the annual GDP of the Philippines.[1]

In the Gulf states, Saudi Arabia being the case *par*

*excellence*, migrant workforces are employed in sometimes slave-like conditions to do much of the country's basic labor, both "unskilled" and "skilled." (Approximately 76% of the Saudi Arabia workforce is migrant according to the IMF.) Yet, for all its repressiveness through arms of direct coercion like its notorious morality police, Saudi Arabia is a remarkably weak state. This imported workforce is vital for the social and political maintenance of that weak state, which in turn serves a key function in the globalized order not only as an oil producer but, crucially, as a control on the world's oil spigot. Far from the Malthusian fears of "peak oil," oil is, in fact, plentiful in the world, in the Gulf region and elsewhere, and Saudi Arabia is a key player in limiting or expanding its production to influence prices. As paths for economic advancement narrow in places like the Philippines, and as industries such as fishing are decimated by changing ocean temperature, acidification, coral bleaching, and other cumulative effects of global climate change, conditions intensify this political-economic shift to migration. In turn, such shifts drive profitable increases in energy demand, low-cost labor through dispossession, and even social and ecological crises themselves. As Melissa Wright observed of the "disposable Third World woman," Filipino and Southeast Asian labor more broadly is viewed — in terms of dislocation and distance but also cultural imagination — as docile and pliant. In the Gulf, male Asian workers are considered additionally useful as "less politically menacing" than local and regional alternatives. Ecological resources become sources of social value, and "human capital" is *naturalized* as closely as possible to the supposed infinite free "gifts of nature."[2]

Now imagine a Global North worker across the globe, likely "middle class." Probably white, but not necessarily so. Place her in California — an increasingly unsuitable geography for mass human habitation. Say she's white-collar — perhaps an office assistant, accountant, or coder. In the 1970s, her labor (and now I am shifting from the generic philosophical "she") would likely have been lower in waged-hours than it is today, and it would have included, in the famous phrase of Arlie Hochschild, a "second shift" of unwaged "free" domestic labor. Cooking, cleaning, care work: the often "invisible" aspects of social reproduction found in the home. Today, our imaginary Californian works longer hours in a "productively optimized" labor process, still for a lower wage than a male counterpart, even as part of her "second shift" is now itself displaced onto migrant labor, including everything from general healthcare services to at-home family care and domestic work to independent contract labor for household maintenance, which can range from food preparation and delivery to, in concentrated urban centers, laundry and far beyond. The extractive circuit produces prodigious amounts of such "disposable" people.

This move toward "outsourcing" domestic labor was already occurring in much earlier periods (the 1960s, 1950s, and earlier), but in the United States it was, at the time, shifted instead to differently racialized gendered labor; a largely black, racialized caste system underwrote white middle-class "normalcy" in the United States. Such a caste system unquestionably continues to this day, even if its racialization has taken on "multicultural" hues. But the augmentation that is key to understanding the extractive circuit is the ever-increasing inputs — to use somewhat crude economic language — to maximize productivity and profitability. The spread of comparative advantage from geographies to the body. This is the case even if it leads to, at best, an increasingly socioeconomically and ecologically tenuous *maintenance* of current overall economic productivity.[3] Put more simply, our imaginary white-collar Californian now works harder and longer, helping to maintain an artificially futile level of production and requiring mass consumption of everything from energy to electronics.[4] And she does so for *less*. Her real wage has stagnated for several decades; even though some of the professions imagined above could place her income above the approximately 50% of Americans who earn less than about $31,000 a year, they don't come close to the levels where twenty-first-century capitalism truly pays off (not even taking into account the grimmer picture in wealth).[5] Her lifestyle is thus supported by financialized debt, which in turn requires her to ever further "innovate" and "diversify" her "human capital." This includes but is not limited to intensive self-maintenance through biochemicals, pharmaceuticals, and other technological interventions to her body and its internal composition and processes.[6] She does *consume* more, but this is not a particularly *pleasurable* form of desire fulfillment. It is rather her very desire to continue to exist that is rechanneled into the logic of capital accumulation.[7]

Similar anhedonic dynamics propel the Filipina domestic worker covering the "outsourced" second shift, only with greater severity. Facing steeper expenses, lower, variable wages, and generalized precarity, her personal structural adjustment program requires digital platforms and devices to access and coordinate the multiple labor "opportunities" needed to get by (let alone to remit income to the Philippines). Pharmacology finances her "flexibility" for maximal, unpredictable hours. Tallied as "consumption," these technologies are rather the fixed capital costs of her "entrepreneurial self." Non-financial damages are "externalities" absorbed across the fragmented "sinks" of her body, no matter how deleterious. In return for pleasures pressed as much as possible into productivity or palliative support, desires not devoured by debt are dollarized and delivered for dicey (personal) or delimited (postcolonial) development. In the most likely outcome of her less than likely, but hardly to be dismissed, *success*, she exchanges the socioecological disasters of her former home for the intensifying, if less severe, ones in her current one. Out of the fire, into the melting frying pan.

Even as a rather small node on the extractive circuit, our Californian is also a component along several different if recognizable lines, for example through the "global value chains" (GVCs) producing her computer or smartphone, from cobalt mines in the Democratic Republic of Congo (DRC) to Shenzhen manufacturing.[8] The unprecedented speed, precision of production, and international division of labor, as well as the ungovernability of these value chains, propels her own speed-up. The food system her (un)well-being rests upon is one form or another of industrialized petro-farming, itself dependent on the labor (and the hyper-depressed wages) of a deeply precarious migrant labor workforce composed largely of undocumented migrants and special H-2A agricultural "guest workers." Its supposed "green" revolutionary efficiency is a figment of resource-intensive industrial agriculture, which leaves the world both malnourished and obese while *increasing* energy costs and greenhouse gas emissions (GHG), not to mention the growing enclosure and dispossession of people practicing far more sustainable modes of agriculture that could feed the world far better today.[9]

Most carbon dioxide is emitted in the process of production. Take her smartphone. The electricity that powers it accounts for only about 16% of the total carbon emissions that span its lifetime as a commodity.[10] Over 79% of the emissions occur across its multiple sites of manufacture and the shipping which connects their global spread.[11] The final production of most phones is in China, whether their end-use destination is the United States, Germany, Japan, or any other country on Earth. And, if we start looking at the other ways phone production exceeds planetary boundaries, in purely ecological terms, we find all measurable boundaries breached.[12] It's not just a matter of carbon dioxide and other GHG emissions but also the processes of resource extraction (mining) itself: excessive freshwater use, eutrophication from biogeochemical flows, deforestation of nearby lands, biodiversity loss from land-use, and others. At the same time, such extraction is dependent, for just one component, on cobalt found primarily in mines in the DRC. The labor in such mines is, almost without exception, either slave labor, child labor, or both. Capitalism-as-we-know-it is dependent on the political subordination of places like the DRC, the designation of its people as expendable, "disposable." Slave labor keeps the cost of cobalt as cheap as possible.

Cobalt is a critical rare metal in high demand.[13] The cobalt from such mines will likely be both refined and processed into phone batteries in manufacturing centers in, again, places like Shenzhen, China.

Globally speaking — although improving within China itself — such "cheap" labor is required to maintain the profitability of that phone. Meanwhile, as we've already seen, the end-use of that phone dramatically increases the conditions of specific forms of "hyperwork" — the speed-up, the exhaustion — in the Global North (and pockets of the North in the South and vice-versa). All manner of new, "more efficient," and "more productive" labor practices across a host of traditionally blue- and white-collar sectors. The very design of those phones — forced obsolescence within approximately two or three years — requires not only an increase in extraction across all these nodes, but is itself a source of profitability.[14] At this juncture, capital must keep burning through more of the biosphere and the human systems inside it to keep up margins.

In other words, the logic of the extractive circuit is one of the most vicious cycles imaginable: at every turn, an *increase* in energy inputs from both ecological and social sources, and, with every increase in energy inputs, an increase in overall inclemency for a global human ecological niche that stretches from rising ocean temperatures and acidification to overall warming, each of which drives a further demand for energy inputs, and so on. The increased consumption demands *in the service* of accumulation require further fossil fuel extraction, further migration into low-wage, high-risk, precarious or informal labor, and even the geo-strategic necessity of different kinds of post-colonial states.

At every node in the circuit, there are two simultaneous and related phenomena: value extraction and nodal exhaustion. Value is extracted *not only* through human labor but also through a series of natural and social inputs. Ecologically speaking, value at the simplest level can be drawn from the "free" use of water or air and other "commons." But it's also the value derived from their commodification, from so-called "externalities" in waste flows and mountains; in complex socioecological processes like industrial agriculture, where not only are soils exhausted but output is dependent on massive fossil fuel, unsustainable pesticide and chemical-fertilizer inputs, or with the "free" exploitation of flora and fauna, as straightforward as logging or as complex as the patenting of DNA strands. Almost every measure associated with anthropogenic climate change — not least fossil fuels — represents a process by which value is added through extraction and exhaustion.

Similarly, the extractive circuit derives value from a panoply of social sources, from the global majorities "relatively surplus to the functioning of capitalism" to the gendered and often unwaged work of social reproduction. In perhaps the most unintentionally radical paper ever written, the late economist Chong Soo Pyun pegged the "monetary value of a housewife" in 1969 through a set of neoclassical and neoliberal methods at $83,807.58 (or $626,410.28 in inflation-adjusted 2021 dollars). Just under 30 years later, the ecological economist Robert Costanza and an interdisciplinary team would argue, "only somewhat facetiously," in political scientist Alyssa Battistoni's words, the value of the world's "ecosystems and natural capital" as, on average, $33 trillion (or $56 trillion, inflation adjusted) per year. As Battistoni notes, feminist arguments regarding social reproduction should serve as bridge concepts connecting the "historical and structural" similarities between labor exploitation and ecological value. At the same time, in those majorities (across both South *and* North to differing degrees), refugees, for example, can be mined for data, exploited for informal economies, or leveraged for geopolitical advantage.[15]

Racializing and/or Orientalizing populations helps render entire peoples and geographies suitable for valuable subordination, disposability, or wholescale abandonment. Dispossession and expropriation are socioecological processes. Value can be derived from the limitations of popular power over capital. Or from the usefulness of social, political, and ecological crises themselves. In a very different (but still market-based) world-system, the DRC should be able to command extraordinarily high prices for its cobalt, especially if mined in more sustainable socioecological ways. This is not *only* an equity or justice issue; wealthier, more powerful states in the Global South can be "resistors" in the extractive circuit. A key part of the *politics* of its reversal. Under different trade and intellectual property frameworks, developing states in the Global South will be better able to quickly adopt renewable energy and other sustainable systems, as well as maintain those sustainable and prosperous practices which already exist. Thus, it may not be necessary for this growth to be *as* destructive as that of wealthy Global North states has been — it could happen, depending on the outcomes of a host of *political* conflicts, without the imperatives of capitalism-as-we-know-it and beyond the bounds of the extractive circuit. However, at this moment, *value* is derived from the DRC's ongoing sporadic armed conflicts and general instability — themselves shaped in part by the extractive circuit — in providing ideal conditions to maintain child and slave labor, environmentally catastrophic extraction, and capture the extraordinary revenues for transnational corporations like Swiss-based Glencore.

## Dolls, Doldrums, Dynamics, and Disease

The key connective tissue in this circuit is "fossil capital," to use human ecologist Andreas Malm's clarifying construction,[16] but carbon is not, by itself, the only input. In the wake of different modes of exhaustion — at the level of formal labor but also at the level of those thrown out of the labor market, at the level of communities, at the level of societies, of states, and of the ecologies we can live and flourish in — one thing is pursued: profit. Neoliberalism's *matryoshka* doll of financialization, international economic governance, risk shifting, state policies, and adjustments in cultural logic helped nurse profitability out of its 1970s doldrums. It did so through the redistribution of labor income to capital; through creating historically unprecedented speed and mobility for transnational capital flows, business-to-business commerce, and firm-level debt/currency creation; and through transforming the social functions of states into profit-generating enterprises, diminishing democratic sovereignty, inhibiting decolonization, and quite a lot more.

After the dawn of the Industrial Revolution in the nineteenth century, atmospheric carbon stood at approximately 280 parts per million (ppm). By the beginning of the "Great Acceleration" in the 1950s, that figure stood at just above 300 ppm, a difference of 20 ppm over seven decades. In 1980, that level was approximately 340 ppm, a difference of 40 ppm in just three decades. As of 2023, concentration is about 420 ppm, a difference of *80 ppm* in just under four decades. Although carbon emissions have been increasing rapidly since the Industrial Revolution, it is no accident that 67% of all such emissions have been produced in these past 40 years or so. *Pace* the Davos set, these emissions track neither population growth nor the rapacious appetites of the expanding "wretched of the Earth." Their path is unimpeded by the proliferation of eco-conscious marketing schemes, "corporate social responsibility," and promises (and non-existent realities) of mystical techno-fixes. They track the return to and *difficult* maintenance of profitability.[17]

As another team led by Steffen concluded, in terms of GHG emissions, ocean acidification, rainforest destruction, aquaculture depletion, global warming itself, and so on, climate change tracks not only cumulative GDP growth (as is widely discussed) but such conspicuous features of contemporary global capital as the increased use of telecoms, non-recreational transportation, and FDI, which moves from almost zero in the 1960s to trillions by the 2010s. Following Polanyi, Steffen's team dubbed this "the Great Acceleration." Such acceleration does not aggregate with population growth; perversely, the relation is inverted. Emissions, material intensity, and other climate measures are causally concentrated where end-point consumption is greatest, as many climate scientists now openly state, among the world's wealthiest.[18] In the global top wealth and income deciles, population growth is lowest or even negative. And as the rate of population growth is curbing globally, climate change continues its exponential pace. And as we've seen, many climate scientists today go further still, like physicist and social ecologist Julia Steinberger, in arguing that "the root causes of global environmental changes are coupled social and technical phenomena, not natural ones. The social system that dominates the world is not 'the economy', i.e., a system detached from time and space; but capitalism, i.e., a historically specific class-based system of extraction, transformation, and organisation of flows of commodities for exchange in markets."[19]

Liberalization promised that for some increase in inequality there would be a return to the high growth rates of the recent past and an overall increase in prosperity. But while the slump from the mid-1970s to the 1980s did relent from lows below 0.5% to rates in the 1990s and later hovering around 2%, it never returned to the *sustained* 3-5% heyday of the 50s–70s.[20] And yet at the same time, *profits* reached all-time highs. There are a number of factors which help explain this: extremely business-friendly regulatory and tax environments, the "labor squeeze" or fall in the share of income going to labor vs. that going to capital, the rise of massive transnational corporations (TNCs) which surpass even the monopolies of the past, globalization, financialization, and more — not least an ever-increasing level of material and energy inputs on a global scale.[21] Far from "decoupling," in many dimensions at a global level we only see intensification. Obviously, as we've seen, these are all interrelated. To use just one example, the rate of atmospheric carbon concentration is *increasing* even as global growth rates *decrease*.[22] Neoliberalism never really "solved" the "productivity crisis" of the 1970s.[23] What it "solved" was a *profitability* crisis, from record lows in the early 70s to record highs today. As Piketty observes, while far from the dizzying heights of the mid-century, an approximate 1% growth rate is not negligible.[24]

It is a component of profitability as well. But when viewed through the extractive circuit — that is, global capital in its most complete, ecologically embedded form — we can see that this current profitability is extremely difficult to maintain in social, ecological, even psychological terms.

It was in 1973 (not entirely incidentally) that the German social theorist Jürgen Habermas first proposed his conception of "legitimation crisis," in which crises — perhaps managed in one sphere — would then spill over into another. Economic crises could become cultural crises which could become political crises, *ad infinitum*. At the time, the ecological was already on the table, as state-managed win-win economic growth could run into the twin problems of finite resources (even if artificially so, as experienced with the oil shocks) but also of the inability of the biosphere to absorb all the concomitant "externalities," including, "in the long run, a global rise in temperature." As the distance between the principles of systemic legitimacy and systemic actuality grow implausibly far, "the legitimizing system does not succeed in maintaining the requisite levels of mass loyalty."[25] Put differently, this is the erosion of hegemony. In this moment, potential responses are hemmed in by the realities of already exceeded planetary boundaries and of social systems already stressed to the point of crisis. It should surprise no one that at this moment — of such extreme and extended extraction, with social conditions radically diverging in so many geographies, and with climate change not simply more *visible* but more *palpable* — a multidimensional legitimation crisis ripples across the world. Ecological-economic crises are *already* social and political ones. In this scenario, strategies of crisis management — increasingly desperate attempts to square the goal of fundamental system preservation with material conditions produced by that very system which undermine such attempts — have little room for maneuver. In the memorable phrase of German sociologist Claus Offe, it is a "crisis of crisis management."

Far from reaching a new equilibrium, such a system,

similarly, has little room for maneuver, although it's not at threat of automatic, mechanical collapse — a perennial, millenarian fantasy of so many across a broad political spectrum. Even with anemic growth rates, every little bit of real capital accumulation requires yet more inputs, more extreme extraction, increased dispossession, and new "sacrifice zones" — completely given over to exhaustion and debilitation. The COVID-19 pandemic, itself a socioecological effect of accelerating capitalist enclosure and zoonotic spillover, proves just how well shocks can be absorbed into the system of macrofinancial capital flows and global value chains of the extractive circuit.[26] Nearly one third of the total wealth that accrued to the United States' 719 billionaires since the 1990s was accumulated during 2020, even as wealth stagnated or declined for vast majorities. Similar patterns can be seen globally. This is the system *working*; profits recover and the brief dip in emissions is reversed, higher in December of 2020 by 2% over the previous year. The shock intensified already existing systemic tendencies while revealing the slim margin of error needed for such a perfectly optimized "real economy." Some recent shortages demonstrate just how taut the logistics of the extractive circuit are. Other shortages, like those of high-grade silicon (i.e., sand, as observed by political theorist Laleh Khalili and internal industry reports alike[27]) for microchips and especially construction, are examples of genuine resource exhaustion.

It is hardly the geographic location of rare earth metals alone, for example, that dictates the location of zones for their extraction. These zones are produced rather through a logic that delicately weaves the power and needs of transnational firms, states, and other strategic actors. Some states may *want* extractive frontiers within their boundaries for a measure of geostrategic leverage. Some local actors (a diverse array of workers, surrounding communities, and social movements) are pitted against others.[28] Extractivism is one of the only paths available to material development at many nodes along the circuit as it is, towards some hope of relief. But it also promises destruction and exhaustion in its wake. While many fully aware of this reality in the Global South are rendered dependent on resource exports, in Pennsylvania, families similarly enroll in the latest fracking initiative or otherwise sign away mineral rights as one of the last remunerative games in town. Capital profits off the mine drainage, the freshwater depletion, the emissions, the social strain, and desperation alike.[29]

## Feeding Frenzy

The frictionless flow of "cheap money" and "cheap energy" — to borrow from Jason Moore and Raj Patel's lexicon — has facilitated the "unbundling" characteristic of twenty-first-century global value chains. These chains are "networks" of small- and medium-sized enterprises, subcontractors, and suppliers under the coordination of a central TNC or "lead firm." Financialized supply chains are structured to allow firms to ignore or skirt local, national, international legal or even physical attempts to restrict the flow of extraction.[30] They are vital for ensuring profitability through maximizing velocity and social and ecological inputs. They facilitate the shift of risk to the actor — whether at-will contractor, off-the-books migrant employee, or indigenous community in a resource-rich area — structurally least able to absorb it. This extends not only into society, but into individuals, their internal biochemical orders, and to the whole ecological niche.

Each of these factors contributes to a logic of absolute profit maximization and, as already described, a *necessarily* ever-increasing *speed-up* — in extraction, production, labor, dispossession, enclosure, and overall niche exhaustion. This is the world of lean-production, just-in-time manufacturing *and* delivery.[31] While most would, with good reason, focus on the human and ecological costs created by these processes — one key factor here is rather what they make possible.[32] The one-day or one-hour delivery, the expedited shipping, the synthesis of business and "leisure" hours: all of this is a lifesaver to the single parent, the double-shift employee in a food-desert, the downwardly mobile 12-hours-a-day professional, the hustling informal or aspirational employee, hoping to claw their way out of generalized precarity. All of these, in the understanding of contemporary law and neoclassical economics, are "services" provided for consumers.

We should see these "services" instead as facilitating the frenzy of these lives, as shifting literal time and energy *not* to these individual consumers, but rather to the needs of an "always-on" capitalism, creating the very crises to which these services respond. They don't strictly fulfill consumption ends; they are also part of *production*. Every moment of life is integrated, profitable, from literal labor hours to the production of micro-units of digital value (via social media and other avenues) in the hours-for-what-we-will.

Such services are dependent on the speed and ungovernability of GVCs which can, to a degree not possible before, dice up a production process into its most minute parts and spread them as far a global distance as comparative advantage dictates, limited, in strictly economic terms, by the current state of communications technology and the price of oil.[33] GVCs function best — that is to say *fastest* and *most profitably* — when the network of small- and medium-sized enterprises as well as more informal "arms-length" arrangements are "governed" by a TNC.[34] Think again of cobalt mining in the DRC. Such arrangements and "governance" allow a TNC like Glencore to be both the largest cobalt mining corporation in the world and to avoid not only Congolese legal accountability but American and British as well.[35] Glencore dominates cobalt extraction through organizing a network of subcontractors, subsidiaries, and informal arrangements. The opacity and complexity involved allows risks — political, legal, environmental — to be largely circumvented. Glencore exerts a form of localized sovereignty over its mining concerns, taking advantage of questionably sourced minerals and maintaining the very labor and environmental practices that it forswears in meaningless corporate "environmental, social, and governance" rhetoric. Its form effectively moves responsibility from the TNC to "the miners themselves" for their own enslavement and abuses. This "governance" is backed by the direct coercion of subcontracted semi-private "state" mine police and private military corporations like G4S.

The British-based G4S is not only contracted by Glencore or the DRC (technically through a local subsidiary); it operates in 90 countries, including running mega-prisons and migrant camps in the UK, despite numerous accusations of serious abuses in the UK and elsewhere. In the US, G4S has been subcontracted by private firms as well as the military, Customs and Border Patrol, the departments of State, Justice, Energy, Homeland Security, and the DEA, in addition to subcontracted work for prisons and police at state and local levels. Private clients include GlaxoSmithKline (initially in the United States, Argentina, Costa Rica, Venezuela, and the UK, expanding to 28 other countries GSK operates in) and Citigroup. G4S was recently acquired by the security conglomerate Allied Universal which boasts the largest security force (150,000 members) in North America. It is the third largest employer in North America and the seventh globally. Similar mechanisms of speed, ungovernability, risk-shifting, and even violence are found at the site of end-use consumption, even in the arguably most powerful state on the planet, as at sites of basic resource extraction in some of the poorest geographies.[36] Whether a logger in Indonesia or a delivery-person in the United States, TNCs rely on similar methods if not to the same degree. At both nodes in the value chain, human and ecological destruction are rampant. It is not physically possible to achieve the just-in-time production and delivery-on-demand described above without burning through fossil fuels and human bodies with merciless efficiency. The *proponents* of global value chains often cite this ungovernability — if, of course, not these expressions of it — not only as fact but as *added value*.[37]

And such massive TNCs — sometimes "headquartered" in and yet untethered to a national economy — also weaken the political power of the end-user, eroding what remains of the "safety net" and social fabric. The form and function of GVCs is fundamentally at odds with basic principles of self-determination and popular sovereignty. The simplest expression of this is the changing relationships in which TNCs shape institutions, including states: "my factories for your reform." While the United States retains its unique position as the world's quasi-central bank and imperial guarantor of global capital "centered on (not in) Washington," this new mode of capital organization facilitates its own versions of "special economic zones."[38] Samsung, for example, can extract concessions in the form of proposed local, state, and federal tax breaks and low-cost labor incentives, on top of existing environmental and social deregulation, in establishing a new semiconductor plant in Texas. Such "reshoring" is boosted by the catastrophic turn to jingoistic great power conflict with China, but it is also not possible without the new socioecological facts-on-the-ground, both generated in the effloresce of the extractive circuit.[39] Many (but not all) in the United States are also increasingly untethered from the national economy, and firms too, frequently subject to the same logic of extraction and exhaustion in socioecological terms.

## Millenarian Burnout

Feelings of exhaustion, taken together — fatigue, burnout, stress, depression, pain — are globally prevalent. Reported levels of extreme stress are only marginally higher in the Philippines (58%) than the United States (55%).[40] Cumulatively, these numbers tick up every year. At every node of the extractive circuit, we find speed, coercion, and the inevitable stressors on the individual. One of the best analytic lenses we have for this intersection of affect, environment, psychosocial pathology, and neurology is the psychology and political theory of Frantz Fanon. The extractive circuit is a "divided society [...] characterized by a predominant nervous tension leading quite quickly to exhaustion."[41] For Fanon, mental "illness situates the patient in a world in which his or her freedom, will, and desires are constantly broken by obsessions, inhibitions, countermands, anxieties." Later anti-psychiatry would posit mental illness itself as freeing, while for Fanon the pathologies were all too real and debilitating. His practice took seriously the need to use psychiatric tools to reconstitute individuals, in dialectical tension with a colonial society that is clearly itself "neurotic," managing pathologies through psychological intervention but towards a clear understanding that the cure lies in radical change in social structure. Although many psychological theories and studies have long explored not only social environments but particularly the effects of violence and trauma, most unidirectionally focus on the individual as the object of study and site of intervention.

The late cultural theorist Mark Fisher called this the "privatization of stress." In Britain, by the mid-2000s, depression had become the most treated disease by the National Health Service. "I want to argue that it is necessary to reframe the growing problem of stress (and distress) in capitalist societies," wrote Fisher. Neoliberalism had compounded such individual focus into an "incumbent" commandment that "individuals resolve their own psychological distress." Instead of "accepting the vast *privatization of stress*," we should ask "how has it become acceptable that so many people [...] are ill?" As Fanon argued of neurotic colonized conditions, Fisher posited that "the 'mental health plague' in capitalist societies would suggest [...] capitalism is inherently dysfunctional, and the cost of it appearing to work is very high."[42]

The "privatization of stress" is a particularly apt phrase: just as one can mine fossil capital to boost petro-farming outputs, one can squeeze the standard "labor power" of a hyper-employed worker while also exhausting her "mind." These are some of the latest frontiers in the long history of transforming ecological inputs (for what is labor but an extension of nature working on itself, as Marx says) into abstract value. The "mental health plague" is the expression of this condition — perhaps, again, in normative economic terms, an externality — but the "privatization of stress," of *stressors*, is a method by which an individual's exhaustion can lower the cost of capitalism's apparent functioning. If a worker takes it upon herself to "resolve" her own distress, or if a member of the vast global surplus populations accepts the "cruel optimism" that they are the source and aspirational solution to their extreme stress, these are real costs saved systemically as individuals squeeze just a little more out of themselves, all while reinforcing the ideologically valuable conviction that the existing system *works*; all evidence to the contrary is on you.

At quick glance, the "mental health plague" might seem a classic case of "First World" problems, but reports from the World Health Organization (WHO) indicate that this kind of depression — and the panoply of affects collected under the rubric of exhaustion — is a global concern. Relatively well-known social phenomena like farmer suicides in South Asia or an individual story like that of the self-immolation of Tunisian street vendor Mohamed Bouazizi, widely cited as the beginning of the Arab Spring, are just aspects of the mental health plague most visible in the media. In recent years, suicide has become the fourth leading cause of death among youth worldwide. And 77% of suicides occur in low- and middle-income countries. Meanwhile, there are the global psychological impacts of climate change itself: the direct experience of extreme weather events, for example, is compared in many studies with PTSD or grief more commonly associated with modern warfare.[43]

This is not to say that pressure is applied equally everywhere. Nor do the mental and physical health crises experienced by Filipino migrants or Californian office workers share the exact same etiology or epidemiology. It is obvious that the stressors experienced by migrant laborers, the permanently unemployed, or even a relatively stable healthcare or logistics worker have different particular causes and attendant problems. But they *are* experiencing what we can call "exhaustion," not only medically but as a relationship to climate, society, and its hegemonic ideologies. There are relays on the circuit — particularly those between ecology and economy — where this exhaustion is very nearly calculable, in the theories of energetics which once so entranced "scientific management."[44] But there are others that are found in individuals, communities, societies, and political systems, attuned and modified to the needs of the circuit in far more qualitative terms.

The "Golden Age" (not really so golden) of win-win shared growth prosperity in the overdeveloped North has long since passed. Almost every drop of fossil capital and what the late political economist Samir Amin called "imperialism rent" is now needed to shore up flagging growth rates. The G7's share of world GDP peaked at 67% in 1988 and, although this is still quite disproportionate to population size, as of 2010 it was less than 50%. Interestingly, this correlates with economist Branko Milanović's calculation that the "citizenship premium" seems to have peaked about 13 years ago. In one of his last articles, Amin comes to a remarkably similar conclusion. Over the past decade or so, "imperialism rent" increasingly flows to oligopolistic monopolies, not in a real sense to "national" economies, nor evenly: "the privileged (high salaries) have become the direct agents of the dominant oligopolistic class, while the others are pauperized." This still spills over to many far outside the upper echelons, including to many who remain tied to the fortunes of firm and nation. And can be seen not only in income levels but in phenomena like vaccine apartheid, in the net energy and material flows that *presently* (not just historically) still pour into the US, EU, and Japan, and the number of coordinating TNCs within these geographies.[45] It is not yet the case that these more direct forms of governance and rule happen, as Rosa Luxemburg wrote of colonial possessions in the early twentieth century, "without any attempt to disguise them."[46] But the "sermonizers, counselors, and 'confusion-mongers'" Fanon identified as crucially intervening in colonizing societies are now working overtime in the metropole.

Where the growth of the firm could be said to be in some

sense in the *interest* of many workers in the "Golden Age" of the Global North, today the growth — or more so the profitability — of the firm is an integral part of the same circuit which also internalizes the further immiseration, exhaustion, alienation, or disenfranchisement of those same workers and communities. The globalized economy in the twenty-first century is so different from that of the 1950s or 1850s as to make national accounting itself difficult.[47] But where the growth of the national economy could be said to be, in some sense, a real, *national* interest in the 1950s and 1960s in places like North America or Western Europe — in the way that it still actually *is* for the majority in the Global South today — most of what can be recorded as growth in the national account today is the disaggregated profits of "headquarters" corporations along a global value chain. It is just a hazy indicator of the gargantuan energy and material inputs and outputs that create the already existing catastrophic ecological conditions that impact precisely those "pauperized" majorities more so (and in some cases *only* so) than their national "compatriots." In other words, for the first time in modern history, there are majorities on both sides of Du Bois's "color line — the relation of the darker to the lighter races of men in Asia and Africa, in America and the islands of the sea,"[48] across the whole of empire, whose most mundane, material interests *align*. There are tremendous ideological barriers to surmount and organizational challenges ahead, but one of the most profound political consequences of this moment in the Anthropocene is that, to be even more precise: for the first time in modern history, the vast majority of workers and communities in the Global North have a mundane, material interest in the wealth and power of people and states in the Global South. Continued colonization compounds their exhaustion; *decolonization* promises the possibility of both their increased power *and* their material benefit.

The rolling legitimation crises of our time are the evidence of just how much ideology in the form of consent is breaking down and just how much direct coercion is ever more the norm, even for groups who were not *previously* subject to colonization. In the face of this coercion, massive increases in repressive apparatuses, personal and social "structural adjustment," and the palpable experience of anthropogenic climate change, the interests of *some* workers and majorities in general (if far from universal), begin to align with the refugee, the South Asian or Latin American peasant, as well as with those *already* racialized and/or colonized in the North. The concrete realities of the extractive circuit place such individuals and communities far closer to the super-majorities of the Global South than "the 1%" or even "10%." In other words, unevenly and to *vastly* different degrees, more and more are in the colony now.

## Colony of the Exhausted

The point here is not simply to invert a Panglossian, Pinkeresque worldview, but rather to see dialectically, as Walter Benjamin suggested. Climate change itself is the most obvious contemporary avatar of Benjamin's "storm of progress," an almost unimaginable techno-social achievement containing within it an almost unimaginable horror. Weathering the storm requires overturning ideals of both progress and regress.[49] The example of Cuban agroecology is illustrative. Beginning in the 1990s, scientists, peasants, and other agricultural personnel launched a series of agroecological experiments drawing on traditional practices and contemporary scientific research in agronomy and ecology. Over three years such projects demonstrated not only enhancements in soil quality and biodiversity, but also increased yields, decreased energy inputs, and decreased labor hours (down to four to five hours per day). The new methodologies would eventually scale up to cover 70% of Cuban domestic food production. Crucially, this case helps dispel the myth that any deviation from the official path-of-progress will be met with a regress into pre-industrial drudgery (or a romanticization of an imaginary utopian past). In contrast, it underscores what I have called elsewhere the "temporal luxury" that contributes to a socioecologically necessary systemic slow down and provides a palpable, extraordinary emancipation possible within a sustainable, flourishing global human ecological niche. As labor journalist Sarah Jaffe summarized, there are profound ecological, social, and economic advantages of even just "less work."[50]

With capitalism increasingly experienced as inherently dysfunctional, the question about living on a burning planet is not how individuals or communities can become more "resilient" to climate change. Rather, the question is, how has this level of degradation become so acceptable? Profits may be at an all-time high, but capitalism-as-we-know-it — capitalism as the organizing principle of our ecological niche — is *difficult* to maintain. As the sociologist Oliver Nachtwey has recently noted, neither financialization by itself nor neoliberalism as a whole was able "to stop the wellsprings of growth from drying up." Measured by emissions and the toll of climate change, every ounce of anemic growth becomes catastrophically expensive. The energy intensity of today's waning growth rate is four times that of the "hyperbolic" growth of the mid-century. The famous "Keeling Curve" — the measure of atmospheric carbon concentration from Hawaii's Mauna Loa Observatory — is getting steeper. In the 1960s and early 70s, with extraordinary global economic growth rates, annual increase was approximately 1 ppm, but as we approach the last 20 years, with lower economic growth rates, it has increased to over 2 ppm per annum and is still increasing. A laser focus on growth can miss how, viewed from both mainstream and heterodox accounts, questions of speed, delay, diverging interests, and, above all, all those socioecological inputs, all those lost efficiency gains, principally are constitutive of and structurally driven by profits. As, again, both mainstream and heterodox analysts find, at least as early as the late 70s to today, pricing in the "externalities" could easily result in wiping out much profitability. Cut costs, exploit socially or ecologically, increase market share, or die.[51]

What is often lost in ecological debates about growth is that growth itself, whether one wants it or not, appears to be largely at an end. While Global South states will continue to grow (even as rates in states like the People's Republic of China or India are in fact *falling*), this is highly unlikely to make up for the collapse in the Global North. In every sense *except automatic collapse*, capitalism-as-we-know-it lives on borrowed time. More centrally to the question of a climate politics is how much time and where it's borrowed from. From a socioecological point of view, "we" are investing ever more resources — ecological, social, individual, political — for paltry returns, a grim present, and a darkening future for the majority of human beings.[52]

Strangely, this turn of affairs is better acknowledged in global business discourse — as the "productivity crisis" — than it is in many common left theories, ecologically-oriented or otherwise. Global North growth rates, as argued by both liberal economists like Robert Gordon and Larry Summers through the "secular stagnation" thesis, or more contemporary Marxian accounts like Aaron Benanav's work on industrial overcapacity and the long economic downturn, will never return to the sustained 4% or greater rates regularly experienced during the "Thirty Glorious Years." It is far more likely that growth will hover at around 1 to 2% or otherwise enter a period of stasis, permanent recession, or some mix of the two.

TNCs — while having a longer history — only became globally dominant in the postwar era, with increased activity mostly coming after the 1960s. Correlation is of course not causation, but the correlation between climate change and the rise of these TNCs — first in industries like fossil fuels and resource extraction, later with the expansion of GVCs particularly in the late 1970s and onward — is staggering. "Acceleration" is aptly put; this is a system predicated not only on increasing material and economic rates as recorded in "Great Acceleration" graphs, but quite literally in its *speed*: the exponential explosion of ecological and economic transformations, facilitated by financialization, fossil fuels, and every socioecological process described, and additionally the experiences, the feelings, of such velocity and violence. In this period, as macrofinancial economist Hyun-Song Shin identifies (and economic historian Adam Tooze cites in analyzing the 2008 financial crisis), "island models" of national economies give way, in terms of both value chains and finance, to "an 'interlocking matrix' of corporate balance sheets," in a *supra-, semi-sovereign* system.[53]

There has never been a time when capital could move more rapidly or more freely — from firm to firm, from geography to geography. It could even, theoretically, move out of bounds entirely — capital could cash out (it could be *worse*, as McKenzie Wark remarks). Today, some 80% of global trade happens through far-flung, high-speed GVCs. Of that, 60% is in intermediate goods and services. In other words, most trade is happening firm-to-firm or within a cross-border GVC. The carbon and materially intense structure of trade is inextricably linked to fossil fuels, "sacrifice zones," and generalized exhaustion.

The "labor squeeze," the "unbundling" of the production process, the expansion of profitable economic activity into every waking hour (and perhaps beyond), the *increasing* rate of atmospheric carbon concentration and every other ecological measure already discussed, the already catastrophic conditions of climate change, alongside increasing mass displacement and enclosure: these are the costs of business-as-usual or even simply slow, less intense mitigation and adaptation. What we have experienced over the past 40-odd years is a simultaneous massive redistribution of wealth and income to the top income decile and a simultaneous and interdependent increase in extraction and resource burn to prop up massive profits and an approximately 1% growth rate across the Global North. Without this growth, however paltry, it would be difficult for capital, in its developed twenty-first-century form of the extractive circuit, to continue, to fundamentally preserve the system. Capital will chew through the biosphere and societies alike in pursuit of an ever more costly maintenance of profitability.

Although capitalism, particularly in this form, promises a

cheap, sleek, efficient path to plenty and prosperity, it delivers instead a costly, privatized system based on impossible inputs in a finite natural and social world — so unimaginably cumbersome and irrational that it requires constant, vigilant, crisis-level maintenance from the scale of the microbial to the human brain to states, geopolitics, and beyond. It delivers all-time high atmospheric carbon concentrations of 420 ppm, the hottest years on record, a climate change-induced pandemic which promises both the continuation and furthering of such epidemiological crises, extreme weather events from wildfires burning Siberian permafrost to Indian flooding and Pacific Northwest extreme heat, and all those political economic, social, and personal structural adjustments that are the daily violence of the extractive circuit. This socioecological violence is a constant norm of this moment for the vast majority of people on Earth, "an atmosphere of permanent insecurity," as Fanon called it. "There is, first of all, the fact that the colonized person, who in this respect is like the men in underdeveloped countries or the disinherited in all parts of the world, perceives life not as a flowering or a development of essential productiveness but as a permanent struggle against omnipresent death."[54] While normative disaster-management literatures preach absorption, quiescence, internalization, Fanon instead proposes externalization, which "implies restructuring the world."[55] What we are promised — through market-based delusions, through techno-mystical fantasies, and romantic reveries — is a return to normal that is hardly possible and not particularly desired or desirable. The extractive circuit describes the zero-sum game of a bifurcated world.

This can be a bitter pill to swallow, in mainstream and many radical politics alike. Many look to growth as the easiest path to achieving substantive social change through "win-win" arrangements. But not only is that essentially no longer possible for a left-wing climate realism, the reality is not quite so straightforward. As one can see from even a cursory glance at the empirical record, it is only within very specific conditions that growth works this way. India, for example, experienced stratospheric growth rates in the 1990s, but its ranking on the United Nation's Human Development Index actually *fell*. In contrast, China experienced similar growth rates, but its rank rose prodigiously. Even though the United States' growth rate returned modestly in the late 1990s, the US fell in both rank and measurement in the same time period. As a host of thinkers (most famously Amartya Sen and Jean Drèze) have demonstrated, growth neither necessarily leads to even a fairly restricted range of qualitatively desirable social outcomes nor do all (including many of the most fundamental) such outcomes depend on a high level of resource use and economic growth.[56]

A case in point is the Indian state of Kerala from about 1960–2000. During this period, Kerala, a populous, diverse state in the south of India, managed to achieve development outcomes comparable to relatively advanced capitalist states with the per capita GDP of one the world's poorest. For example, in 1995 there were 31 million Keralans living on a median income of $292, somewhat less than, for example, Sudan or Cambodia. In contrast, US per capita GDP was $29,980. However, social outcomes were shockingly high in comparison: literacy stood at 91% (vs. 54% for Low-Income Countries [LICs] and 99% for the United States); life expectancy stood at 71 years (vs. 56 for LICs and 77 for the US); infant mortality at 13 per 1,000 births (vs. 89 in LICs and eight in the US); and so on. "Thus," as political economist Patrick Heller writes, "despite per capita income levels that are well below the average in India and other low-income countries, Kerala has achieved social and human development levels that approximate those of the first world."[57] Kerala was contemporary Portugal on the budget of literally 1/100th that of the United States.

This was not, as we will return to, some image of a grey socialist egalitarianism but a necessarily vibrant cultural and political life. Kerala achieved this level of social prosperity *not* through economic growth, but rather through "effective public action."[58] Although Kerala is *technically* a market economy, it has to this day "Five Year Plans" and many other classic features of Communist planned economies, including state or other public ownership of over 50% of the sector.[59] Crucially, insofar as Kerala has financial control within the broader federal state of India and further still within the legal structures of the global market, democratically accountable state institutions also dominate the financial system in addition to a wide network of cooperative financial institutions.

Through these features, alongside fairly unique horizontal participatory structures, there was massive state intervention in labor markets — including high minimum wages and protections for organized labor. Alongside broader social transformations in education and women's equality, many key sectors are dominated by the state: electricity, housing, water, transport, etc. The same lack of growth that was once deemed a fundamental flaw of the "Kerala model" of development can begin to look, particularly when we're thinking *outside* the Global South, like a feature. At the scale of the city and social movements, there are influences of the Keralan model in places like Jackson, Mississippi, led by the ecosocialist and Black liberation movement Cooperation Jackson.[60] The Keralan model — alongside those many other ideas and experiments — can be scaled as part of very different systems in the *Global North*.

And yet Kerala cannot be abstracted from the extractive circuit, from the global economy, or even its extreme precariousness within a now neofascist India. Kerala still has many of the salutatory social outcomes described above (the best in all of India even with India's, and now Kerala's, prodigious growth), but it is increasingly wracked by economic, political, and ecological forces. Kerala's pursuit of heavily managed capitalism in one federated state left it highly vulnerable to recent extreme weather events like floods and mudslides, both through the limits of its management and through the impossibility of escaping anthropogenic climate change. Even recent successes such as its management of the pandemic have been short-lived. Kerala was praised internationally for having one of the best COVID-19 strategies in the world through its remarkable healthcare system and networks.[61] However, after holding out for several months with one of the lowest infection rates found anywhere, Kerala was hit with a massive second wave as citizens who work abroad and in other Indian states returned. Following Indian liberalization in the 1990s, Kerala was well-integrated into the world market — its domestic agriculture increasingly geared towards export crops; its highly educated but underemployed population increasingly seeking foreign work, particularly as professionals in the Gulf states; remittances becoming a major and increasing portion of the total federated state GDP.[62] These factors made the infection surges inevitable (as they did in almost every isolated success story; Kerala's health system still managed to keep death rates lower than average). And, just like the Philippines, with each turn of the socioecological screw, the extractive circuit churns, producing "cheap labor," pushing further into ecological devastation for profitable advantage, turning social crisis into opportunity.

Accounts can often make it seem as if capital hovers about the Earth in almost ethereal form. But the extractive circuit is not a metaphor. It works through real people, specific geographies, economically strategic areas organizing, linking, and connecting our global human ecological niche. The granular level I began with in my paradigmatic example is the very real, material workings of this system. Imperialism does not vanish in such a system. States (in a wide variety of forms) still play a crucial if dramatically changed role. They are *instrumentalized*, rendered more brittle, more permeable through their increasingly coercive nature, even in the core. Colonization is more generalized. But whether one is committed to somehow keeping the system in motion in perpetuity or to holding on until a "cash-out," the extractive circuit, and its maintenance, one piece of it or another, currently has your fealty. It is the exhausted world — the climate — capitalism built.

# Chapter 3
# Climate Lysenkoism;
# Or, How I Learned to
# Stopped Worrying and
# Rescue Class Analysis

*If you've done six impossible things this morning, why not round it off with breakfast at Milliways, the Restaurant at the End of the Universe!*
— Douglas Adams, *The Restaurant at the End of the Universe*

*Accumulate, accumulate! That is Moses and the prophets! [...] Accumulation for accumulation's sake, production for production's sake: by this formula classical economy expressed the historical mission of the bourgeoisie and did not for a single instant deceive itself over the birth-throes of wealth.*
— Karl Marx, *Capital Volume 1*

## Climate Lysenkoism?

On July 31, 1948, Trofim Lysenko addressed a pivotal conference of the Lenin All-Union Academy of Agricultural Sciences (VASKhNIL). Over the course of the 1930s and 40s, VASKhNIL had served as the central stage on which many key debates in the natural sciences took place, none probably more well-known than that between Soviet geneticists and the "Michurinist" (named after the Russian then-Soviet horticulturalist Ivan Vladimirovich Michurin) school headed by Lysenko.[1] Lysenko is perhaps best known as the infamous Soviet agronomist and biologist under Stalin who spearheaded a rejection of Mendelian genetics. His speech, "The Situation of Biological Sciences," took up much of the first day of the conference and set out an agenda denouncing Mendelian genetics as "bourgeois" science in favor of a kind of synthesis of Lamarckian theories (alongside ideas from Ernst Haeckel and others) subjected to the increasingly rigid, dogmatic Soviet interpretation of Marxism.

In many fields, science in the Soviet Union — particularly in early ecology and theories of climate change, which we will return to — was ahead of its time. And Soviet geneticists had in fact not only been researching but applying genetic research, particularly in agriculture. However, as Lysenko set out at the start of the conference, theories reliant on mutation (for example) were dismissed on grounds of philosophical idealism, while acquired heredity and other aspects of Michurinism were in keeping with the "great" teachings of Stalin and the Stalinist interpretation of Marx. Indeed, the conference and the elevation of Lysenko were not only given Stalin's blessing but proceeded at Stalin's charge and behest. Scientific opponents of Lysenko were committed Marxists and viewed Lysenko's attack on them and the broader field of genetics as of significant detriment to the Soviet project, to Marxism as a whole, and to scientific understanding through a reality-defying methodological misunderstanding and misapplication of historical materialism.[2]

Although there is more to the story, put simply, Lysenkoism has come to be shorthand for the one-sided subordination of adjudicating empirical phenomena (and their theoretical explication) to political and doctrinal expedience.[3] It is unsurprising that liberal system preservationists will opt for apolitical technocratic hopes tied to technological fantasies. But while the right increasingly turns to right-wing climate realism, a new form of Lysenkoism is becoming omnipresent. In the Global North broadly, and the resurgent Anglophone left in particular, the relatively recent (and welcome) revival of Marxism and socialism has been accompanied all too frequently by what I call "Climate Lysenkoism." Climate Lysenkoism covers a broad range of self-ascribed "left" and "Marxist" perspectives that subordinate both natural scientific and historical realities to a quasi-mystical technophilia and an ahistorical romance of the mid-twentieth-century Northern nationalist welfare state.

The charge of Lysenkoism as an epithet can hardly be restricted to the Cold War imperial view that developed, "advanced" capitalist states pursued a "value-neutral" brand of scientific inquiry, while it was only Marxist and other anti-colonial, anti-imperial movements and states that "politicized science." Lysenkoism reaches far beyond the historical case. As the biologists Richard Levins and Richard Lewontin put it:

> Dialectical materialism is not, and has never been, a programmatic method for solving particular physical problems. Rather, dialectical analysis provides an overview and a set of warning signs against particular forms of dogmatism and narrowness of thought. It tells us, 'Remember that history may leave an important trace. Remember that being and becoming are dual aspects of nature. Remember that conditions change and that the conditions necessary to the initiation of some process may be destroyed by the process itself. Remember to pay attention to real objects in space and time and not lose them utterly in idealized abstractions. Remember that qualitative effects of context and interaction may be lost when phenomena are isolated.' And above all else, 'Remember that all the other caveats are only reminders and warning signs whose application to different circumstances of the real world is contingent.' To attempt to do more, to try to distinguish competing theories of physical events or to discredit a physical theory by contradiction is a hopeless task.

Not only do Climate Lysenkoists plunge right into these errors, they mutilate methodology and theory at the same time.

Climate Lysenkoism is a subvariant of so-called "ecomodernism" (which, as I return to in Chapter 5, is neither really ecological nor modernist). This is an understandably status quo and often right-wing approach to climate change that emphasizes markets, high tech (read: technology only as understood from the POV of Northern capital), endless economic growth as defined roughly within the parameters of GDP, and a cosmo-utopian vision for the *preservation*, not of a sustainable human ecological niche but of, as we've seen with business-as-usual and right-wing climate realism more broadly, as much of the current concentrations of wealth and power as possible. Climate Lysenkoists depart from their capitalist partners (sometimes quite literally partners, as with the industry thinktank The Breakthrough Institute) *only* in arguing that the triumphant march of Promethean progress is "fettered" by market mechanisms.

The more "scientific" strand of this bizarre constellation is animated by technological wonders and wraps denunciations of climate science, environmentalists, and other leftists — from Marxists to anti-colonialists, eco-feminists, and anti-racists — in the bright shiny packaging of a vision of the American "way of life" for all. Is economic growth a problem? No, (biophysically improbable) green growth is not only the solution, it's already here! Difficulties in energy planning vexing you? No problem! Existing and next-generation nuclear energy that is *always just about to* break through will untie all our knots. (It won't.) The Green Revolution in agriculture no longer feeding the world and clearly contributing to everything from greenhouse gas emissions to the nightmarish matrix of zoonotic disease like the COVID-19 pandemic — and still binding ever more desperate farmers from Uttar Pradesh to the United Kingdom into impossible debt? Rubbish! Agriculture has never been more sustainable, and any minor problems observed merely require the intensification of industrial food production. Are the material costs of transportation in the 24/7, always-on, just-in-time world of contemporary capitalism, of the extractive circuit — from emissions to land-use to biodiversity loss — too high to maintain? No radical transformation needed! Carbon neutral fuels are here to save the day. (They're not.) Planetary boundaries already exceeded in possibly six of nine dimensions? Those are just rough heuristics invented by anxious climate scientists and pushed by perfidious activists. In fact — against every scrap of evidence — there are no ecological and biophysical limits at all. (There are.)

As Douglas Adams once joked, "If you've done six impossible things before breakfast, why not round it off with breakfast at Milliways, the Restaurant at the End of the Universe?" Adams' comedy is about the absurd in the social and the physical: impossible time-travel, impossible observation of universal destruction while enjoying fine dining, impossible always-already booking, impossible meeting of people across all time and space, the impossibility of meeting yourself (due to embarrassment), and the impossibility of paying an impossibly gargantuan bill. Climate Lysenkoism — which consistently puts scientists and activists in its crosshairs and, in the name of fighting capitalism, always manages to argue for it — is only funny in the most tragic way as the "left flank" of right-wing climate realism. First time tragedy, second time farce.

The more political strand of Climate Lysenkoism conjures a "Marxism" that is a stripped-down version of the evolutionary, deterministic, and "progressive" distortion of Marx's undogmatic historical materialism even more egregious than early twentieth-century European social democratic politics.[4] In order to make such a political model look even remotely plausible today, to *deny* the concrete realities at hand, advocates turn to both strands — denial of scientific and historical realities for theoretical convenience and imagined political expedience. This is not the climate skepticism of yesterday. The real denial today is these technological and political *fantasies*.

Climate Lysenkoists ignore and sideline the global nature

of production and distribution, of movements and struggles. Imperialism, in its economic and material dimensions, is dismissed. There can be no question that the attack on welfare states in the Global North (and imposed from the North onto the South) constitutes a ruling class project. However, reverting to a romantic idealism, the Climate Lysenkoists assume that we can simply "go back" to the supposed ideal of the mid-century — an insidious nostalgia common in liberal and right-wing nationalist thought. They fail to ask whether such arrangements are applicable to climate change, or even *desirable*. They willfully turn a blind-eye to the real political opportunities generated in this socioecological moment — what I have described already as the greatest material foundation for international solidarity in modern history.

While elements of Climate Lysenkoism can be found in many parts of the Northern left, particularly paradigmatic cases are the works of Canadian technology enthusiast Leigh Phillips and American geography professor Matthew T. Huber.[5] Although the threads can be difficult to disentangle, Phillips is more representative of the pseudo-scientific strand, while Huber, the political. Phillips's articles and book *Austerity Ecology* are a "left"-tinged unholy marriage of Malcom Gladwell-esque counter-intuitive (and power-comforting) *bon mots* and Steven Pinker's laughable accounts of historical violence and progress (which stand up to neither mathematical nor historical scrutiny). Meanwhile, Huber's *Climate Change as Class War* — in its magical thinking, lazy research, historical and scientific illiteracy, and even policy imperatives — is barely distinguishable from the recent mountain of "crisis of liberalism" texts like Yascha Mounk's *The People vs. Democracy*. Phillips is the contrarian bomb-thrower, ready to make the eco-socialist case for private jets, yachts, cruise ships, and infinity pools. While Huber, befitting his professional status, couches his arguments in the comfortable collegiality of the academy. But he is no less enthusiastic in his historical fabrications, projected political expediency, pathological projection, and outright bizarre misrepresentations. Just-so stories about technology are sutured onto the flimsiest histories (when history is discussed at all). Marxist concepts are deployed as talismans loosened entirely from meaning and methodological coherence; a radical politics is promised but all that's delivered is a transubstantiation of center-right programs.

Talismans, transubstantiation, unholy marriages, *faith*. We are forced, to paraphrase Marx, to "take flight into the misty realm of religion"[6] in order to make sense of this phenomenon. While the Climate Lysenkoists masquerade as a swaggering scientific, Marxist, and technological splash of cold water, what we really have is the dismissal of even the most critically engaged science, a "Marxism" abstracted into a simple nationalist welfarism, and a definition of technology only the most gung-ho venture capitalists could approve of. This is not techno-utopianism; it is techno-mysticism.

The thread that ties all the hard right turns in nominally "left" thought — from this climate techno-mysticism to anti-trans "gender realism" to increasing support for police and prisons[7] — is the attempt to restart the exact (failed) politics of European welfare states in social, economic, and ecological conditions which render them impossible, all while ignoring the new, far more radical, and genuinely international politics and possibilities opened up in this conjuncture; possibilities that are more desirable and have a genuine chance to go so much further than a romantic "return" to the so-called "Thirty Glorious Years" (which were not so glorious for most). It is no accident that Phillips mocks anti-police activism, celebrates anti-Islamic sentiment, couches traditionalist natalism in languages of left emancipation, or gives cover to the scientifically baseless and xenophobic COVID "lab leak theory."[8] Nor is it surprising that both are enthusiastic promoters of the explicitly red-brown (Strasserite) *Compact* magazine project.

In trying to accomplish so many impossibilities all at once, Climate Lysenkoists desperately turn to Norgaard's "implicatory denial"; they deny the extent of radical social transformation an emancipatory climate politics and program requires.[9] Unwilling to countenance the degree to which the unprecedented nature of material conditions in this moment are fundamentally incompatible with their theoretical models, they end up embracing ever more conservative fantasies that seem to comport with their historical ideals (whose own shortcomings are obscured). They *repress* the material realities that demand a socioecological project largely unexpected and unaccounted for in previous theory, with unanticipated radical directions, needs, temporal intensities, and desires. As in psychology, these realities return as symptoms — Norgaard talks about the rhetorical reflex to the defense of progress, unidimensional myths of the Enlightenment, and the defense of a fantasy of the American or European "way of life."

As Norgaard observes in Norway — often held up by such thinkers as a kind of ideal European "social democracy" — in the face of the inexorable decline of the welfare state, in seemingly full comprehension of the economic and ecological *information*, many turn to comforting conservative illusions — the "traditional" fantasy, the "safety" of the patriarchal law-and-order society, disgust with new forms of desire and expression. The synthesis of high-tech fantasy with romantic kitsch is a hallmark of conservative modernism. In extending Norgaard's analysis, Andreas Malm posits that recent fantasies of geoengineering, for example, are a return-of-the-repressed, a *deus ex machina* which performs a double repression: first as the sublation of denial into symptom, and then obsessive attachment to that symptom. Only here it is not the individual neurotic washing her hands, but a whole social formation holding on to "normalcy" through fantasies of technological magic, which "align with the bleak realities of capitalist society, of which their geoengineered world will be an extension and intensification."[10] The endless feedback loop of repression, sublation, symptom, and further repression — understood on the psychological plane or simply as material impossibilities — drives Climate Lysenkoism to its inevitable marriage with conservativism.

Malm argues, by way of Adorno's *Negative Dialectics*, that "a bourgeois society will choose total destruction, its objective potential, rather than rise to reflections that would threaten its basic stratum." Even the IPCC is not so conservative.

Other forms of left techno-optimism do not always share this conservative bent. Accelerationism in its contemporary form is often predicated on the reality of the end of growth, and of radical social liberation. Xenofeminism takes up technological strands in socialist feminist thought to think through technology as a way to subvert social norms. Climate Lysenkoism's conservativism stands apart.

The critique of Climate Lysenkoism is not a return to "scientism" nor an argument that nothing within scientific inquiry needs critical engagement; there are many places for necessary intervention — not least in political and economic lacunae and assumptions. Sociobiology, the rambling discourses of New Atheism, and similar scientistic causes are among the absurd attempts to apply the methods of the natural sciences to social questions. (It is unsurprising that the arch, ahistorical "materialism" of New Atheism has proven to be a funnel to reactionary politics.) Many climate scientists, seeing where their analyses begin to fray at the nexus of the social and the ecological, often actively seek out radical social and humanistic thought. However, to dismiss questions of limits, boundaries, risks, probabilities, technological challenges, and so on, is to do precisely the opposite: to *dismiss* the realities that a left-wing climate politics must face — and which shape the social ground of political possibility in this moment. In actuality, it is increasingly clear that climate science in particular *is* internalizing many of the lessons of Marxist and radical social critique, and that any genuine historical materialism must adopt, as McKenzie Wark has argued, a more comradely relationship with science and scientists.[11]

While the actual Lysenko was animated both by theoretical conviction and pragmatic considerations (the need to radically increase food supply in the Soviet Union), programmatically

Lysenkoism was instituted under the aegis of Stalin's near-total power. (And apparently, after the practical underperformance of Lysenkoism, Stalin was inclined to dismiss both the man and the program. This was only delayed by his death.) Climate Lysenkoists, by contrast, have no such excuse: their lodestar is capital; their comrades are existing industry and its spokespeople; their utopia is this miserable, exhausting world extended *ad infinitum*. It is Macronism dressed in red.

If the right — certainly in practice, and increasingly in rhetoric — is moving towards right-wing climate realism, this "left" formation is moving *away* from radical political or ecological realism. Climate Lysenkoists posture that their program is a final break with capitalist realism, from the ideological constriction of neoliberalism. But they are not boldly storming the barricades of ideology. They are among its last victims, red-washing its most basic tenets — there is no alternative, a rising tide raises all ships — as they unload inchoate socioecological upheaval and affect into the arms of right-wing climate realism. They are a new breed of "market Stalinist" making their homes in major organs of the Northern left, like weevils eating grain from within. Only in excising their techno-mysticism, their political idiocy, and their implicatory denialism can we make way for the real possibilities and politics of left-wing climate realism.

## Six Impossible Things: The Science of Climate Lysenkoism

In 2019, the usually taciturn Jeff Bezos gave a remarkably candid presentation on Blue Origin, the massive commercial space firm that Bezos now heads after stepping down as Amazon CEO. In the near hour-long presentation, Bezos admits that limitless growth — which is the foundation of Amazon's and his own wealth — is incompatible with any meaningful notion of sustainability.[12] Long-term problems, like flagging productivity growth and inadequate energy to continue to prop up an Amazon-like enterprise, will be "solved" by opening up a new frontier. He doesn't even try to counter the widely acknowledged truth that existing technologies *could* facilitate "good life within planetary boundaries."[13] Rather, he presents his case for what constitutes a life worth living in existential terms: "Do we want stasis and rationing, or do we want dynamism and growth?" Bezos's answer is obviously the latter. Not just for Bezos. All of "us" (I don't think any attempt was made at First Contact with all us global plebeians) want what Bezos says we want. Dynamism and growth are *exciting*; stasis and rationing, a dystopian nightmare.

It is unsurprising that the wealthiest capitalists on Earth, like Bezos or Elon Musk, embrace so-called "ecomodernism". The 2015 Ecomodernist Manifesto principally promotes a program predicated on continued, infinite, capitalist development in which "accelerated technological progress will require the active, assertive, and aggressive participation of private sector entrepreneurs, markets, civil society, and the state" to facilitate a "good, or even great Anthropocene."[14] It is yet another program that facilitates the maintenance and enhancement of existing wealth and power — another version of right-wing climate realism, this time dressed in the business-casual of Silicon Valley. Bezos does not have to contort himself with arguments about "green growth" as with liberal system preservationists, technocrats, and, well, Climate Lysenkoists. Bezos doesn't have to mislead anyone about magical vs. existing tech, about absolute decoupling or efficiency. Expressed in hyper-capitalist terms, these arguments actually have a coherence, albeit vague. There are no promises of avoiding mass death, eco-apartheid, and the host of horrors we've already examined. Of course, Bezos paints a rather overly rosy portrait of capitalist technological progress, but, then again, so do the nominally "anti-capitalist" versions of this argument. Bezos' vision for the spacefaring civilization does regurgitate some classic Breakthrough Institute conservationist-style talking points about an Earth returned to some kind of idyllic state — "Earth ends up zoned residential and light industry," Bezos says. "It'll be a beautiful place to live, it'll be a beautiful place to visit, it'll be a beautiful place to go to college and do some light industry." Common ideological rhetoric about children and grandchildren is deployed. It is companies like Amazon who will do the zoning and governance, of course, like a twenty-first-century East India Company.

The capitalist case for ecomodernism is rational, if perhaps far-fetched. It *can't* be the case that "stasis and rationing" are desirable since that deviates from their theoretical and ideological convictions. We *must* agree that "dynamism and growth" are the *only* paths to a "good, even great Anthropocene." Most people experience the famous "treadmill of production" as destructive and exhausting. "Rationing" conjures images of breadlines, austerity, gray nightmares. As a right-wing political program, so-called "ecomodernism" makes sense; as a path to a *mass* sustainable human ecological niche, it is absolutely incoherent. As Armin Grunwald, an intrepid German physicist who specializes in technological assessment and systems analysis, remarks: "The ecomodernist approach grounds on premises which are not based on knowledge or experience, but rather on mere belief in technological advance."[15]

However, what rationing actually means is *redistribution*. With all the limitations of polling stipulated, this is globally incredibly popular. Even in the United States, majorities support the redistribution of wealth, a long-term secular trend, if accelerated modestly by both the 2008 financial crisis and the COVID-19 pandemic. As of 2020, US support generally exceeded 60%. Surveys find international consensus across the board to even greater degrees, with the United States at the lowest end.[16] Stasis *sounds like* stagnation. But it can also mean sustainability, free time, different modes of technological development, artistic pursuit, and a veritable cornucopia of human freedoms. Rephrased, it also means, quite simply, rest. Not only are such positions popular; they are *necessary* for genuine mass ecological flourishing.

Often, particularly in political and economic research,

this is framed (and appropriately so) as a question of political palatability or, even more importantly, power.[17] However, there is increasing consensus within interdisciplinary climate research that inequality — both within national boundaries and between countries — is a *driver* of climate change in biophysical terms.[18] This is not only the result of the widely acknowledged hyper-consumption of the wealthy. Increasingly, even non-radical studies look at the impact of capital's investment power and influence over the shape and development of production and political control. Malthusian arguments — increasingly popular among the Davos set and even the basic premise for Disney's *Avengers: Infinity War* films — that the real danger is the dark huddled masses developing higher standards of living don't stand up to empirical scrutiny. Reducing intra-state and international inequality is a fundamental need for climate mitigation and adaptation — not only in moral terms, but in physical and political ones as well.

One would imagine that this would be music to the ears of any socialist and, frankly, anyone interested in a flourishing life for the vast majority of actually existing humans on a boiling planet. And yet Climate Lysenkoists like Phillips take the Bezos and Musk position: even prospective socialist relations simply produce *more*. More "stuff," more "growth," more "progress" — each term accepted in its capitalist meaning, simply redecorated as "socialism." Unlike Bezos, Climate Lysenkoists have to tie themselves in contradictory knots: absolute decoupling *must* be possible; a good — if, in Bezosian coloring, bland — life *cannot* be possible for all within planetary boundaries; "dynamism" cannot be sacrificed, so "miracle" technologies must solve the problem.

You can start with any item on Phillips' list of potential technical marvels or any of his models of "technosolutions" past. Take his infatuation with synthetic fuels, synthetic jet fuels in particular. Maybe these keep "us" moving *just like today* — and faster. Sadly, while industry and various governmental and intergovernmental bodies have been promoting this for years, even the most enthusiastic (non-industry funded) research marks the technologies as nascent and dubious: "Though our analysis evaluates several performance variables for accelerating the deployment of SAF [Synthetic Aviation Fuel], our results should be considered preliminary." Findings across technologically enthusiastic researchers still almost without exception posit the existing technology as lacking "maturity," establishing promising grounds for "future research," requiring clarification of even nominal environmental improvements, and necessitating "significant advancements [...] to ensure emerging SAF routes collectively drive toward lower carbon intensity."[19] As a recent comprehensive review concludes:

> Given the current state of the art in the presented disciplines, none of the cited technologies can alone reach the sustainability targets for green aviation as mentioned in the European Flightpath 2050.[20]

Those targets are a 75% reduction in carbon emissions. And SAFs fall significantly short of that measure. Existing methods can sometimes net *higher* carbon emissions than even fossil fuels in lifetime assessments, especially due to the extraordinarily high energy costs in their production.[21]

Another *highly enthusiastic* study finds the best theoretical scenarios producing synthetic fuels with *worse* "global warming potential" and "ozone depletion" than fossil fuels. Still another finds — even in the best theoretical scenario — almost comically worse outcomes in "respiratory effects, smog formation, acidification, and eutrophication." A lifecycle assessment conducted by a team of Tunisian chemists and engineers on synthetic fuel in a real-world model predicted that it would generate not only a greater carbon footprint and increased global warming potential (alongside all the above categories) but a pronounced *social* cost to "workers and society," not limited to "forced labour, gender wage gap, and health expenditure." Meanwhile, another study promoting the use of SAFs noted, "decreases in demand and improvements in energy intensity" would be necessary, while "the interactions with food security, local communities, and land use are enormous hurdles for such a ramp-up and come with their own increasingly difficult trade-offs."[22] Not the music Phillips' ears want to hear. For example, in a best-case scenario, SAFs would still emit more carbon in 2050 than aviation did in 2020, and would need approximately 19% of the world's agricultural land for their production.

According to Phillips, to acknowledge these limitations is to be "Malthusian" or "anti-worker." (Workers in Tunisia, particularly women, don't count, I suppose.) The best-case analyses all acknowledge that SAFs are still in serious need of further research and development. The obvious, immediate need is in stopping emissions and other socioecological impacts, which means reducing demand, which requires additional provisioning, especially to those most in need (say, working class migrants far from home), while cutting off the most egregious production and consumption (say, private jets). For Phillips and other Climate Lysenkoists this is heresy. "There is no point in history where we can say: 'Okay, we've enough now. Let's stop,'" Phillips writes. "It follows that the progressive argument must be to say yes, we do want more stuff, and to argue against those calling for less stuff."[23]

Surely, you ask, he doesn't mean private jets, luxury cars, and McMansions — consumption goods of dubious use and far out of the reach of even the best-off proletarian? But he does.[24] To argue for less of anything — even if it's less of the massively wasteful and destructive toys of the uber-wealthy — is to potentially deprive future workers of these anti-social luxuries. Indeed, any restriction on "stuff" is anti-humanist "eco-austerity"; such restrictions are neoliberalism painted green. Phillips revels in straw men: he compares high-end capitalist luxury goods like bespoke furniture with "the free plastic Elsa doll from Disney's *Frozen* accompanying a Subway Fresh Fit Kids Meal"[25] to argue against "elitist hypocrisy" and for the Gospel of Stuff. Of course, it never occurs to him that *both* are capitalist "crap." An apt motto for Climate Lysenkoism would be, *"Nothing too shitty for the working class.*" For all the bluster about neoliberalism, Climate Lysenkoists are market fundamentalists at their core: what capitalism makes reflects what we want. Supply and demand. That there is practically no basis for any of this — scientific or social — is an understatement to say the least.

Phillips is fond of facile comparisons. He regularly compares the challenge of carbon emissions to that of chlorofluorocarbons (CFCs).[26] Easily solved through some regulation over the groans of industry. Total global cost estimates for implementing the Montreal Protocol are roughly $37 billion in losses to business and industry.[27] The CFC industry in the United States was worth $8 billion and concentrated largely in a single company, DuPont Chemicals. CFCs weren't trivial, but they were not the backbone of the global economy, unlike fossil fuels. Estimated decarbonization costs — which I will return to later — range from $16–200 *trillion* over the next 20 years. The IPCC has a conservative short-term benchmark of $31.2 trillion. All these estimates include trillions in stranded assets and other immediate, unrecoupable losses. This isn't just an apples-to-oranges comparison; it's apples to atom bombs. Not only does Phillips seem incapable of understanding the qualitative difference; he also misunderstands the order of operations. The CFC ban came *before* replacement technologies. "Substitutes for most CFC uses were not developed, alternatives to aerosol products containing CFCs were created in response to the ban on non-essential aerosols in the US and the cutback in the EEC."[28] If anything, the main lesson from the Montreal Protocol — for the North in particular — would be to "keep it in the ground," reduce demand, and decarbonize with actually existing technologies and techniques first.

Another passion of Phillips is the agricultural Green Revolution, from its "synthetic fertilizers and pesticides" to later genetically modified seeds.[29] Phillips' enthusiasm is tempered only in highly attenuated market considerations; limitations are economic only and should never be seen as a result "of the technological advances." This places Phillips to the right even of the World Bank, the UN, and certainly the IPCC, who all acknowledge serious inherent social and ecological shortcomings of the technologies themselves. For the Climate Lysenkoist, there's never a mixed bag. There is only progress "to infinity and beyond!" But the reality, as one UN rapporteur put it, is that in the Green Revolution "productivity was not measured in terms of human and environmental health, but exclusively in terms of commodity output and economic growth."[30] This led to a host of paradoxical phenomena: "hidden hunger," for example, in which pure caloric intake increases but health outcomes worsen and malnutrition spreads. Within this there is the still stranger phenomenon of "globesity" — the increasingly global prevalence of overweight individuals who are simultaneously chronically undernourished. And of course ecological devastation — to the climate as a whole and to agricultural systems. In the words of the IPCC, the "green revolution" led to "excessive use of agrochemicals, inefficient water use, loss of beneficial biodiversity, water and soil pollution, and significantly reduced crop and varietal diversity."[31]

The food system produced through the Green Revolution has transformed the agricultural sector from an "energy producing" sector, yielding more calories than inputs as late as the early twentieth century (even accounting for labor), to a system which expends the equivalent of 15 calories of largely fossil-based energy inputs for every one malnourished food-calorie.[32] Today agriculture contributes some 10–12% of emissions in direct farm production and approximately 34% taking in upstream costs like the petrochemical inputs to those magical "synthetic fertilizers and pesticides" that Climate Lysenkoists like Phillips conveniently forget to mention.[33] IPCC wording is classically conservative, but even they note, for example, a 100% increase in water use for irrigation alone; an 800% increase in the use of nitrogen fertilizers, etc. Without enumerating every aspect of its broadly accepted failures, it's enough to note that the "green revolution" has undercut its one questionable achievement: yields of its nutrient-poor staples are slowing or decreasing while resource use and "externalities" in their production increase.[34] This trend has been obvious for some time. As a groundbreaking 1985 book put it:

> The high-technology monocultures increase the vulnerability of production to natural and economic fluctuations. The plant varieties developed for the green revolution give superior yields only under optimal conditions of fertilizers, water, and pest management. They have been selected to put most of their energy into grain rather than vegetative parts, and the resulting stout dwarf stems make it easier for weeds to outgrow them, making herbicide use mandatory. Their reduced root growth increases the plants' sensitivity to a shortage of water. Irrigation buffers the crop against the vagaries of rainfall but increases the farmers' sensitivity to the price of fuel. High-nitrogen fertilizers and the growth-stimulating effects of herbicides make the plants more vulnerable and attractive to insects. The use of fertilizers offsets local variations in soil nutrients but makes fertilizer prices part of the environment of the roots of plants. And monoculture removes diversity as one of the traditional hedges against uncertainty. Despite modern agricultural technology, crop loss to pests has not been reduced since 1900 and probably is increasing [...] Modern agricultural technology results in environmental deterioration.[35]

The irony is that this extended quotation is from probably the most famous work of *Marxist* biology ever produced: *The Dialectical Biologist* by Richard Levins and Richard Lewontin. As Levins and Lewontin emphasize, the ecological (and social) qualities are inherent in these technologies. A Marxist analysis does not negate an empirical finding, but it aims at different goals and asks different questions. Thus, they argue that unsustainable industrial agriculture "must be replaced by a radically different system of production." Both practical and theoretical investigation should pursue "a gentle, thought-intensive technology in which the object of research is not to find new inputs but rather to find ways to reduce inputs." They encourage syntheses in line with this goal of the best understandings of contemporary scientific inquiry with indigenous practices and new social arrangements; more or less exactly what even the most cautious contemporary literature encourages.

This does not, though, comport with the unidimensional understanding Phillips applies to technology. While he has never met a theoretical technological wonder that isn't simply "progress" — when it comes to questions of clear social costs like farmer suicide in India, he suddenly discovers obfuscatory complexities. Phillips claims that the phenomenon either didn't occur or that it can't be attributed to his "superior", futuristic, (and highly profitable) tech.

The first claim is patently false.[36] Bizarrely, the study he cites for the second claim confirms increases in farmer suicide rates by 18% in 1998 and 10% in 2002, with a compounding rate of 2.5% from then till 2006, when the article was published. Phillips also claims the study argues that "focusing on genetic modification and ignoring the real causes — as Nagaraj puts it: an 'acute agrarian crisis in the country — and the state policies underlying this crisis' — is a dangerous distraction." Nagaraj's paper makes no claims at all about genetic modification. The quoted section is, in fact, about media attempts to victim-blame individual cases on "moral" failures (alcohol use, etc.). The paper does point out a connection to Phillips' beloved cotton cash-crop though, as well as to the provision of seeds, fertilizers, and credit. Phillips tries to pawn off any shortcomings of "miraculous" technological achievements exclusively to an ill-defined, entirely external neoliberalism.[37]

In reality, the technology is shot-through with *and* exasperated by the capitalism.[38] According to another study on farmer suicides:

Beginning in the mid-1980s, synthetic fertilizers showed their first signs of being overwhelmed by the state's decaying soils, as well as by their own inherent limitations. Without further warning, yields began to stagnate while input requirements stayed the same or even rose. Although in some cases yield actually declined as input requirements grew, for more farmers, an ever-increasing amount of fertilizers had to be purchased simply to sustain even stagnant yields. At the same time, pesticide use became ever more necessary, both as the result of weakened soils, and of the inherent vulnerability of HYV seeds to pests and disease.[39]

Most of the literature agrees that the crisis cannot be attributed to debt alone; it was ecological, technological, *and* economic systems, all embedded and inseparable from each other. The new seeds are dependent on a constant and increasing supply of fertilizers, pesticides, machinery, and more. The increased use of those energy-intensive technologies causes soil degradation and water depletion and increases greenhouse gas emissions. Climate change further suppresses yields and increases costs, and even more inputs are needed. Capital and debt run through every process; one tiny loop on the extractive circuit. The technological "miracles" built a highly fragile system even public provisioning could only partially solve.[40]

As countless studies on the Green Revolution note, sustainable and nutrient-rich crops (like millet in South Asia) were displaced and destroyed, alongside much of the knowledge of their cultivation.[41] Ironically, for all their bluster about "anti-Malthusianism," the Green Revolution was sold as an answer to Malthusian fears of a Southern population explosion and anti-Communist fears of Maoist peasant rebellion.[42] It succeeded by taking advantage of the waning of Nehruvian socialism. But Climate Lysenkoists never let a little history — or human suffering — get in their way. The machine must keep running.

And to keep it running with the thinnest possible veneer of ecological plausibility, we turn to Phillips' greatest love: nuclear power. The early retirement of active nuclear plants is almost certainly a mistake, and there is even some possible merit in pursuing lifetime extensions for them. However, there is little evidence that building new nuclear plants can be a major contributor to a genuine energy transition or even net positive.[43] The reason that early retirement is probably unwise is actually the same reason that we shouldn't pursue much new nuclear: upfront *material* costs and emissions — from uranium extraction to production to physical construction — are extreme (as are retirement costs and its overall water consumption). All energy projects[44] have significant material usage — one reason that nearly every plausible study includes demand-side reductions. But for nuclear, any positive is achieved after significant delay; in the meantime, material costs come immediately. Climate Lysenkoists like Phillips (but as frequently their comrades in the nuclear and fossil fuel industries) chalk this up to red tape and the need for long-term and public investment.

In his 2015 book, Phillips places his hopes on China since, as he correctly points out, the PRC has proven capable of complex construction in record time. This reality is as visible in 2015 as it is today — from bullet trains to skyscrapers to the astounding construction of advanced hospitals in about two weeks during the COVID-19 pandemic.[45] He foresees Chinese nuclear buildout reaching 88 gigawatts (GW) by 2020 and 400–500 GW by 2050. However, under probably the best possible conditions for feasibility in the world, Chinese nuclear construction lags — not for lack of trying or unfavorable conditions but because of nuclear's inherent complexity, difficulties to build, and staggering upfront material costs. In 2020, Chinese nuclear capacity had increased, but only to approximately 51 GW, well under the already reduced 14th Five Year Plan target of 58 GW. Meanwhile, Chinese renewables consistently exceed targets: the 13th Five Year Plan called for 680 GW of renewable energy but actually delivered 1,063 GW. While Chinese nuclear energy remains fairly flat, 80% of new energy installation in China in 2022 was from wind and solar. Both analysts and officials anticipate the PRC easily exceeding the 14th Five Year Plan's 2030 target of 1,200 GW of renewable energy production well ahead of schedule, and likely reaching 1700 GW by 2030. And nuclear targets continue to be revised down; the 14th Five Year Plan aims at 70 GW of nuclear. The explanation is simple: building nuclear, even under near-perfect conditions, is massively wasteful, difficult, and expensive. The 14th Five Year Plan confirms what research both inside and outside China has already suggested: resources are far better spent on storage — "low" tech as well as high — and improving grid transmission.[46] And raising renewable goals, and implementing fossil fuel phaseouts, efficiency measures, and demand reduction. China has engaged in basically a real-time experiment, and the results support rather more mundane measures than the Climate Lysenkoists imagine. (And China is an exceptionally useful case for questions of cost and feasibility, as we'll see again in the case of carbon capture and storage (CCS)).

Even some of the rosiest pictures, like those from the International Energy Agency (IEA), predict nuclear as a miniscule portion of energy production across all sectors, maybe maintaining its 10% current share (and even then with major caveats). The IPCC Special Report on Global Warming of 1.5 °C sees nuclear power as shrinking to about 4–5% proportionally in all 1.5 °C scenarios and absolutely in the relevant, immediate low/no overshoot scenarios. Furthermore, there is significant evidence that a both/and approach is not possible. Nothing supports Phillips' call for a "massive worldwide buildout" at all costs, especially considering that even supportive estimates still find nuclear falling short.[47]

The Climate Lysenkoist, though, does not know any limit. Instead of admitting he might be a smidge wrong, by 2019 Phillips started focusing more intently on "4th Generation" and "Small Modular Reactor" (SMR) technologies. Phillips foresaw these technologies solving any perceived problem with older nuclear. But even the IEA's 10% is predominantly smuggled in through the existing fleet, some backups built in the developing world, and SMRs — even though they admit they're barely, well, real. Phillips imagined several pilot programs would already be in operation. "NuScale Power, based in Portland, Oregon, has a 60-megawatt design that's close to being deployed" — as of writing it still hasn't come to fruition. What evidence there is on SMRs shows that they may have even more drawbacks than conventional nuclear.[48] "4th Generation" sounds nifty but, in reality, is mostly repackaged old, failed tech that, even if it works, is — yet again — not useful as climate strategy due to time and cost.

Even Phillips admits that 4th Generation — and the true golden goose: fusion — are a long way from deployment. As is observed across the board, and even by most advocates, nuclear — even in these supposedly revolutionary forms — simply isn't ready. New Chinese revisions reflect this reality; the real goal of the 14th Five Year Plan is that nuclear remains at approximately 4% of the energy mix; ideal-case scenarios see no difference for nuclear, just a more rapid drawdown of coal.[49] Genuine advances already made in renewables, simple grid upgrades, and storage make this already feasible. The PRC is hardly an example of "folk politics" or "small is beautiful" — although, again through trial and error, China's megaprojects are complemented by massively distributed, spread out, local networked projects, a legacy of Chinese economic planning that is underappreciated.[50] No matter how exciting or indeed technologically astounding — take China's extraordinary expansion of rooftop solar — nothing less than a mystical projection of the bleeding edge will cut it. "They are impressed by the flashiness of 'advanced' science (the more molecular and expensive, the more impressed they are)," to adapt Levins and Lewontin again.

Like the Communist Party of China (CPC) almost certainly does, I wish that fusion was here, and I don't oppose research that may in the future produce *real* wonders. But there is no excuse beyond swaggering Promethean arrogance to pursue the shiny impossible instead of the actually realizable, the urgently necessary, and the *more* desirable. Phillips doesn't believe there really are any limits, though. Planetary boundaries, he insists, are hazy outlines, not hard limits. A "heuristic" whose flexibility only bends upwards. But Phillips has it backwards, as always. In addition to the widespread adoption, integration, and expansion of these heuristics across climate research, most *critical* revisions express concern that the boundaries' rough quantification means that they may not be *upper* bounds, but rather that climate policy needs to be oriented towards a "safe operating space" well *below* such measures.

Amazon is perhaps the perfect extractive-circuit firm — nearly no aspect of its business operations comports with a sustainable social or ecological niche. Even by its own fuzzy accounting, it produces some 71.54 million metric tons of carbon dioxide per year (as measured in 2021). We've already seen how carbon accounting can be tricky. But Amazon's sprawling scope and the multiple value chains it coordinates illustrate this better than production and consumption numbers do. "Scope 3" emissions — "the result of activities from assets not owned or controlled by the reporting organization, but that the organization indirectly affects in its value chain"[51] — account for more than half of Amazon's reported emissions. However, Amazon creatively avoids accounting for approximately 60% of its scope 3 emissions by only counting its own products. Even by the dubious standards of CSR, this obfuscation is notable. Amazon doesn't disclose how much water the datacenters of its most profitable division, Amazon Web Services, consumes (again conspicuous even among other mass corporate water consumers). Google — approximately 9% of the web services market — uses over six billion gallons per year for server cooling. Amazon is 33% of that market. Amazon's notorious "last-mile" outsourcing is responsible for egregious labor conditions and dangers to public safety. Its speed — from sourcing to shipping — is inextricable from labor exploitation and social annihilation as well as massive emissions and material throughput use. The Big Data systems that are part of what make it appear such an efficient distribution allocator are inextricable from its gargantuan energy consumption or their racialized surveillance, to name just a few aspects. Yet, Climate Lysenkoists like Phillips envision socialized Amazons as the cornerstone of a "good Anthropocene," preserving its speed, data, and size. Socialized Amazons or Walmarts are the green Gosplans of their imaginary future.

Of course, like the rest — and this barely scratches the surface of Phillips' absolutely baseless assertions — the idea of a socialized Amazon that performs *exactly* like today but for the public social and ecological good is impossible. Well over six impossible things; Phillips should be thrilled with his reservation at the Restaurant at the End of the Universe. It is actually remarkable how much science fiction comes into these pictures, *Star Trek* in particular.[52] As a genre enthusiast and *Star Trek* fan, I find this somewhat amusing. All Climate Lysenkoists seem to see is utopian technology, but the canonical premise of the fiction (even in the original 1960s high-utopian version) is that pursuing the kinds of technologies beloved by Phillips led to deep inequality, eugenic wars, environmental degradation, and approximately a century of social and ecological collapse.

Bezos apparently unironically loves *The Expanse* novels and personally intervened to continue their TV adaptation — even though these novels trace a trajectory of private spacefaring which leaves capitalist Earth in steady decline, communist Mars forced into a constant state of military vigilance, and a vast colonial and anarchic outer belt providing much of Earth's basic service while itself struggling for even the most basic necessities. In Bezos's future, Amazon can really be the East India Company. Musk has named his proposed Mars-exploring spaceship after Adams' fictional Heart of Gold, seemingly missing the joke — the Heart of Gold runs on an "Infinite Improbability Drive" fundamentally impossible to control.

The joke is at least as old as Goethe, and likely much older, given that *The Sorcerer's Apprentice* borrows liberally on accounts of the Golem from Jewish folklore.[53] Marx adapts Goethe's *Sorcerer's Apprentice* in *The Communist Manifesto*. For over a decade I've taught Marx to probably thousands of working adult students, and when we get to those sections of the *Manifesto*, most will instinctively try to make the text match a one-sided picture of Marx as condemning capitalist progress and technology. I push them first to see broadly a position like the Climate Lysenkoist's; Marx is effusive about capitalist progress and technology. However, then we see that this one-sided view also fails to capture Marx and, more importantly, Marxist methodology. Many interpreters focus on Marx's clever switching of the apprentice for the sorcerer himself. The bourgeoisie is the sorcerer in Marx's rewrite, and has unleashed forces he cannot control:

> Modern bourgeois society with its relations of production, of exchange and of property, a society that has conjured up such gigantic means of production and of exchange, is like the sorcerer, who is no longer able to control the powers of the nether world whom he has called up by his spells.

However, just as many commentators miss the point of including the story at all. In seeking to simply replace human labor with "magic," the sorcerer's apprentice conjures uncontrollable forces and machinery. The *Manifesto* is one of Marx's most productivist and stagist texts and yet, in adapting this story, it inverts the actors but *not* the fundamental message. The task now falls to the subordinate or dominated class to "break the spell," to put the genie back in the bottle, in Goethe's words:

> To the lonely
> Corner, broom!
> Hear your doom.

Similar dialectical appraisals are found throughout Marx. He sees in technology a promise of progress perverted to a reality of destruction, just as the human becomes a mere "appendage" to the machine in the *Manifesto*. Bringing us back to industrial agriculture, Marx argues:

In modern agriculture, as in the urban industries, the increased productiveness and quantity of the labour set in motion are bought at the cost of laying waste and consuming by disease labour-power itself. Moreover, all progress in capitalistic agriculture is a progress in the art, not only of robbing the labourer, but of robbing the soil; all progress in increasing the fertility of the soil for a given time, is a progress towards ruining the lasting sources of that fertility. The more a country starts its development on the foundation of modern industry, like the United States, for example, the more rapid is this process of destruction. Capitalist production, therefore, develops technology, and the combining together of various processes into a social whole, only by sapping the original sources of all wealth — the soil and the labourer.[54]

Marx was constantly revising positions and arguments over his life as befits a necessarily open-ended methodology, but how one can come to arguments like Phillips' from these sources beggars belief.

Phillips' "Marx" is a linear proponent of Enlightenment progress. Phillips's "Marx" is methodologically positivist. The actual Marx called positivism a "shit" philosophy, worse and more guilty of idealism than Hegel. Phillips' account of scientific progress would make even Karl *Popper* blush. Phillips — perhaps piqued that the PRC is stubbornly ignoring his fantastical projections — manages to simultaneously call for universal respect for "the scientific method" while resurrecting the notoriously racist and xenophobic COVID "lab leak" pseudoscience.[55]

Yet he also blames "parts of the academic left" — he cites Thomas Kuhn and Paul Feyerabend as somewhat surprising examples — as promoting pseudoscience. Phillips defends the "battle of evidence" as the only path to knowledge but seems unconcerned about scientific consensus and empirical evidence whenever it challenges one of his positions (which is almost always). Phillips' cosmic techno-mysticism subjects available scientific data not to critical scrutiny but to *a priori* assumptions about the boundless possibilities of technological progress. I don't intend to open an entire argument about the philosophy of science,[56] but rather to underscore that this is the very definition of Lysenkoism. Instead, Phillips drapes Marxish language and selective quotations around a profoundly *conservative* vision of keeping things *exactly* as they are, just theoretically expanded to benefit some nebulous abstraction he calls "workers." This in spite of the fact that every position just so happens to coincide with those put forward by industry, by finance, by tech. This is because the non-magical trajectory of Promethean so-called "ecomodernism" is fundamentally about the continuation of Actually Existing Capitalism — at any cost.

Does it matter to the Climate Lysenkoist that the epidemic of farmer suicide is now present in the Global North? "The US farmer suicide crisis echoes a much larger farmer suicide crisis happening globally: an Australian farmer dies by suicide every four days; in the UK, one farmer a week takes his or her own life; in France, one farmer dies by suicide every two days; in India, more than 270,000 farmers have died by suicide since 1995." Probably not. As we'll see more closely in the next section, the only "workers" who exist for the Climate Lysenkoist are cultural projections of burly, industrial labor. These actual people don't really matter. Phillips certainly hasn't revised any of his positions for them. American farmer suicide looks a lot like South Asian farmer suicide.[57] He rightly sees a kind of misanthropic tendency in deep ecology, but his "environmentalism of the rich" is more hateful and lethal.

Reflecting on the obesity/malnutrition paradox, Lauren Berlant asks in language eerily similar to Fisher's discussions of mental health and ecology:

How else, then, to understand the intersection of the long history of poor people's shorter lives and the particular conditions of contemporary speed-up — people working harder and longer just to keep afloat? [...] the poor are increasingly less likely to live long enough to enjoy the good life, a good life whose promise is a fantasy that justifies so much exploitation?

The word that best captures the reality of living in Phillips' or Bezos' *fantasy* world of exhilarating and exciting dynamism — is exhaustion: "Working life exhausts [...] At the same time that one builds a life, the pressures of its reproduction can be experienced as exhausting." What, Berlant wonders, might be the possibilities of a *slower* life? Or, as their later work suggests, relinquishing the cruel optimism of those fantasies?[58]

Nonsense, replies the Climate Lysenkoist. We need more and we need it faster:

> The socialist says: through rational democratic planning, let's make sure innovation arrives so that we can move forward without inadvertently overproducing. And move forward we must, in order to continue to expand human flourishing. So long as we do that, there are in principle no limits. Let's take over the machine, not turn it off![59]

Jeff couldn't have said it better himself. But for the vast majority, the machine, the extractive circuit, is *exhausting*.

## Concrete Denials: The Politics of Climate Lysenkoism

In June of 1920, Vladimir Lenin penned one of the most famous phrases in the Marxist canon. Responding to a published analysis of conditions in Germany at the end of World War I that, in the name of Marxism, depended on a rigid, dogmatic projection of transhistorical first principles, Lenin wrote that this mode of analysis "absolutely evades what is most important, that which constitutes the very gist, the living soul, of Marxism — a concrete analysis of a concrete situation."[60] When it comes to the *politics* of climate change, what is missing in even some of the best radical thought is precisely this *concrete* analysis (even if concrete — one of the world's most common building materials — is correctly understood as ecologically disastrous). Left-wing climate realism builds from and contends with the unique qualities and challenges — historical, temporal, social — unique to this political moment. Climate Lysenkoists like Matthew T. Huber prefer to dwell in a world completely divorced from any such realities.

The late Mike Davis contrasted "structural" with "conjunctural" understandings of class politics to help illustrate the quandary. A purely "structural" or abstract account of class imagines a transhistorical ideal, relying on a nineteenth-century paradigm of industrial workers' strategic agency at the "point of production." However, even a cursory review of history finds that "class capacity" — i.e., agency — "arises conjecturally, as activists reconcile both in practice and in theory different parallel demands and interests." The conjunctural, "limited to historical stages or episodes is ultimately transient," but that is its strength, not its weakness. It "denote[s] the intersection of unsynchronized histories"; it synthesizes "grievances and aspirations" in "specific crises."[61] While the former provides a kind of dated if powerful theoretical model, the latter begins to answer that most dogged of contemporary Marxist and other radical political questions: who, today, is the subject of mass politics? Discussions of the subject can get mired in what is often misrepresented as an impossible metaphysical or linguistic morass, but put more simply: the subject is the who or what of politics; not only the Marxist idea of class but, for example, the liberal individual, or conceptions of "the nation," or the vague "people" of populism, or the *demos* of radical democracy, or even, as Hegel argues, Reason — *Geist* — itself.

In the wake of the collapse of the Soviet Union and Francis Fukuyama's (much overblown) supposed "End of History,"[62] some even theorized post-subject politics.[63] This kind of disillusion has no theoretical and even less empirical necessity. Rather, the question at hand is, in Seyla Benhabib's left-liberal formulation: "how and by whom, [can a] general interest [...] be recognized and articulated?"[64] Her question differs little from Lenin's "who, whom?" formula. And some would note that we need to follow Ellen Meiksins Wood and argue that as important as that interest is the "strategic power" and "capacity for collective action" in a given historical situation.[65] To this we must add the *desire* for things to be *radically* otherwise. This is not merely some scholastic addendum in service of contemporary affect theory (although affect theory has much to teach here). It is not only Marxist theoreticians like Lukács (among many others) but Marx himself who sees this widespread and increasingly radical *dissatisfaction* with the whole of society as vital; who sees that general interest, capacity, agency, and power requires "the desire for social transformation" and "class antagonisms."[66]

There's a temptation to leap here from the twenty-first century back to the nineteenth — as if politics are not subject to the changing conditions that reshape and restructure the mode of production. If capitalism is at the center of so many analyses, across so many fields, as the *chief* driver of climate change — natural, social scientific, and humanistic — isn't a left climate politics then simply synonymous with class struggle? The answer is frustratingly dialectical: both yes and no. Even among those who acknowledge the unique nature of this socioecological conjuncture, those with the most probing and lucid analyses, when it comes to the specific *politics* of climate change, it is all too common to reject the discomfiting and unfamiliar realities.

Climate Lysenkoists aren't even following Marx's own thought, let alone any subsequent elaboration. As the political ecologist Murat Arsel argues about class analysis today: "it bears repeating that fidelity to Marxist class analysis does not require a rigid two-class framework, the erasure of genuine cross-class formations, or the possibility that additional conceptual tools might need to be developed to respond to the exigencies of climate change."[67] Huber's "theory" ignores nearly every insight from conjunctural analysis, an ignorance particularly deleterious to climate politics. Huber's "class analysis" — in the *kindest* terms — uses the most bowdlerized class theory to dismiss inconvenient scientific findings and an imaginary projection of "the working class," from which he reverse-engineers, against all evidence, his entirely dubious technological claims.

A broader, non-dogmatic Marxist analysis — and, crucially, a global understanding — following Fanon, a necessarily "stretched" one that accounts for those realities and the *conflicting* interests — the conflicting *feelings* — they generate, is vital for a genuine left-wing climate realism. The political and scientific distortions of Climate Lysenkoists are actually illuminating: they demonstrate with dumb alacrity everything confounding about the all-too-common attempts to simply staple transhistorical political paradigms onto the acute realities of climate change.

The always thorny questions concerning class, subjectivity, and agency are confounded for Climate Lysenkoists in two particular dimensions. First by empirical trends which do not concur with the nineteenth-century model noted by everyone from Eric Hobsbawm to Kathi Weeks, from Ruth Wilson Gilmore to the quasi-anonymous *Endnotes* collective. These authors (among others) address in different ways conditions we will examine more closely: a humanity which, far from economistic and deterministic predictions, is, as Davis summarizes, comprised mostly of "surplus populations," who largely experience the "jobless growth" of extraction, disposability, and the "vice of war and climate change."[68] A chorus of authors (and a larger chorus of evidence) demonstrates how the broadest predictions of a global urban proletariat (on the mid-nineteenth-century model) to its most pointed definitions (agency as based in certain workers' strategic location in relation to the "point-of-production") is *at best* useful only as a rough rubric. The reality of social change, whether reformist, radical, or revolutionary, is always — even when unified into a *subject* — coalitional in nature.[69]

To flesh out this picture, first, it is clear that the industrial proletariat never became a majority in every society, or globally.

This did not even take place within the imperial capitalist centers of Europe — with one exception. As Adam Przeworski notes, "in Belgium, the first European country to have built substantial industry, the proportion of workers did break the magic number of majority when it reached 50.1 [percent] in 1912."[70] At the most basic level, this underlines the inter-class syntheses that have characterized *all* working class and radical movements. Even a passing glance at history confirms this. All twentieth-century *mass* movements of significant reform (from women's suffrage to American Civil Rights) were heterogenous coalitions drawn together into a common project. The case is the same for revolution (the Bolshevik Revolution, the Chinese Revolution, the Cuban Revolution, a world-wide wave of anti-colonial revolutions, the "last great revolution," the Iranian Revolution, etc.). Regardless of how one views these, none would have been possible without complex, cross-class blocs. Even the very Euro-American welfare state politics that so enthrall Climate Lysenkoists reflect this reality.

But even more damningly, the reality — denied by Climate Lysenkoists as much as the discomfiting findings of climate science, engineering, and even unequal global exchange[71] — is that actual world demography in terms of class and "relative surplus population" produces even starker numbers. Out of the global working-age population of just under five billion, fewer than two billion could be considered "workers" under even the loosest standards. Overall, the relative surplus population is some 2.47 times the "active army" (including the actively unemployed). Broken into developmental categories, the least developed countries have a ratio of 8.9 to 1, middle-income nations 2.4 to 1, developed world 1.37 to 1. And this is only increasing. As the authors of this study — the late Marxist labor studies specialist David Neilson and developmental sociologist Thomas Stubbs — study grimly note:

> a clear majority of the world's labouring population is now relatively surplus to the functioning of capitalism. Of particular

concern is that engineering increasing demand to stimulate the redeployment tendency is no longer a solution because it implies the escalating consumption and destabilisation of an already materially depleted and ecologically destabilised planet.[72]

In far more vivid detail, and closer to home, Gilmore notes, in the case of California during the massive expansion of its carceral apparatus, the "overall trend is for labor force growth to exceed employment growth by about 4%." Several forces come together — some specific to the time, like the particular mode of market restructuring during globalization; and some more general, such as trends towards efficiency "producing a growing relative surplus population — workers at the extreme edges, or completely outside, of restructured labor markets, stranded in rural and urban communities."[73] This is entirely in keeping with Neilson and Stubbs's finding that even in the United States, the ratio of surplus population to active army is 1.25 to 1.

The future of mass politics — ecological or otherwise — lives or dies here, as Davis concedes, while at the same time cautioning against "imprudent [...] abstractions" like Hardt and Negri's "multitude" which "simply dramatize a poverty of empirical research."[74] Indeed, works like *Empire* and other early analyses of globalization tended to make sweeping, largely abstract claims with little grounding in empirical reality. Climate Lysenkoism's "class politics" is defined by an equally intense empirical and theoretical poverty — at *best*, to borrow sociologist Michael McCarthy's apt formulation, a "class abstractionism."[75] Empty calls by Climate Lysenkoists to "reclaim the future" are undercut by an (empty) romance of the near past and a blindness to the far more radical possibilities of class and subjectivity in this conjuncture, a conservative imagination unable to grasp the emancipatory movements, too constricted to see the far more radical horizons that have vast appeal in this moment.

In the name of a perfectly abstracted "working class" hovering above actual history, Huber dismisses inconvenient elements of climate science as the guilt-ridden, pathological projection of the class fraction that includes most scientists: the "Professional Managerial Class" (PMC).[76] At the same time, every element of Marxist theory that could be interpreted to support an uncomplicated understanding of capitalism as fundamentally progressive is conveniently emphasized.[77] "Legitimate" politics are reduced to union bargaining — particularly at the industrial point of production — and liberal democratic elections. Therefore, just as with so much liberal thinking, fantastic technological solutions and materially impossible or garbled positions like unlimited "green growth" are deemed necessary. With niggling problems like science and history out of the way, the path is supposedly cleared to working class power through liberal institutions. A bright, green future is promised which looks suspiciously like a carbon copy of our catastrophic, exhausting present.

Every aspect of this story is false, fabulated, or fanciful. It is also barely fair to call it "Marxist," even in a dogmatic sense, given that almost none of its self-appointed "orthodoxy" bears any deep relation to Marx or the vast majority of subsequent Marxist research and politics. There are occasional lucid moments — familiar arguments about the centrality of capitalist production to climate change, about the futility of individualized solutions. But it is this entire model — this whole story — that is of interest here. Setting aside for the moment nearly a century of research and theory from Marxist-feminists, eco-Marxists, world systems thinkers, and countless others, this way of thinking is at odds not only with the specific conditions of the contemporary socioecological conjuncture; it's at odds with reality, and even itself.

In order to get around, for example, problems in class history and theory, Huber suddenly remembers far more expansive definitions of class generated by heterodox Marxist geography and Marxist-feminism, that include, among others, unwaged domestic workers, peasants, and the ever increasing "vast informal proletariat," which, in the words of Davis, "is a wholly original structural development unforeseen by either classical Marxism or modernization pundits."[78] However, in the effort to bang ill-fitting nineteenth-century pegs into unprecedented twenty-first-century holes, Huber subordinates all such formations ("Populations in the household or the informal economy") beneath the tiny and globally *decreasing* fraction represented by the industrial working class, particularly small in Huber's chief case, the United States.[79]

He spends little time on existing climate struggles and movements since these are composed of "professional class-based NGOs/academics" and "marginalized classes." (Incidentally, this roughly describes the leadership and mass base of the Russian Revolution.) This distortion belies an ignorance of the complex realities of existing movements and especially "ecological distribution conflicts." Huber bemoans that on a list of climate justice organizations only 30% are unions (largely in line with the International Labour Organization's estimate of 42% unionization as of 2022) but, in critiquing Ariel Salleh's eco-feminist analyses, he excludes "the vast majority" which are "peasant and farmer unions."[80] The sleight of hand is constant; strike numbers and examples of militancy are conveniently bolstered by those who Huber then fundamentally discounts as non-strategic or unimportant. Huber not only subordinates all social relations beneath the formal wage relation, he limits them to essentially valorized concentric circles emanating from the "true" industrial worker.[81] That real accounts of surplus populations demonstrate, as Gilmore shows, that within such numbers are "internally dynamic but structurally static racial hierarchies" is largely irrelevant to Huber.[82] Even Wood — whose polemics Huber hangs much of his class theory on — doesn't go quite so far: "class relations are not reducible to production relations."[83] Huber's PMC guilty consciences promotes "irrelevant" marginalized populations; these are of course the very working poor and others who far-from-unorthodox thinkers like Davis and Hobsbawm view as the *sine qua non* of any politics under current conditions.

Numbers are only one part of the story. Power is the other. Huber writes:

> it is workers' *strategic* location at the point of production which gives them tremendous power to disrupt capital's profits at the source. By using the power to strike, workers have historically forced major concessions from capital — not only to pay workers more, but also to transform workplace conditions more broadly.[84]

Or, perhaps more grandly, as Vivek Chibber declares: "Workers, therefore, are important for a strategic reason, which is that they are the agent, and the *only* agent, that has a structural place within the society that can bring the power centers to their knees."[85] In this way of thinking, workers in service sectors and especially the ever-growing numbers in informal sectors (let alone many forms of social reproduction) do not have this *precise* "lever." Service sector workers *might* force concessions through workplace action but, for Huber, this is weak at best. Not so, he argues, of industrial workers: "in the context of climate change and other ecological crises, such industrial sites remain the belly of the beast. Workers have the power to disrupt these industries at their core through the power of the strike."[86] The logic dictated therein is obvious. The fundamental strategy is to focus on the organization of labor in the workplace, from where grows a more radical workers' movement and radical politics expressed through strikes and elections. However, the spokespeople Huber quotes never argue for anything remotely like an actual left-wing climate realism. "The refinery strike lasted for several weeks before a deal was reached. The deal did not 'halt capitalism's assault on the planet,' but it might have reminded industrial workers of the power they have ceased to use over the last several decades."[87] This is not an argument about ecological politics; it is a romance about a slumbering giant.

This kind of logic has long been ridiculed — and rightly so —

within the Marxist tradition itself. Marx himself never once made an argument remotely like this. Even as well-known a theorist and supporter of "The Mass Strike" as Rosa Luxemburg argued against "the fanciful form of reasoning" that "the trade-union struggle was the only real 'direct action of the masses' and also the only revolutionary struggle."[88] This, she noted, was not remotely how actual historical movements played out, but was just a principle of syndicalist faith. And presaging a century of coming critical theory and global Marxism, she presciently argued that this principle also underwrites the seemingly opposite evolutionary social democratic position. And although I don't think recourse to Marxology is necessary, it is quite easy to observe that both Marx and Engels had ambivalent views (at best) about this kind of union-only or union-first strategy.[89] Marx did not argue that it was workers' strategic location at the point-of-production that lent them a lever or power to "disrupt [...] industries to their core." Marx at his most positive hoped that mundane struggle in the workplace might help spark a broader consciousness of potential power for the working class as a whole.

Even within his limited historical purview, and indeed his patriarchal and Eurocentric myopias, Marx was never as silly as these self-appointed "orthodox Marxists." Marx vacillated on these questions and many others in relation to changing historical conditions and ongoing refinement of his critique of political economy. But he anticipated later critiques of exactly the kind of politics put forth by Climate Lysenkoists. To pivot again to Przeworski's analysis, "economic interests" alone — analyzed from the point of view of history, theory, and political economy — do not lead to "socialism." And, one can add by extension, to any sort of genuine left-climate politics. Indeed, the pattern he notices for this kind of politics is to "protect profits from the demands of the masses because radical redistributive processes are not in the interests of wage-earners."[90]

Reflecting on labor in the United States closer to our moment, Gabe Winant notes that it was (and is) the task not of the labour movement but of the socialist movement to bring into contact with one another the various struggling fragments — those who are organized as carriers of the relations of production, those not organized in any manner, and those engaged in struggles that do not correspond to any broadly conceived system of production...[91]

Revitalized, radical labor struggle is nascent and unlikely to take the forms it did in the mid-twentieth century. Today, class struggle occurs across a range of terrains, not simply or even centrally the workplace. The privileging of this one specific site is not only not particularly ecological, it's also not remotely Marxist. Worse, it has little basis in the historical realities of large-scale social change.

In fact, one might say that the Climate Lysenkoist doth protest a bit too much. Constant references to Marx, claims to hew to "orthodox" Marxism, and so on, belie this non-Marxist framework.[92] On its own this would not be a problem, but wedded to its particular fantasies it dictates a politics divorced from the global nature of current production, laser-focused on returning to the supposed Golden Age of the postwar period. The real model in Huber's work is the openly non-Marxist framework of professional union organizer Jane McAlevey. While there's much to commend in McAlevey's work, it's fundamentally blind to the historical forces that are so vital for climate, or really any radical politics *tout court*. As labor journalist Alex Press puts it in reviewing her work, "unions are defensive forces: history suggests that alone they are insufficient to combat rule by the rich, much less the wage-labor system."[93] Or as Davis notes: "the old working class [...] to put it crudely, has been demoted in agency, not fired from history."[94]

Huber's political theory aims at sectoral bargaining, robust legal protections, and the reinstitution of the welfare state. However, as Oliver Nachtwey observes, examples from countries like Germany — with even more robust legal protections, sectoral bargaining, and a dramatically larger welfare state —

have not only experienced similar union decline but a pattern much like the United States, where major existing industrial unions are not only bound ever tighter to their respective firms but are more than happy to endorse two- or even three-track employment structures, proudly protecting the best-off workers in concert with capital, while writing off the needs of newer contract workers and tertiary migrants and other informal arrangements.[95] Huber ignores intense debates across radical political history and theory (Nachtwey cites the early work of Franz Neumann, among others, including Marx himself) which observed in real time how legal formalization and institutionalization often goes hand in hand with decreasing militancy. Recent radical US militancy — as with 2018's wildcat teacher's strikes or the United Auto Workers' current 2023 auto-sector strike — reflect cross-sector alliances, broader social upheaval and radicalization, untethering from firm and nation, and even disintegrating legal guardrails (many teachers' strikes took place in "right-to-work" states, for example). Davis, observing now-clear theoretical and historical "blind spots and misdirection" in Marxist thought, also points to the impossibility today of trying to "rely on a single paradigmatic society or class to model the critical vectors of historical development."

I return to Davis again and again not simply because of his magisterial and wide-ranging work, or that I think he has everything exactly right, but because Huber hangs Davis like a saint's medallion to sanctify his "theory." While it is Davis who, observes in historical detail that "class capacity" — even delimited to the American and European accounts — is not found in some logic of the strategic point-of-production but across a host of terrains and institutions where the difficult organizing work of synthesizing the broadest sense of class "grievances and aspirations" occurs. Huber claims Davis sees the New Deal (policy) as "the high-water mark" of American class politics; in fact, Davis's "high-water mark" is about the militant *power* of *unorganized* workers in the early New Deal *period*. Huber disdains all talk of "overconsumption." Davis cites "overconsumption" as a key problem in the United States, particularly among the growing middle class which — as with Nachtwey's analysis of twenty-first-century Germany — includes members of the formal industrial proletariat.[96] Huber dismisses what he, following Phillips, calls "the politics of less," while Davis, who genuinely engages in ecological research, argues for a "sustainable equality" driven by "public affluence," displacing individual "consumption" and defined through "great urban parks, free museums, libraries and infinite possibilities for human interaction."[97] Huber is fixated on wages and votes. Davis points out that American labor militancy (in both the 1930s and the late 1960s) often focused on the tyranny of the workplace — from health and safety conditions to radical demands for autonomy in design. Huber sneers at talk of imperial drain from South to North. Davis — and nearly all serious Marxist research and certainly all serious climate analyses — sees these as pivotal. Northern "development" is predicated on the continuing extraction of debt from the South.[98]

Davis's entire account of "late Victorian holocausts" is predicated precisely on the international nature of capitalist production, consumption, and colonialism. Davis is at pains to point to the fundamentally international nature of capitalist production and warns in another of his last writings:

> as excited as I have been about the leftward evolution of a new generation and the return of the word "socialism" to political discourse, there's a disturbing element of national solipsism in the US progressive movement that is symmetrical with the new nationalism. We tend to talk only about the American working class and American radical history [...] in what sometimes veers close to a left version of America Firstism [...] Socialists should stress the urgency of international solidarism at every possible occasion.[99]

This "left" America Firstism is precisely the formation that Huber is building and that's currently promoted by bastions of "left" Anglophone thought — a mirage "left" to work hand-in-hand with climate apartheid.

## The Labor of Carbon Capture

Labor action and industrial workers, unions and other classic organizations of the left — from Shenzhen to Los Angeles — have a vital role to play in left-wing climate realism. The age of hippies vs. hardhats (always a touch overblown) is long gone.

In Huber's schema, it's at least seemingly the opposite. Since it is only strikes and elections that matter, only workers at the point-of-production and a broader electoral movement, he reaches for technological fictions like a good liberal technocrat. In order to "manipulate" (Huber's phrasing, not mine) the workers he imagines, he assumes we need every technology possible to make the magic of green growth possible.[100] Huber's fundamental political strategy is based on kick-starting capital accumulation and hoping the lifeboat trickles down. For example, Huber emphasizes the necessity of carbon capture and storage (CCS). The problem is that CCS — despite significant investment, research, and development — doesn't exactly work. Recent studies focused on the American electricity sector find trivial, zero, or even increased carbon emissions from various negative emission technologies (NETs).[101] *Pace* Huber, they are highly *profitable* but of questionable ecological utility.[102]

CCS is a relatively tame case on the spectrum of techno-fantasies. And yet it is a telling example both ecologically and politically. First, as many studies conclude, even in the best-case scenario, CCS — alongside technologies like direct air capture (DAC) — are not actually effective carbon emission mitigation technologies.[103] Some of the most positive assessments note that "e-mobility and heat pumps reduce the climate impact more strongly per kilowatt hour of electricity used. Hence, from a climate perspective, the implementation of these technologies should be prioritized over the hydrogen-based CCU technologies."[104] The question is not about technology in the abstract, but what counts for technology, where, and by whom.

The very article Huber cites actually gives priority to environmental engineer Claire White's advocacy of existing low-carbon, scalable, alternative materials: "it's possible to make cement-like products using other substances instead, she noted, including recycled byproducts from other industries, such as steel slag, fly ash from coal-fired facilities or certain types of clays."[105] Such alternative materials, as we'll see, are part of the quasi-utopian possibilities of a left-wing climate realism. CCS is introduced by way of a throwaway comment referencing a 2016 paper. That paper (which Huber did not read or chose to ignore) is not about CCS, but about exploring whether cement itself reabsorbs carbon. A single line at the end mentions the possibility of CCS to add to that absorption.[106] While the argument for CCS in cement is stronger than in other areas, it is still a picture of continuing at current levels of emissions or even — as with power generation — *increased* emissions. Almost all serious research comes to the same conclusions: "as this study shows, there is no NET (or combination of NETs) currently available that could be implemented to meet the <2 °C target without significant impact on either land, energy, water, nutrients [...] 'plan A' must be to immediately and aggressively reduce GHG emissions."[107]

Sympathetic studies find the technology underwhelming. In one of the most recent comprehensive reviews of existing and potential CCS technologies in cement production, the engineering team of Dwarakanath Ravikumar, Duo Zhang, et al. ultimately found "that the $CO_2$ emissions from OPC [Ordinary Portland Cement] production are greater in CCU concrete than in conventional concrete" production. Similarly, recent reviews of CCS in China and Canada by mechanical and energy system engineers Sheng Li, Lin Gao, et al. find that "current $CO_2$ capture technology in coal-fired power plants cannot be applied broadly due to its high energy consumption," and, furthermore, even theoretical large-scale technical innovations would merely catch up with *already existing* technologies in ordinary renewables like wind and solar. Like so many others, they find it a dubious investment of material resources in comparison with renewables. They would have liked it to work, they are plain in saying, but renewables have simply been far more successful in terms of mitigation potential and resource use and are available on a much more realistic climate mitigation timeline. In other words, the best use case *if* CCS ever does work is not in the immediacy and intensity of current climate challenges, but in some future use cases. Similarly, oceanographer David Ho, who sees an absolute *future* necessity of carbon-dioxide removal (CDR), argues that focusing on its near-term deployment, before "society has almost completely eliminated its polluting activities," is "pointless."[108]

The political calculus is even more damning. Huber claims that CCS utilization is blocked by "the freedom of capital."[109] CCS's actual investment, development, and propaganda campaign is vigorously pursued by the fossil fuel industry itself. Firms like Shell, ExxonMobil, and BP are *currently* pumping billions, and soon trillions, of dollars into CCS, well above the $4 trillion the IEA dubiously suggests is needed. The reason for this is obvious: efforts to invest in CCS or DAC right now are a convenient way of *extending the life of fossil fuels*. Indeed, internal BP memos argue that CCS will "enable the full use of fossil fuels across the energy transition and beyond."[110] In the narrower case of CCS and concrete, we find key multinational firms — like Heidelberg Materials, the German-based multinational corporation (MNC), which is the world's largest producer of construction aggregates and one of the world's largest carbon emitters, or Japan-based Mitsubishi Heavy Industries — similarly investing billions.[111] Capital is not fettering the possibilities for CCS development. Capital — fossil and extractive capital in particular — are *all in* on CCS and any other technological "bridge" that can maintain profitability and keep business-as-usual rolling. It is betting on technologies like NETs to bring climate transition closer to its timeline. It is deep decarbonization, among many other measures across the whole spectrum of climate change, that threatens capital. The idea that CCS is somehow being held back by capital, waiting for the gentle push of sectoral trade unions, has no relationship to reality. Capital is generating technologies in *its* interest — technologies with trivial if any current merit, technologies to buttress the ever more difficult job of keeping up the waning global productivity growth rate.

Climate Lysenkoists are not just willfully ignorant of the "concrete situation"; they are willfully ignorant about concrete itself. "*Production!*" screams the Climate Lysenkoist, with no sense of how production is happening. Huber is not merely *wrong*. He distorts this conjuncture — from historical realities to climate science and technology — to fit his already fundamentally flawed theory. In doing so, he buttresses the power of capital. Huber actively campaigned *against* New York's Build Public Renewables Act because it didn't include CCS, nuclear, and the "all-of-the-above" position championed by fossil capital and Climate Lysenkoists.[112] This is not ecological *or* class politics. It's Nationalist Welfarism with American Characteristics.

Climate Lysenkoists need these facts not to be true in order to complete their just-so story about neat, progressive history and their desperate "class abstractionism." There is no reason to dismiss workers, including Northern industrial workers. The question is about *real* interests, real grievances. Huber claims CCS works because this is the position coordinated between the International Brotherhood of Electrical Workers (IBEW) and industry management. He can't countenance the reality that for all the power of a concept like abstract universal interest — or his "proletarian ecology" — it's "objective" only in the most distant meta-theoretical realms, where real politics go to die. In reality, workers — and non-workers — can have clear, *rational* interests which do not align with radical change of any sort, let alone with left-wing climate realism.

This is a question not of dismissal, but of strategy. It is

probabilistic, not deterministic. There are thousands of movements already in action. Are these enough? No. Is a more *crystalized* subject formation — indeed a class formation, if you will — needed? Yes. However, a left-wing climate politics has to build from existing formations *and* from a broader global majority (in both North and South) that is most likely to reject the world of exhaustion, not to bind themselves to the oars of the armed lifeboat.

In an analysis of any historical moment, particularly one of deep, fundamental crisis, one finds contradictory, overlapping, and often irreconcilable interests. As W.E.B. Du Bois observed in *Black Reconstruction*: "the proletariat is usually envisaged as united, but their real interests were represented in America by four sets of people: the freed Negro, the Southern poor white, and the Northern skilled and common laborer." He famously notes the *intra*-class divergence within "the working class" during the Civil War. As he argued:

> It must be remembered that the white group of laborers, while they received a low wage, were compensated in part by a sort of public and psychological wage. They were given public deference and titles of courtesy because they were white. They were admitted freely with all classes of white people to public functions, public parks, and the best schools. The police were drawn from their ranks, and the courts, dependent upon their votes, treated them with such leniency as to encourage lawlessness.[113]

In the real conjuncture, politics is not drawn from the theoretical abstract, but from the real subjectivities engendered. Not merely subjective interest, but also structurally conditioned *affective* experience or orientation. In the , this wage (psychological and literal) was, as Du Bois put it, a "bribe" which sundered the class project.[114] The project of emancipation, of abolition, could not suffer the dogmatic assertion of objective abstract class. Like *all historical social change* — certainly on the scale of climate politics — it assembled and crystalized, at least for a time — a heterogenous social subject, like Fanon's anti-colonialism. Or one could phrase it as Gilmore does:

> Unions that represent low-to-moderate wage public sector job, which have a high concentration of people of color as current and potential members, might join forces with environmental justice organizations, and biological diversity and anti-climate change organizations, and immigrants' rights organizations, and others to fight on a number of fronts against group-differentiated vulnerability to premature death — which is what in my view racism is. And if that's what racism is, and capitalism is from its origins already racial, then that means a comprehensive politics encompassing working and workless vulnerable people and places becomes a robust class politics that neither begins from nor excludes narrower views of who or what the "working class" is.[115]

The door may be open for any to join the project of abolition, but it does not wait for social consensus. It needs only enough consensus to form a coherent political subject. As Jodi Dean perfectly quips, "anyone but not everyone can be a comrade."[116] The world of the extractive circuit is rife with contradictions, much like previous moments of intense social struggle, in organized labor as much as anywhere else.

You have cases where industrial unions like those gathered in Kerala's Standing Council of Trade Unions (SCTU) are in conflict with agricultural unions like the Nation Fisherman's Front. You have cases where industrial unions like the National Union of South African Metal Workers (NUMSA), the largest trade union in South Africa and famous for developing one of the paradigmatic "just transition" models in its 2012 "Climate Change and Class Struggle" seminar (yes, really), are pitted against *other* industrial unions like the National Union of Mining (NUM). You have cases where mining unions like the Bougainville Mining Workers' Union (BMWU) in Papa New

Guinea or the Rössing Mineworkers Union (RMU) in Namibia fought tooth and nail in concert with environmental social movements. Or cases like those described by political ecologist Stefania Barca surrounding Italy's Taranto steel plant — the most polluting in Europe — in which three separate unions with three separate frameworks all came into conflict. And even in the US energy sector, you have a case like that of the United Electrical, Radio and Machine Workers (UE) rallying to a climate agenda in concert with activists from the broad-based Sunrise movement. UE is *considerably* smaller than the IBEW, but it's a far more promising site of attention if you are looking to build a foothold among unionized electrical workers in the United States or, indeed, if you are interested in galvanizing American labor. Not long after the initial collaboration between UE and Sunrise, UE and the Democratic Socialists of America (DSA) set up the "Emergency Workplace Organizing Committee" (EWOC) as an alternative class organization, a "support desk" for labor and other movements. Such alliances are far more useful to the current moment, but Huber ignores them because they don't fit his pre-fab, transhistorical model.[117]

This is hardly an exhaustive list, but it represents cases illustrative of actual working class socioecological formations. The idea of an empirically unified working class defined in the narrow abstract sense is patently false. Sectoral position, while far from determinative, weighs heavily on ecological orientation.[118] Ideological orientation, downplayed by Huber, is actually vitally important (if still not determinative). Painting all environmental organizations as ineffective PMC mobilizations is not only untrue; such organizations — and social movements more broadly — are key to generating militant ecological struggle.[119]

Left-wing movements are not hobbled by "centering marginalized peoples" — as has become *doxa* among so many Anglophone "Marxish" thinkers. They are *strengthened*. Such populations, as we've seen, are the *sine qua non* of any future left politics. The unions and broader working class formations that worked in concert with informal workers and placed issues of social reproduction and women's liberation at the forefront, who viewed the shop floor as one coequal site of struggle across society and states, were both the most radical and most effective. NUMSA's plan is not simply a green jobs program: in addition to calls for a swift transition to renewables, it identifies capital accumulation as the key driver of climate change; it emerges from and emphasizes this broader social and political struggle. This is true not only in the Global South but in the North as well. And the most important working class union ecologies and left-wing unions in general were explicitly internationalist, not just in rhetoric but in practice.

However, each of these cases faces a fundamental impasse. NUMSA — still the largest union in South Africa — has been viciously attacked by transnational capital, the right-wing government, and NUM at different times and in different formations. It is now on the back foot, despite being massive, militant, and Marxist. The UE/Sunrise alliance holds promise, but only as a sliver of a counter-hegemonic bloc. And then there are the unlikely alliances across the world that Murat Arsel has dubbed "the environmentalism of the malcontent." As Alyssa Battistoni argues, a social-reproductive focus changes how we see both workers and capitalism:

> We should see workers performing the work of social and ecological reproduction, whether for wages or not, as workers who can form part of the political force for left climate programs. Yet the disjuncture between the sites of militant labour struggle and the sites of capital accumulation [...] is likely to produce new contradictions as the former build power.[120]

The discomfiting reality is that Climate Lysenkoists simply do not share her ecofeminist commitment to undoing the logics of extraction and exhaustion. Huber's "manipulation," his bribe, what Barca derides as "jobs blackmail," reveals how, as Barca puts it, "labour's ecomodernism is a conservative strategy, built around the defense of production."[121] A serious ecological focus would not put industry-friendly tech ahead of actual mass necessity; a genuine focus on organized labor would engage the new structures of global value chains and production networks — including informal and social-reproductive and non-workers — in thinking socioecological politics and its challenges.

Huber seems unconcerned with how changes in the structure of global production and demographics have rendered "traditional" strikes far less effective.[122] Strikes with more reach, like the 2018 US teachers' strikes, took the form of broad social protest aimed at state legitimacy, as opposed to the "classic" model of work stoppage.[123] Meanwhile, the largest traditional strike in human history, the Indian general strike of 2020 — which brought together most major unions, center and left-wing parties, industrial and agricultural workers, numbering some 200–250 million people — didn't amount to much.

Climate Lysenkoists' methodological (and ideological) nationalism[124] is at odds with historical realities of labor and politics *and* climate and technology, where full pictures are only clear with an analysis that, just like the extractive circuit, crosses and links so many boundaries and borders.

## Just as Radical as Liberalism Allows

When thinking about *power*, strikes are part of political ecological struggle, but they are hardly the key or only source of power. Extreme cases like the RMU or BMWU are particularly illustrative of real political struggle that the professionally demure Climate Lysenkoist would like to avoid: unions working in concert with environmental and other civil society organizations were met with the full power of state and private violence and responded in kind (as is conveniently written out of many just-so stories of the origins of American labor).

In facing the same enemy — the Anglo-Australian Rio

Tinto multinational metal and mining firm — the RMU became part of the Namibian struggle for independence.[125] The BMWU itself took up arms and guerilla warfare both against the firm and also towards independence.[126] As we've already seen, this kind of confrontation — which characterized the nineteenth and early twentieth-century rise of the American labor movement as well — is increasingly prevalent in the metropole. For example, violence against native Americans is well-documented around both social and climate struggles. As Gilmore suggests, we should regard mass incarceration and policing as the violent subordination of the growing masses of surplus populations. And as we've seen, this kind of colonial relation, if still attenuated, is more omnipresent. Police, internal security forces, and private security forces are increasingly prevalent in the violent repression of protest — including both labor and environmental action — across North America and Europe. Sometimes the same security forces that keep workers like those at Rössing "in line" — that keep the uranium cheap for the techno-mystics who so covet it — are those who round up union organizers and environmentalists under the guise of "anti-terrorism" or more general "law and order" policies in the metropole.[127]

It Is not that labor action is now irrelevant or that violence is the principal mode of struggle. It is rather that something like a left-wing climate realism will work across a host of terrains ranging from social protest and movement to labor action to electoral and state politics to violence and more. Logistical weak points in global value chains are susceptible to sabotage by organized social forces including but not limited to labor movements. One of the most powerful levers in a world of frictionless capital flows is denying it a place to land. This can involve everything from novel experimentation to complex transnational coordination. We will return to strategy and tactics, but the idea that fossil capital will bargain away its existence in a legally ordained agreement, or even submit to a legislative or administrative edict without bringing the weight of state and private violence to bear, is absolutely insane.

Huber is little interested in popular uprisings, spontaneous social protest, and other such social phenomena outside the workplace, the electoral sphere, or the extension of the leadership of the workplace *over* other struggles. McAlevey dismissively calls such phenomena "pretend politics." From Occupy Wall Street to Black Lives Matter, these are trivial exercises in a kind of activist *jouissance*.[128] Similarly, Huber sees little strategic hope in the NoDAPL, Keystone XL, and other pipeline protests. These, he argues, are only moderate, defensive successes. Yet his political model (which has not generated even defensive ecological successes) is *precisely defensive*. As we've already seen, the attack on trade unions in the neoliberal age was part of a massive clawback to revive profitability, but it was one that took place on many terrains. Where riots broke out — indeed broke away from strikes — Marx saw promise. Such moments are the building blocks of politics and a potential political subject for Marx, Fanon, Gramsci, Luxemburg, Du Bois, and so many others. These moments are the beginning of something broader, a political consciousness taking shape. There is no magic trick — whether the supposed strike lever or Blanquist revolutionary volunteerism — that defines the potential power here.

One need not endorse a quixotic, totalizing revolution to see how the model of labor-firm, or even labor-sector, compromise Huber celebrates is ill-equipped for a zero-sum climate politics that may not necessitate "full socialism" but is just as radical as reality demands. As the developmental economist Cédric Durand puts it, "a smooth transition beyond carbon is no longer an option. There is no Pareto-efficient way of eradicating fossil fuel use in a timeframe compatible with the prevention of climate disorders. A zero-sum or even negative-sum game is in play."[129] Huber postures as the hard-nosed class warrior but actually champions an impossible, technologically ameliorated capitulation. The contemporary moment of "unbundled" value chain and supply chain production makes even a relatively fixed sector — like electrical utilities — difficult to organize.[130] But it also opens up new political possibilities that are connected across the supply chain, reflecting genuine new potential and genuine new *power*.

Or rather, one might say, left-wing climate realism must look to those groups, blocs, masses, and class fractions taking shape in rather different forms as divided by the socioecological moment.

Just as Stalin dreamed of catching up to a fantasy of American modernity, Climate Lysenkoists dream of a fantasy-land mid-century "Golden Age" whose unique historical and ecological conditions no level of CDR will ever be able to recreate. Worse, they fail to consider that something *much better* is both possible and desired — including by a majority of Americans. We've already seen that, as much as it was predicated by the compact between labor and capital to renew capital accumulation, it was equally dependent on the injection of fossil capital in the form of oil, cheap food in the form of petro-farming and the so-called "Green Revolution," cheap labor from racialized and feminized populations, massive draining of resources from the Global South, geopolitical competition from the Soviet Union, post-war rebuilding, and more. But as Aaron Benanav points out, in terms of labor itself, Americans settled for a set of protections weak by global standards (North or South). Meanwhile,

In European and wealthy East Asian countries [...] postwar labor market institutions were not designed by left-wing governments but by right-wing politicians who emphasized the importance of national-imperial identities, the formation of male-breadwinner households, and the maintenance of relatively fixed workplace hierarchies. In return for accepting corporatist arrangements, male heads of households received substantial job protections: unlike in the United States, regularly employed workers were not hired and fired at will.[131]

It's easy to see why the Heidelbergs and the Mitsubishis, the

Shells and the BPs, the Bezoses and the Musks, are so enamored with the techno-mystical world of so-called "ecomodernism." Theirs is a clear-eyed, right-wing climate realism. Their vision — limited in scope to a chosen few outside their class — makes perfect sense as a conservative project to maintain or even enhance the status quo long into the future. They might massage futuristic messaging, but they don't actually live in a fantasy world. In contrast, Climate Lysenkoists have to retreat to the "mist-enveloped worlds" of techno-mysticism and political naïveté, as (unconsciously or not) handmaidens to right-wing climate realism.

## Once More with (Class) Feeling

For all the lip-service to a working class agenda, Climate Lysenkoism doesn't actually think much of workers, even those placed at their prized tip of the spear. For all the bluster about "Marxism" and scientific analysis, this vision of "class" politics is predicated on a fundamentally *cultural* projection of "the worker." *Real* workers are industrial laborers, predominantly male, and *dumb*. This is no kind of left climate politics or Marxism. Huber correctly understands class formation to be anything but automatic. However, he turns patronizing to solve this; political education is not an actual process by which subjects connect grievances, passions, social immiserations, and so on with their actual social and ecological conditions into a form of consciousness. Workers won't understand "complex" science. They need "easy-to-understand material gains."[132] The only way this dichotomy makes sense is if one assumes that connecting complex analysis to material gains is beyond the ken not only of workers but of people in general.

As C.L.R. James astutely argued, this was a key error of Toussaint Louverture in the Haitian Revolution: "It was dangerous to explain but even more dangerous not to explain." The full scope of the Haitian Revolution at that moment was nothing less than total independence, but Toussaint did not trust the people with the risky truth. At the end of an arduous conflict with the Spanish in 1800, Toussaint feared that the Haitian masses would not understand the ends of abolition and independence; but even more, he did not want to present the masses with the discomfiting truth that more perilous struggle and loss lay ahead. The officer with the lofty Enlightenment ideals would not countenance such risk. It was the more politically astute — if less formally educated — Jean-Jacques Dessalines who would broach the reality: "the war you have just won is a little war, but you have two more, bigger ones. One is against the Spaniards who will not give up their land [...] the other is against France who will try to make you slaves again as soon as she has finished with her enemies."[133] This tension raises deep questions of political education, strategy, and more to be addressed in the next chapter. And no matter how well James's critique fits, it's preposterous to compare the petulant politics of Climate Lysenkoism with a towering figure like Toussaint. Beyond the tensions already cited, the connective tissue is not fear of demobilization but — bringing us back to Norgaard's analysis of the flight to conservative safety — *fear* of the people themselves, to whom they serve up a fantasy of conservative stasis, a promise of positivity to quell their own managerial anxieties and insecurities.

Indeed, pulling a page from the self-help and positive psychology playbook, Climate Lysenkoists (in tune with many currents) will only countenance "positive" messages where the numbers run up.[134] The only other option is "negative" messaging of "dangers," "disasters," and "traumas." None of this holds up to even mainstream, let alone radical, psychological and sociological scrutiny.[135] The full affective register is dismissed, along with a goal of material, political subjectivity. A left-wing climate politics must take seriously how those conjunctural formations are always riven — not only by already-given, *material,* non-class identifications, structural locations or social positions, segmentations, and fractions, but also by equally *material* affects, feelings, passions, and emotions all vital to political possibility. It is not "positive" messages about social gains; it's "social desire" for the world to be otherwise. "Negative" affect doesn't have to be fear-mongering or eschatological apocalypticism. Where is the *hatred* for the class enemy? Where is the joy in collective action? Where is the anger or disgust with contemporary conditions? Where are the "infrastructures of feeling" (Gilmore), the "affective infrastructures" (Anderson, Dean, Bosworth), the "structures of feeling" (Williams, Thompson)?

Not only is affect central — if often underplayed — in political theory (think Machiavelli's discussion of the love and fear a people might have toward their prince, and the fear or love a prince might have of his or her people, or Hobbes' discussion of anxiety and his underexplored desire for "commodious living" as the central connective motivation for his theoretical commonwealth to form[136]), it is far from alien to Marxism. If anything, Berlant is understating the case when they write:

> Marxism has a long tradition of interlacing descriptions of the present across relations of ownership and control, the reproduction of labor value, and varieties of subject position with the affective components of labor-related subjectivity. It is not claimed that subjects *feel* accurately or objectively historical — this is why the concept of ideology has to be invented — but this tradition has offered multiple ways to engage the affective aspects of class antagonism, labor processes, and communally generated class feeling the emerges from a zone of lived structure...[137]

The long history of Marxist writing on the subject is already theorizing affect. Immediate feeling must proceed into a more formally articulated grievance which must relate to material reality. Berlant has in mind Marx himself, but also Lukács, Benjamin, Adorno, Jameson, and Raymond Williams.[138]

If one is suspicious of so-called "Western Marxism" (a category error I address briefly in my conclusion), one can easily add revolutionary figures to this list. Lenin on how "leaflets," reflecting the miserable reality of working life, work on and through "passions" which in turn provoke further writing from workers and *all* dominated social elements — the passion of industrial workers is affected by "the exposure of the evils in some backward trade, or in some forgotten branch of domestic industry" and indeed "all cases of tyranny, oppression, violence and abuse, no matter what class is affected." Restricting this to the "economic struggle" alone is not only an error but "harmful and extremely reactionary in practice." In practically Spinozist terms, understanding and feeling are two sides of the same coin.[139] A pattern even more descriptive of Mao's emphasis on "class feeling," "speaking bitterness" about individual material experience, engendering both outward antagonism and inward bonding, but similarly, such "that no one is loved or hated without a reason."[140] In this, "CCP leaders appreciated what Western social scientists are only beginning to understand: emotions cannot be dismissed simply as a residual, irrational domain of consciousness. Rather, emotional gestures and utterances hold a unique capacity "to alter the states of the speakers from whom they derive."[141]

Or Luxemburg, for whom "class feeling" arises in and from understanding the "intolerable" life of "social and economic existence"; radical political practice requires "the most adroit adaptability to the given situation, and the closest possible contact with the mood of the masses."[142] And of course Fanon's discussions around "affectivity," whose examination is key in the next chapter. None of these positions are *exactly* the same, but each demonstrates in different ways the necessity of affect and the full affective register for mass politics. They also all perform critiques of *failed* affective politics. Across this broad spectrum of relatively early thinkers, affect is not just some "autonomic activity," in Berlant's phrasing, but a specific relation to historical conjunctures.

Today, one is most likely to encounter affect in Marxist thinking through a series of related concepts: "infrastructures of feeling," "structure of feeling," and "affective infrastructure," although this is hardly an exhaustive treatment. Gilmore elaborates the concept of "infrastructures of feeling" by arguing,

> In the material world, infrastructure underlies productivity — it speeds some processes and slows down others, setting agendas, producing isolation, enabling cooperation. The infrastructure of feeling is material too, in the sense that ideology becomes material as do the actions that feelings enable or constrain. The infrastructure of feeling is then consciousness-foundation, sturdy but not static, that underlies our capacity to recognize viscerally (no less than prudently) immanent possibility as we select and reselect liberatory lineages — in a lifetime [...] as well as between and across generations.[143]

Drawing on Raymond Williams' "structure of feeling," particularly in its mode of the accretion of affect over time and across space, Gilmore recasts the Black Radical Tradition as an accumulation of structures of feeling, which underlines "the productive capacity of visionary or crisis-driven or even exhaustion-provoked reselection." Feelings — "the least coherent aspect of human consciousness" — can "enable or constrain" for Gilmore.

Williams' "structure" or "structures of feeling" is itself notoriously difficult (as well as constantly cited, along with Benjamin and Fanon, as a connection between historical materialism and contemporary affect theory). "Structures of feeling," for Williams, capture "a kind of feeling and thinking which is indeed social and material, but each in an embryonic phase before it can become fully articulate," but which "exert palpable pressures and set effective limits" long before they are fully understood. "The term is difficult, but 'feeling' is chosen to emphasize a distinction from more formal concepts of 'worldview' or 'ideology' [...] specifically affective elements of consciousness and relationships: not feeling against thought but thought as felt and feeling as thought." Structures of feeling are "social experiences in solution" — in between formal social structure and "specific articulation in material practice." Structures of feeling characterize a given time in a broader, more fluid way. Williams developed the concept over a long period of critical dialogue with Paul Gilroy (as Gilmore notes) and E.P. Thompson. Thus, by its later articulations, it contains the idea of "differentiated structures of feeling to differentiated classes."

The example Williams uses is the English Civil War, in which two structures of feeling can be identified in literatures, "though neither is reducible to the ideologies of these groups or to their formal (in fact complex) class relations."[144] Thompson took up the political implication in emphasizing that

> class happens when some men, as a result of common experiences (inherited or shared), feel and articulate the identity of their interests as between themselves, and as against other men whose interests are different from (and usually opposed to) theirs [...] [as a] structure of feeling.[145]

Here we might begin to think through palpable pressure and embryonic affect, of the inchoate affects of socioecological climate, of the novel affects attributed to climate directly, but also much more broadly of the affective matrix of exhaustion.

Theorizing the "affective infrastructures" of "pipeline populism," the American geographer Kai Bosworth notes the limitations of an amorphous account of "the people" as the political subject of climate change, looking at the various groups that gathered — and then fell away — around the Keystone XL and Dakota Access pipelines.[146] In examining these movements, Bosworth notes how "emotion emerges from political-economic contexts and material landscapes, non-deterministically conditioning political struggles."[147] However, the transient, semi-spontaneous eruptions of "the people" do not confirm concepts like Ernesto Laclau and Chantal Mouffe's transient "chain-of-equivalence" or "collective wills," but rather point toward the need to understand how affect can inform and be differently organized into a heterodox Marxist class subject composed (following Gramsci, Hall, and Marx) of myriad relations and social groupings and blocs produced by capital in this moment.[148] In all of those Marxist accounts, this limitation is cited as the danger of affective politics — emotion, feeling, affect alone is simply manipulation without material grounding, cause and effect, any sense of structure or differentiation. It slides easily to the right. "Left populism" posits affect as purely discursive. In contrast, for theorists like Berlant and Sianne Ngai, affect is historically and materially situated.[149] And Bosworth's "affective infrastructure" is of course material too: "however complex, the afterlives of the blockade channeled everything from vengeance and trauma to humor and solidarity [...] I was surprised when I saw some of the farmers and ranchers I knew from KXL meetings at the more radical DAPL blockade."

Across significant conceptual differences, all these classic and contemporary understandings of materially grounded affect — often calamitously ignored on the left — require heightened attention in the case of climate politics. Affect is not crucial because, as some theorists posit, the free flow of affect is liberation in itself.[150] Nor does affect, any more than social structure, organize or produce political subjectivity on its own. "Grievances" and "aspirations" are fundamental to any understanding of class. Rather, as with Fanon's analysis of decolonization or Du Bois' of the Civil War, the particular temporal intensity of our moment cuts across those clean *theoretical* lines of social division; it fractures actual social life within them. "Class," "classes," "hostile camps," "historical blocs," provide a rubric for "stretched" mass political subjectivity in the socioecological conjuncture — its "who, whom?"; its strategic capacities and agencies.

In any moment, the unique configuration of a mode of production defines the terrain on which *political* subjectivity is possible, on which actual struggles take shape and might connect across already given, material, non-class identifications, structural locations, or social positions and segmentations. But to adapt Fanon's language again, the "immediacy" and

136

"urgency" of climate politics bring the affective elements more radically to the foreground. "Generalized affectivity" is as much of a guide to political organization for a left-wing climate realism as social position. It's not simply feelings of exhaustion that course through the extractive circuit; it's the potential for a burgeoning disgust with the world of business-as-usual, detachments from its good-life fantasies, or the seductive safety promised, albeit with subordination, by existing wealth and power, by capital writ large, in right-wing climate realism, aboard the armed lifeboat. The extractive circuit not only sets the stage on which its own "gravediggers" are possible, it *structures*, engenders inchoate disaffections, desires, disgusts, attachments, and grievances.

We must attend to these inchoate, sometimes at best pre-political affects, as with Ngai's discussion of disgust — whose "intense and unambivalent negativity [...] seems to represent an outer limit or threshold of what I have called ugly feelings, preparing us for more instrumental or politically efficacious emotions."[151] Such affects mark the *potential* to draw an increasingly stark line of political conflict. What this means in more ordinary terms of politics, strategy, organization is discussed in the next chapter. For the moment, we focus on the articulated, the nascent, and the inchoate disaffections with the world of business-as-usual, on the shifting planes of desire, which is also a terrain of struggle. Desire not in place of "negativity" but as its complement. Desire not for *fantasies* of business-as-usual futures, dressed in any political color, or "cook shops of the future," "bribes," or blackmail but, as the queer theorist José Esteban Muñoz writes, "these pictures of utopia (a term that is used in later comments Adorno makes in the dialogue) [that] do the work of letting us critique the present, to see beyond its 'what is' to worlds of political possibility, of 'what might be.'" Such pictures give greater form to the latent dreamworlds truly fettered within the present.

"The education of desire," Thompson calls it, rehabilitating William Morris and the salutary form of utopianism which

Marx argues is usefully pedagogical. "When it succeeds, it liberates desire to an uninterrupted interrogation of our values and also to its own self-interrogation [...] the recourse to utopian writing signifies exactly the desire to make a breakthrough, to risk an adventure, or an experience, in the fullest sense of the word, which allows one to glimpse, to see, or even to think what a theoretical text could never, by its very nature, allow us to think, enclosed as it is within the limits of a clear and observable meaning."[152] Not the free flow of directionless desire or desire as simply narrow quantitative increase, but an understanding that: "Desire may actually indicate choices or impose itself as need [...] for to suppose that our desires must be determined by our material needs may be to assume a notion of 'need' itself already determined by the expectations of existing society. But desire also can impose itself as 'need' [...] the form of more open choices between need."

Intervening here might begin with the most prosaic facts, *limits* even, but within them are a spectrum of elements which don't draw a precise blueprint but begin to give form to multiple *disaffections* with the world as it is, that speak to the desire for things to be otherwise.

## The Minor Paradise of a Sustainable Niche

During the waves of social upheaval that followed the 2008 financial crisis — the movements of the squares, the Arab Spring, the various iterations of Occupy — Walter Benjamin's proposition, "Marx says that revolutions are the locomotive of world history. But perhaps it is quite otherwise. Perhaps revolutions are an attempt by the passengers on this train — namely, the human race — to pull the emergency brake," appeared, apparently independently, in street graffiti in multiple languages across the world.[153] Pulling the brake is quite different from shifting to reverse. The rupture allows something *new* to emerge, to capture possibilities unrealized. Something different.

Lateral — to use Berlant's phrasing. It also speaks to the desire to *stop* the machine, to slow down, the desire for rest, relief, for the inversion of exhaustion.

In the simplest formulation, the *project* of a left-wing climate realism is to carve out a sustainable global human ecological niche capable of supporting the flourishing of some 7–9 billion humans. An ecological niche "is a term for the position of a species within an ecosystem, describing both the range of conditions necessary for persistence of the species, and its ecological role in the ecosystem."[154] The term delimits the project. Jason Moore's argument that capitalism "organizes nature" is helpfully translated here. What capitalism organizes is this ecological niche; Malm's fossil capital crucially underwrites the "Great Acceleration" of our niche. The niche helps us understand that climate mitigation and adaptation is not about "all of nature," but rather about promoting and securing a specific set of "conditions necessary" which are always simultaneously also about the broader niche ecosystem.

The niche is, by definition, Anthropocentric, but that does not mean a logic of Baconian domination; it simply means that humans have already demonstrated and will continue to demonstrate in any outcome our agential capacity in unique ways within this niche far beyond the capacity of other biotic and abiotic elements. Such a niche is necessarily global; humans are a dominant planetary species. The concept is also global, though the specifics for any given state or geography are "both affected by the environment, and at the same time affect the environment for other species."[155] It is also global in the sense of the *necessary* convergence of global material throughput use, of removing as many sites, nodes, zones — social and ecological — as possible. Carving implies a transformation in the metabolic relationship between society and ecological niche from a fear-driven domination to an ongoing reconciliation.[156] Ecological maintenance concerns, against romantic imaginaries, do not vanish with the horizon of climate mitigation and adaptation. They are an active and continuous intervention to promote specific conditions; as the very concept of the Anthropocene suggests, humans as a geologic force are here to stay.

Already we are off the roadmap of many of today's heated debates about technology, development, modernity, and more. This project is *unexpected*. It captures more of the original humor and joy of the meme of fully automated luxury communism (FALC) (or the even better fully automated luxury gay space communism) than the actual (impossible) attempts to articulate FALC as a remotely plausible reality. It captures the wild personal and collective *life*, materially and temporally rich, that is missed, obscured, or moralized in so much "degrowth" thought. The project of what I am calling the "minor paradise" of the sustainable niche does not lie between the two. It is a project whose "urgency" and "immediacy" (to use the same Fanonian terms we will return to in articulating the subjectivity, agency, and strategies of left-wing climate realism) mark a waystation (neither a beginning nor an end) largely perpendicular to either.

In purely quantitative terms, even the highly restricted IPCC posits a rough estimate of mitigation and adaptation costs around \$31.2 trillion in the immediate term.[157] This is not just spending or investment — a great deal of what is being measured here are stranded assets from the fossil fuel industry. Such quantitative accounting is highly complex, particularly depending on whether it is only energy that is being discussed or other aspects of climate change and planetary boundaries. As the environmental policy scholars (and IPCC authors) Arthur Rempel and Joyeeta Gupta note, common estimates range "between \$16 trillion and \$200 trillion" depending on a number of factors, particularly the downstream effects of rapid energy transitions or the costs of mitigation and adaptation in other areas.[158] They use the unique conditions of the COVID-19 lockdown to help model additional effects in areas like financial markets, where they observe 30–40% losses in relatively diverse portfolios, demonstrating a further immediate loss of wealth in likely rapid decarbonization scenarios.

Pegging a number, particularly a dollar sign, is even more difficult when moving from the costs of transition to the socioecological parameters possible in the near term for a sustainable, flourishing niche. Attempts to do so admittedly are always working on rough sketches and with the near-universally agreed upon problem of using GDP as an even remotely satisfying benchmark. Even so, there is surprising convergence in disparate studies. Examining necessary socioecological outcomes and energy use, scholars from across the world come up with remarkably similar quantitative pictures.

A European interdisciplinary team of data scientists and physicists modeled several potential goldilocks scenarios, where energy consumption (inclusive of production, transport, and building  costs) results, in a large but imperfect global convergence, with maximum energy use (post transition) capped at around 30–40 GJ/per capita. This translates, again with imperfection, to a basket of social goods (housing, education, healthcare, food, transportation, etc.) and expenditure incomes in a range from $10,000 to approximately $66,000, with a standard deviation converging on $26,000.[159] Meanwhile, Chinese scholars (including Dajian Zhu, one of China's leading ecological policy advisors) argue for a similar basket of goods and an income of roughly $20–30,000.[160] This calculation involves not only energy but material footprint and planetary boundaries *tout court*. This goal was set into policy in the 14th Five Year Plan, which is also the first to abandon explicit GDP growth targets in favor of broader "indicative" economic measures, in tune, as Zhu remarks, with the "ecological civilization" goal of "development that is both high quality and low-carbon."[161] (Which will only prove possible if China flexes its almost globally unique power to strand assets — wipe out wealth — at will, particularly its rebounded coal plants, no matter how efficient. In turn dependent on geopolitical dynamics, centered on American de-escalation, and described further in the next chapter.)

Both analyses accept "the contradiction between economic

growth and environmental protection."[162] However, the former is approaching a sustainable niche from the perspective of a developed Northern economy, while the latter is looking at climate from a still-developing point of view. The European team is looking to "degrow," but they call into question some of the more radical claims of degrowth. Meanwhile, the Chinese team sees the "ecological civilization" model as an alternative, greener path to development, not only for China but as a potential model for the whole Global South, emphasizing the necessity of South-South cooperation for technology transfer and "leapfrogging."[163] In tandem, this is something like a global model — not the only one, to be sure, but a genuine one to carve out a sustainable human ecological niche.

As members of the European team quip, discussions of limits are pilloried as returning to caves, but the caves are essentially social housing with

> highly-efficient facilities for cooking, storing food, and washing clothes; low-energy lighting throughout; 50 L of clean water supplied per day per person, with 15 L heated to a comfortable bathing temperature; they maintain an air temperature of around 20 °C throughout the year, irrespective of geography; have a computer with access to global ICT networks; are linked to extensive transport networks providing ~5,000–15,000 km of mobility per person each year via various modes.[164]

The "cave-dwellers" all have mobile devices (forced obsolescence having been abolished and cleaner extraction and production implemented) and considerably reduced workloads. But even this detailed quip — or the more grandiose pronouncements of the CPC's "Pursuing Green Development and Promoting Harmony between Humanity and Nature" — still leaves us in the world of gray policy. In the final chapter, I will return to these approximate numbers, which techno-mystics decry even though they are majoritarian improvements in places like the

US and UK. But for the moment I want to linger on how these models and plans for thriving life within planetary boundaries point beyond a set of policy prescriptions and development goals; how this is pulling the emergency brake; how the technocratic language obscures the vivid, quasi-utopian — even if far from perfect — lateral expression of the sustainable niche.

I have borrowed the phrase "minor paradise" from my longtime colleague Rebecca Ariel Porte. A minor paradise is not a utopia; it stands in contrast to Arcadian and Golden Age fantasies (one of many places in which the "degrowth" and "techno-mystic" views converge) and even with "paradise" as an ordinary concept. It is not Milton's "lost paradise"; it is rather an "earthly paradise" always imperfect in contrast to its *imaginary* prelapsarian Christian cousin. "An image of earthly paradise discards perfection for the more limited — and the more dangerous — proposition — that the world might be differently arranged and more vibrantly [...] to make the best of flawed materials," writes Porte. The imperfection is the mark of its non-utopian nature; the limits are definitional: there *may* yet still be realms of absolute freedom (or as Benjamin reminds us, the hell on Earth that we call status quo), but the minor paradise is not any end. It is where history is always beginning, in multiple voices. Minor paradise is a splinter in the eye separating "the way things are" from "other courses history might have taken," and "things as they might be." To think of a minor paradise is to take account of "missed opportunities, failed potential, counterfactuals, what could have been and wasn't, what might be and isn't."[165] The splinter in your eye is the best magnifying glass, as Adorno remarked. Boundaries, limits, walls — paradise is not always *good* (think national borders and colonies vs. gardens), but it *can* be the overflowing abundance, rummaged from histories and imaginations, technologies and practices, within enabling constraints.

So much literature that has to do with ecological policy and politics either desperately stalks about for a sound way in which "the way things are" can continue, or concedes terrain with deflating theoretical language: *de*growth, *alternative* hedonism, *decent* living, *sufficiency*, and so on. Sometimes the concepts can be partial and incomplete, other times highly illuminating (Katie Soper's work on alternative and post-growth pleasure is exemplary), but the posture and imagination are often still *defensive*. Thinking of the sustainable, flourishing ecological niche as a minor paradise frees us from thinking of it synonymously, impossibly, as some kind of ultimate utopia of the absolute sublation of all ills, but also shifts the affective register. The "caves" joke is an example of the humor that Nicole Seymour notes is relatively rare in formal climate work and activism. But even there, the framing is, "well it's not *that* bad." Instead, we might fully voice that mass climate adaptation and mitigation is *better*. The Chinese government documents — vacillating between the dry prose (no matter how accurate) of official planning and the overblown rhetoric of Party slogans (no matter how true) — are hardly an improvement. "Ecological Civilization" may sound nicer than "balancing global material throughput use towards equality and sustainability." But not by much. These all fall short of conveying the respite, the relief, the abundance, the wild, radical possibilities of the sustainable niche.

This is not just rhetorical, another "bribe" or "manipulation" to garner popular enthusiasm. It is the same brutal honesty — recall Dessalines — which cannot avoid the stark, potentially frightening truth about zero-sum political struggle. Of course, any counter-hegemonic project needs a *counter*. But here the counter does not concede ground — not to the bad faith accusation of the "politics of less," and not, in earlier eras, to the burgeoning ideologues of capitalism and colonialism. In Marx's *Manifesto* or Fanon's *Wretched of the Earth*, the *j'accuse* is turned on its head: You claim that communists wish to do away with property? Yes, we do, your property, so that all may thrive. You claim that communists wish to abolish the family? Well, if by that name you mean *your* idea of family, or the conditions under which working class families are already decimated and decaying? Why yes, we do. What we are generating is *better*: new and old forms of kinship, liberation from domestic bondage, and social reproduction — better examined and explicated by later Marxist feminists from Silvia Federici to Sophie Lewis.[166] The colonizer loudly and proudly proclaims the barbarity of the native and the civilizing mission of colonization; Fanon retorts that it is only the decolonized who move beyond barbarism, beyond these meager limits of the European ruling class.

The minor paradise of the sustainable niche is both an expression of *desire* and a counter to the provocations of small, impoverished, exhausted imaginations. Gray but necessary policy proposals for building new social housing (where needed) and retrofitting existing supply give way to recuperating the proverbial hanging gardens of Babylon through contemporary "hortitecture" or "agritecture" in urban/peri-urban farming and vertical forestry. As the architect Võ Trọng Nghĩa observes, Ho Chi Minh City is starved for green space and sustainable building.[167] Vietnam — a unique outlier in the dataset compiled by Steinberger et al. for "A Good Life for All Within Planetary Boundaries" — meets the most social goals while transgressing only one planetary boundary.

Most "Green Architecture" is high-end, luxury housing — "green" apartheid housing. Nghĩa, though sometimes engaging in that kind of work, has specialized in low-energy, low-cost, high-quality-of-life housing — from modular, long-lasting, energy efficient bamboo structures in the Mekong delta to the "Farming Kindergarten" in Đồng Nai. Many such projects involve plant-covered facades and draped balconies; local, low-carbon, and long-lasting recycled materials; style and vegetation to fit social expectations and local ecosystems; green facades, forestry, and rooftop farming that provide passive cooling, thermal insulation, and capture air pollutants and water, a vitally important adaptation given Vietnam's ever-increasing flooding.[168] The kindergarten is a public school directly adjacent to a shoe factory; factory workers' children live in the entirely passively cooled and heated structure. The roof gardens —

which are accessible to children, workers, and members of the surrounding community — not only contribute to the site's beauty but also produce significant produce. Wastewater from the factory is recycled into the agricultural plots. Even within existing conditions, costs were $350 per square foot, low even by Vietnamese standards, despite the site's notable (and internationally acclaimed) design quality. The project is meant to be a prototype; Nghĩa and other Vietnamese architects, engineers, and planners hope to use similar syntheses of vernacular techniques and cutting-edge research to retrofit Ho Chi Minh City. Nghĩa and others emphasize the social and ecological benefits of these projects, but also note how they fit into a slower, saner imagination of the world.

*Image 1: "Farming Kindergarten" in Đồng Nai.*

Such design paradigms exist and are possible the world over with varying materials and setups based on local socioecological conditions.[169] Their scaling is held back by many sectors of capital — which profit off of accelerating energy use for building, air-conditioning, and more, as preached by the prophets of techno-mysticism. This is not some fantasy scenario; neighboring Singapore is already well on its way towards these kinds of transformations.[170] The hyper-capitalist city-state applies strong regulatory control over key areas like housing, construction, and transportation. Such building and planning is neither a purely Southern nor Northern phenomenon. You have cases in Europe, like the reclamation of a defunct Coca-Cola factory into a social housing project in Wienerberg, Austria, complete with green facades and paths, a school, common areas, urban agriculture and forestry, among other features.[171]

*Images 2 & 3: Ho Chi Minh City and its theoretical appearance through green retrofit.*

Or you have cases which combine material recycling, vertical and horizontal urban agriculture and insulation, water recycling, walkability, the blending of greenery and photovoltaics, social housing, work and recreation spaces, and spatial organization for social and ecological sustainability.[172] The most recent IPCC reports — even within their conservative frames — see these kind of features and their "synergistic" integration as fundamental to ecological and social mitigation and adaptation in one of the classically "hardest to abate" sectors.[173] Already, in places like Kampala, Uganda, urban and peri-urban agriculture produces 50% of the city's food needs, contributing as well to food security, sovereignty, and sustainability.[174] Similar numbers are found for cases like Tamale, Ghana, and Ouagadougou, Burkina Faso.

*Image 4: Urban and peri-urban agriculture and forestry in contemporary Kampala.*

Vitally, this kind of possibility is not limited to some pure urbanization theory of sustainability. It can be part of what Max Ajl calls a "planet of fields."[175] However, for the moment, what is more important than statistics and examples is the vision, the very real possibility, of a far *better* life, a *richer* one. The *aesthetics* of the spaces in the classical sense of *aisthetikos*, the sense perception, the *feeling* they generate.[176] They begin to outline the idea of a varied socioecological project which not only comports with actual ecological need using actually existing technologies, but does so in line with a social flourishing that transcends desperate attempts to maintain the exhausting status quo. Without romanticizing the past or projecting "civilizational cones," we can say that these kinds of spaces draw on roads not taken or quashed in the march of capitalist progress.

One can look to Red Vienna — so frequently cited by contemporary socialists as a model — and forget that it already contained not only many of these design principles but socially revolutionary structures that far outstripped eventual capitalist development: "workers' dwellings were incorporated with kindergartens, libraries, medical and dental clinics, laundries, workshops, theaters, cooperative stores, public gardens, sports facilities, and a wide range of other public facilities."[177] Green space, integration into existing environments, the use of novel and vernacular forms, urban agriculture existed throughout the system; even decorative and green facades and roof gardens were realized in some buildings. These were to provide *rest, respite, relaxation, recreation*. This was the Red Vienna that so terrified contemporary Austrian economists — the original neoliberals.

Red Vienna is hardly alone in its partially actualized or theoretical projects. One could look to a seemingly opposite case like Balkrishna Doshi's celebrated 1980s housing projects for India's "Economically Weaker Sectors" — i.e., peasants, workers, the poor in general. "Aranya Low-Cost Housing," for example, provided a set of materials and an integrated urban plan and infrastructure for basic needs (water, electricity, etc.). The plan minimized solar heat exposure and relied out of necessity on extreme material and resource efficiency, but without sacrificing aesthetic and other pleasures. The plots cluster densely around a central "spine" for social and cultural spaces, workshops and industries, parks and shops, "connected by green pedestrian pathways."[178] Instead of trying to tear down and replace the improvised slums that surround many South Asian cities, Aranya *counted on* the improvised and impressive ingenuity with which slum-dwellers could use local and reclaimed materials. Doshi provided further plans for balconies, terraces, and more that could be built with the materials available, but these were left for the residents to adopt, adapt, or reject. The project was neither slum nor suburb; the hope was that residents would turn the connected plots into rich, vibrant houses, while the broader lineaments would provide the space for social conviviality. And this is precisely what happened: residents from nearby Indore slums quickly realized the initial experimental examples into a dazzling variety of homes, wildly colorful, with integrated functional and ornamental greenery, alongside parks and social centers.

In spite of its promise, Aranya — originally planned for some 8,000 units — was never completed. As India liberalized, the once vast state support for large-scale planning vanished and was replaced by financialized private and public-private development. Funding dried up and was replaced with loan instruments. Suddenly, many residents found themselves saddled with debt for basic services and material banks. And representatives from the newly unleashed private real estate industry informed residents of the rising value of the homes. Ironically, the project had worked in some ways *too well*. Upmarket buyers were eager to buy into the miniature city created by the supposedly backwards slum-dwellers. Most residents sold. By the time Doshi was receiving architectural awards for the project, all but 20 of the original homes were gone; in their place new, standardized "towers of glass and metal which heat up quickly in the Indian weather and therefore have to be air-conditioned every minute of the day" sprang up. A vision of luxury for all — defined not by capitalist wealth but by social abundance and a calm "rhythm of life" attuned to people's, not capital's, needs — gave way to the disastrous "Housing for All" program of the neo-fascist BJP government, which follows an all too familiar pattern of displacement and gentrification in favor of India's new, highly concentrated wealth.

In both these cases, need *structured* desire, necessity *educated* desire. Necessary restraint did not translate to asceticism or simply less but provoked a far more exciting — dare I say, dynamic — imagination. This is not only the case with these examples. Susan Buck-Morss writes of the "missed opportunity" in early Soviet planning "to transform the very idea of economic 'development' and the ecological preconditions through which it might have been realized."[179] Late 1920s "paper architecture" laid out plans for anti-productivist "green cities" — just as early Soviet science was already exploring climate change and what the socioecological nexus might mean; these green city plans already included concepts like wind and solar power and aimed for reconciliation as opposed to the "mastery" of nature. Ideas of *relief, respite,* and *rest* were emphasized not simply as recuperation for the "exhausted" workforce, but implying a "radical criticism" of the new, industrialized speed-up of the dawning Stalinist era, which mimicked capitalist development.

The architecture critic Kate Wagner contrasts the radical imagination that was once the purpose of "paper architecture" into what she calls "PR-architecture." "Paper architecture, of course, still exists, but its original radical, critical, playful (and, yes, even erotic) elements were shed." Instead, "PR-architecture" is dreamed up by marketing and public relations teams, shedding those qualities to "look good" on social media, "a substance-less, critically lapsed" imaginary for a "substance-less, critically lapsed media landscape." One of her examples:

Architect Bjarke Ingels's "Oceanix" — a mockup of an ecomodernist, luxury city designed in response to rising sea levels from climate change. The city will never be built, and its critical interrogation amounts only to "city with solar panels that floats [because] climate change is Serious."

But Oceanix did get Ingels and his firm, BIG, a TED talk and circulation on all of the hottest blogs and websites. Meanwhile, Ingels had been in business talks with the ultra-right, then president of Brazil, Jair Bolsonaro.[180] Techno-mystical anti-utopias — which promise the present unchanged — are exhausting and boring.

The official "1+n" Chinese policy papers call for ambitious plans "to accelerate the development of urban rail transit, bus lanes, bus rapid transit, and other forms of large-capacity public transportation and strengthen the development of bike lanes, pedestrian walkways, and other facilities for slow urban transportation systems," such that 70% of all Chinese transit needs are covered by electrified mass transit and — in contrast with the ecologically disastrous US electric vehicle push — to limit individual electric vehicle transport to 30% or less. Unmentioned is air travel. Realistic assessments argue for limiting air travel to all but the most necessary cases. All the above are certainly admirable proposals, depending on details, but they too exhibit constrictions of imagination.

Given the unfeasibility of replacing jet and shipping fuels anytime soon, some are planning the return of a seemingly dated piece of technology: the airship. While there has been a quiet rekindled interest in airships for several decades, in recent years teams of engineers from around the world have devoted serious research to their resurgence, modernization, and development (and with interest from some small sectors of capital).[181] Airships as they exist already work; they produce 75–90% less emissions than standard freight or commercial aircraft; potentially more within the next several years. Although such calculations do not account for the full lifecycle assessment, the lower-end is based on real-world flight measures, with well-known technologies. Non-industry funded models already find similar (80% and projected higher) claims. Airships are both the detritus of history and the stuff of steampunk imaginations the world over. The minor paradise of the sustainable niche meets *Final Fantasy*.

There are questions, to be sure (fewer airships running on cleaner, safer, but rare helium or more working with far more volatile, even when "green" hydrogen?) but most revolve around *speed*. Airships are more versatile but slower than airplanes. Buried in the comments on the draft of the 5th report from the IPCC's Working Group III, one reviewer objects to the claim: "In aviation, no serious alternative to jet engines for propulsion has been identified." This is not correct, they note, and go on to remark:

Airships are not mentioned. They have large potential especially in freight transport (big energy efficiency advantage compared to aeroplanes) and they can use many types of heat engines and fuel cells, and in addition they are especially suitable for solar power use.[182]

A back and forth ensues between the commentator (who is identified only as "Alan" and presumably is American climatologist Alan Robock, the only Alan in the authors list) and an unnamed editor: "will try to reflect these points — but it seems clear that commercial passenger air travel will be dominated by jet engines for decades to come." Just as states could order fleets of empty aircraft to return to the skies, empty of passengers, during the pandemic, radical political intervention *could* achieve the drawdown of jet fuel and the proliferation of airships. "*What about the workers?*" the Climate Lysenkoists protest disingenuously. They will be desperately needed to guide the just transition in aviation.

In this tiny exchange, we see capitalist realism, even within an intra-scientific discussion, trumping anything approaching social and ecological realism. Indeed, Alan presses his case after more pushback: "Some useful suggestions which will be separately evaluated and would be worth incorporating e.g., need more discussion of the impact of just-in-time and the opportunities of relaxing it to cut energy use and emissions." Alan makes this argument one other time to little avail. The airships are not mentioned, and just-in-time gets one highly attenuated inclusion. Green capitalist hegemony imagines today's vast global value chains running at their current, socioecologically devastating pace or, through ever more extraction, further accelerating. (And green *capitalists* want the airships for themselves.)

Policy plans will discuss *quanta* of clothing but not

the extraordinary aesthetic possibilities of clothing made to last. Contemporary "luxury," greenwashed, "sustainable" clothing is often as drab as "fast fashion." Planned obsolescence is not only in electronics but in clothes; as Esther Leslie notes, Marx observed this already in *Capital III* where he describes the increasing production of "shoddy clothes," whose "great benefit" (Marx puts this in scare quotes) for the "consumer" (again) is that they wear out 3–6 times faster.[183] Leslie imagines sustainable fashion as a kind of post-capitalist Etsy. Others

*Image 5: Examples of Stepanova's everyday fashion designs.*

explore the possibilities in repair and reuse. And still others promote the rich possibilities of adapting traditional techniques and ornamentation. Early Soviet experience shows examples of these, driven by economic necessity to what today would be called ecological or sustainable. "The Soviet government supported the do-it-yourself subculture by organising special sewing and construction training courses for adults and pushing people to keep using their existing clothes," writes David Ferrero Peláez,[184] but the early Soviet period also saw quasi-utopian designs that were never fully realized. The constructionist designer Varvara Stepanova designed not only her famous unisex uniforms but exceptional fashion-oriented clothing which would only "fall out of use, not because they start to look funny when the market generates novel fashions, but rather because the conditions of *byt* [everyday life] will have changed, necessitating new forms of clothing."[185] Fashion would be liberated by breaking *out* of the trajectory of existing capitalist production. Like the half-realized plans of socialist housing or post-colonial agriculture, these would mark socioecologically necessary interventions in gendered and racialized hierarchies and divisions of labor.

Here again we see a different rhythm, a different conception of time, at play. Not just *possible* fashion but *better* than fast and "luxury" fashion. This temporality and its further aesthetic expressions are addressed again in the final chapter. All of these examples guide us toward the *feeling* of the minor paradise of the sustainable niche while remaining well within planetary boundaries.

Rempel and Gupta point out that the vast stranded assets "must remain on the balance sheets of resource-rich actors to equitably build the path towards [a] truly inclusive future." Among these actors they count "rich shareholders, multinational firms and investors predominantly from the North who have accrued mammoth profits from commercialising fossil fuels over the last decades."[186] The reason for this is not purely "equity" concerns; if costs are shifted to poor actors or to states in the South, incentives are amplified for ecologically catastrophic accelerated resource extraction and for fossil fuel reinvestment. What they are calling for is, in essence, a new model of risk.

The world of the extractive circuit operates on modes of risk management that seek to maximize profitability to the absolute point of systemic breakdown — something much on display since the advent of the COVID-19 pandemic. Ulrich Beck famously introduced the notion of "risk society" to describe the supposedly post-class world developing in the late 1980s. Risk society (now writ globally and universally) is less about the distribution or circulation of capital and more about the distribution of risk. Beck saw environmental risk as particularly indicative of the kind to which individuals across social strata must respond.[187] While separate from the developing theories of risk and uncertainty by economists like Friedrich Hayek and Frank Knight, the effect of risk-shifting is similar. Individuals are responsible for the delicate balancing act of risk calculation. Beck saw this as potentially empowering, much like the neoliberal and behavioral economic paradigms. The economic and psychological models are almost certainly more familiar, in experience and in risk management literature.[188] True unknowability for a thinker like Hayek (i.e., in no way calculable by probability) means that planning is fundamentally impossible, only the market can respond. Meanwhile, business practices like lean production, just-in-time, and corporate reengineering — developed, as we've already seen, in the mode of the extractive circuit — are framed as universal goods, risks worth taking.

Although Beck might balk at these conceptions, he too embraces a mode of anti-politics, with remarkably similar models and lines of thought.[189] Attempts to account for this in environmental terms are largely still syncretic across these various concepts. Concepts of risk, hazard, vulnerability, and resilience have evolved across the last three IPCC reports, synthesizing Beck with individual coping models, psychological resilience, national security, and financial theories.[190] The 2022 issue of the *Journal of the Institution of Environmental Sciences* on "Environmental Risk" — including authors who have written about absolute catastrophic and existential climate outcomes — pushes back to a degree, but ultimately it reaffirms many aspects of the risk society model conjoined with these new "insights."

In contrast, much of the direct climate science literature demands a radically different risk model. The concepts of "safe operating space" and "planetary boundaries" discussed earlier both provide this. As Steffen et al. wrote:

A planetary boundary as originally defined is not equivalent to a global threshold or tipping point [...] Even when a global- or continental/ocean basin-level threshold in an Earth-system process is likely to exist, the proposed planetary boundary is not placed at the position of the biophysical threshold but rather upstream of it — i.e., well before reaching the threshold. This buffer [...] not only accounts for uncertainty in the precise position of the threshold with respect to the control variable but also allows society time to react to early warning signs…[191]

In other words, boundary values are not set *at* some kind of knowable limit, but rather uncertainty must be included by creating a "buffer well before reaching the threshold." This is essentially a synthesis of "scientific knowledge of the Earth system" with the "precautionary principle," but even the authors admit the social limitations of this conception, which complicate the picture.

Here again, we can look to largely marginalized or suppressed risk "models." For example, there is a concept in Jewish law of *chumra,* or a fence around the law, which argues for legal interpretation that ensures reasonable restriction beyond the bare minimum — such that there is as little possibility that the threshold is crossed, but also not unduly onerous, such that the "safe operating space" of law underwrites flourishing and not ascetic burden. The gray policy models briefly sketched — European and Chinese — provide an empirical sense of where that *chumra* might lie.[192] Marxian and other radical frameworks can further elucidate potentials or complexities within planetary boundaries, reveal lacunae, contextualize and critique to elaborate a secularized *chumra* the minor paradise of the sustainable niche needs. Well within planetary boundaries and well within the structured necessity — the educated *desire* — for reversal, relief, and respite from exhaustion, for an "epoch of rest."[193]

The broad contours of an agenda for mitigation and adaptation are clear: rapid, planned decarbonization and transition to renewables concomitant with as swift and total restrictions on fossil fuel extraction as possible.[194] Decommodification of basic social goods that we already know are more efficient and effective when publicly provisioned. A vast shift of resources — and sovereign power — to the Global South, not simply out of moral duty, but out of rational necessity concomitant with a move away from economic-growth paradigms and toward "growth agnosticism," human development, and redistribution. An end to further dispossession and enclosure, and a turn to sustainable, agroecological food production. Greater restrictions and popular guidance of capital flows, and greater freedom and facilitation for human migrations. And, underneath all, a release from the vicious cycle described before; a world of greater individual, social, and political security, greater *temporal* luxury, greater if different material freedoms, and greater human flourishing. This is the world we begin to glimpse in the minor paradise. It offers an end to dogmatic debates about *more* or *less, growth* or *degrowth*, or scholastic investigations into "real socialism" or "managing capitalism" (when the real conversation is about socioecological relations, distribution, wealth, and profit). The sustainable niche speaks to the exhaustion found at every node of the extractive circuit and to the heterogenous desires expressed in its quasi-utopian, very possible fragments; the *better*, even luxurious terms of addressing socioecological necessity.

To some, such an agenda and portrait appears as a dystopian, horrifying nightmare. But to so many others — a significant potential mass majority — it arrives as blessed relief. Left-wing climate realism — the politics of exhaustion — turns in part on this distinction.

In a world defined by social realities that defy nineteenth-century models and an unforeseen ecological project, "who," as Davis asks, "will build the ark?" *Pace* God, Moses, or Ezra the Scribe, many arks are possible. The question is more precisely who will build *this ark* — the ark of a global human ecological niche capable of sustaining the flourishing of some 7–9 billion humans? This is not a straightforward question. The "ark" of the sustainable niche as minor paradise depends not only on contemporary discontent, not only on new and potential capacities and power, but also on that equally shifting plane of social desire. Or, as Berlant puts it, on "attachments to good life fantasies." For some, this is the easy question of fighting to hold on to lives (and land) that are already facing today's manifold socioecological violences. For others, it is the more difficult question of *detaching* from either the "cruel optimism" of exhausted life as we know it, or the security — no matter how attenuated, limited, or subordinated — of the armed lifeboat.

If the minor paradise of the sustainable niche requires minimizing socioecological risk while reversing the global conditions of exhaustion, the *politics* of left-wing climate realism requires that we embrace *far more political risk*. Murat Arsel ends a reflection on broadening class analysis in terms of climate change: "In a recent commentary, Battistoni has argued that Malm's 'How to blow up a pipeline?' is not so much a discussion of 'how' but 'why'. It could be added that 'who will blow up the pipeline?' is also an important question awaiting an answer."[195] Who is the subject of left-wing climate realism? Who will build this ark? Who will blow up the pipeline? Who will pull the brake?

# Chapter 4

# The Exhausted of the Earth

*I'm sick and tired of being sick and tired.*

— Fannie Lou Hamer

## Against Resilience

In 1971, Aaron Antonovsky, an Israeli medical sociologist, led a small team conducting a survey with over 1,000 participants concerning how women cope with the effects of menopause. A question on the survey asked, almost as an afterthought, whether the women were concentration camp survivors. In reviewing the findings, Antonovsky was astonished. "How the hell can this be explained?" he exclaimed to colleagues. What he had discovered would prove foundational not only to Antonovsky's career but to an entire new field of research. Of the 287 women who reported that they had survived the camps, over two thirds qualified in the category of "breakdown" — still suffering from "the horrors," as he termed it. Unsurprisingly, this was a vastly higher number than for the women who had not experienced the camps. "What is, however, of greater fascination and of human and scientific import," argued Antonovsky "is the fact that a not-inconsiderable number of concentration camp survivors were found to be well-adapted [...] What, we must ask, has given these women the strength, despite their experience, to maintain what would seem to be the capacity not only to function well, but even to be happy?" The answer was nothing less than a set of psychological dispositions that produce an understanding and acceptance that external stimuli reflect a coherent world; that one has the internal resources to meet any demands from these stimuli; of an optimistic disposition that such demands are "challenges, worthy of investment and engagement." With these it might be possible for a person to withstand life reduced to the absolute degradation and deprivation of the camps and still remain functional by existing social standards. Antonovsky would eventually call his science "salutogenesis," but what he had really discovered is what we now call "resilience."[1]

Earlier, in 1955, the psychologists Emmy Werner and Ruth Smith began a vastly influential 40-year longitudinal study of children on the island of Kauai. Initially the investigation focused on how structural conditions such as poverty affect both pregnancies and subsequent childhood experiences. Although Smith and Werner found conclusive evidence that structural and environmental factors were strongly associated with negative outcomes, they were, in an eerily similar story, shocked to discover that about one in four of those exposed to several, severe structural risks "developed, instead, into competent and caring young adults." Smith and Werner dubbed this group, "the vulnerable, but invincible." Werner later reported how their observations confirmed Antonovsky's initial propositions: a set of beliefs, attitudes, and capacities — what Antonovsky called a "sense of coherence" — could facilitate "health" under even the most adverse circumstances.[2] It was these — the well-adapted Auschwitz survivor and the vulnerable but invincible child — who would become the ideal types of resilience; the docile inhabitants of this exhausted world.

"Resilience" appears some 3,970 times in the IPCC's 2022 *Impacts, Adaptation, and Vulnerability* report. Many of these are citations to natural scientific studies building on the 1973 ecological definition: "resilience determines the persistence of relationships within a system and is a measure of the ability of these systems to absorb change of state variable, driving variables, and parameters, and still persist."[3] Much research under the ecological resilience rubric accurately describes assessments about, say, a given urban environment's inadequate deployment of resources for climate adaptation; or a technical evaluation of a coastal region's threshold for ecological stability; or in modelling geophysical climate change probabilities. Resilience in these senses is not some mirage; a monoculture ecosystem is far less technically resilient than multispecies diversity. An agroecological and agroforestry system is far more resilient to extreme climate events — and far more efficient — than industrial petro-farming. Refinements of this definition cast resilience as "the capacity to adapt or transform in the face of change in social-ecological systems, particularly unexpected change, in ways that continue to support human well-being."[4] Even with such augmentations, the term is acknowledged as murky and its social application highly debatable. Transformation often slides back into preservation or modest modification even when the broader research calls for nothing less than radical change across all aspects of society. However, such uses are only a part of those 3,000 or so IPCC references.

Resilience, "resilience theory", "resilience science" are interested in how "stressors" affect specific systems, in how systems can *persist*, and simultaneously, in social and individual capacities, *absorb* ever greater risk, crisis, trauma, and stress.

Resilient ideals are now ubiquitous. Between 1970 and 2021 some 81,000 academic articles were published focused on resilience, more than 80% of which were in the last two decades. As a concept, resilience barely registered at the mid-century. "Resilience science" is found in risk modelling, vulnerability assessment, disaster management, sustainable development, urban planning, physiology, epidemiology, security, health, and more.[5] Antonovsky's "sense of coherence" and variations have become a common bridge concept between social-psychological and ecological definitions. This is particularly visible in the vast "grey literature" between government agencies, think tanks, and consultancies.[6]

In its common use, resilience is easy to understand. It is the capacity of ecosystems, individuals, communities, or societies "exposed to hazards to resist, absorb, accommodate to and recover from the effects of a hazard in a timely and efficient manner, including through the preservation and restoration of its essential basic structures and functions."[7] Resilience is therefore about risk-shifting, minimum resource levels, and "bouncing forward." Resilience emphasizes some of the stickiest, socially destructive ideals of our time: the hardy survivor, the endlessly flexible and adaptable worker, the self-reliant community, all of whom continue to function within even the most corrosive socioecological conditions and deprivation. This is part of why resilience is so beloved by policymakers. In a crisis-ridden world, it counsels quiescence and parsimonious austerity. Even in its most generous formulations, it looks for just how little some unit — a body, a region, a population — might need, while avoiding the possibility of significant *external* change entirely.[8] Resilience is a management strategy and apology for the status quo, for global capitalism with all its constitutive social and socioecological relations. In resilience thinking, chaos, disease, and stress are omnipresent and often unavoidable — *naturally*. Resilience thinking teaches the absolute limit of risk or stress that can be shifted onto individuals and communities, like a Victorian viceroy counting calories for coolies. And simultaneously, it sighs that should such a limit prove too much for these poor souls, it is a failure of internal capacities. Nothing could be done; they were perhaps, in the phrasing of Ruth Wilson Gilmore, a "disposable population" to begin with.

Resisting, absorbing, accommodating, and recovering — all socially passive and politically inert — rely, as two resilience specialists summarize, on the cultivation of "optimism, intelligence, creativity, humor, and a belief system that provides existential meaning, a cohesive life narrative, and an appreciation of the uniqueness of oneself."[9] This is quite literally the prescription of ideology. As Theodor Adorno once quipped: "there is humor because there is nothing to laugh at." Optimism in a world that is failing; intelligence in knowing it is the best because it is the only possible world; creativity in adapting to that world; a belief system and cohesive life narrative that affirms the world as it is and asserts the value of each and every individual even as it prepares many for mass death.

Resilience thinking is always *reactive* to exogenously described disasters, shocks, and stressors. Even when it is preached prophetically as prophylaxis, it ignores the empirical realities of phenomena endogenous to "the present dominant socioeconomic system [...] based on high-carbon economic growth and exploitative resource use." Sociologist Sarah Bracke reads resilience through Berlant's cruel optimism: "an attachment to resilience [...] effectively prevents us, as individuals and collectively, from going there. Here resilience becomes a symptom of the loss of the capacity to imagine and do otherwise, and cruelty is one of the more politically cautious names for such a condition."[10]

Of course, it is a political reality that emancipatory movements — and left-wing climate realism — act from a place of "significant stress or adversity,"[11] but resilience as a principle sublimates a temporary challenge into a goal itself. Performing or enacting resilience becomes the cruel proof of strength, of commitment, of rugged self-reliance. Attachment to the ideal of resilience only maintains a world which demands it. The achievement of resilience marks the horizon of "success." *Failures* of resilience — individual or collective — demand an inward turn and reckoning. Where did individual capacities or "social support" systems fail? Positive commandments of unbridled optimism, personal adaptation, and meaningful affirmation — which are found not only in resilience theory but in positive psychology, cheap exhortations to mindfulness, and many of the faddish, self-help pseudosciences of the neoliberal era — are completely incoherent with the catastrophic climate change that is already here. They are not just apolitical, but *anti*-political. Resilience is the all-consuming preparation for life (or death), as Walter Benjamin once wrote, in a hell "which is this life, here and now." For some, resilience is the categorical imperative of business-as-usual; it is crisis managers buying time. For others, resilience is *exhausting*.

## The Exhausted of the Earth

Against resilience then, against the atomizing prescriptions to *internalize* stressors, hazards, crises, and other structural phenomena, the political possibility for a left-wing climate realism depends on *externalization*. That is to say, politicization. Frantz Fanon's *Les Damnés de la Terre* — "The Wretched of the Earth," taken from the verses of the *"L'Internationale"* — are today's *Exhausted* of the Earth. Climate change is not the byproduct of contemporary capitalism; your exhaustion and that of the global human ecological niche are fuel for the fire. Our niche has a case of the Mondays. This life, this civilization, is, above all, *exhausting*. Business-as-usual promises only to accelerate your and this world's exhaustion; it can afford to take a leisurely, piecemeal approach. Neither you nor this world can afford that. Neither can you nor this world wait for "the revolution." But relief from exhaustion promises a new, freeing centrality of social and ecological reproduction and the most dramatic reconstitution of political economy in human history. Neither you nor this world can abide by liberal admonitions to propriety, to civility, to patience, or compromise. Exhaustion is not some rhetorical gesture, discursive fiction, or new theoretical fantasy. Exhaustion outlines the historical bloc, the mass political subject of this conjuncture.

Spread out a map of the world and push pins into every location that is figuratively or literally on fire. Just as these are zones of extraction, exploitation, expropriation, these are zones of exhaustion. And like wildfires, they proliferate. Connect each pin with a wire and suddenly you see the outline of the world of exhaustion, the extractive circuit, capitalism in its full socioecological expression. As we've seen, the extractive circuit quite literally crisscrosses the world. In a necessarily expanded understanding of value extraction, the extractive circuit extends from geophysical realities to psychosocial "optimizations." It organizes a global human ecological niche for maximal profitability — no matter how difficult to maintain and at whatever cost.

By recent counts there are well over 3,000 "ecological distribution conflicts" in the world right now.[12] The concept of an ecological distribution conflict attempts to capture the incredible range of social conflict — in terms of class, race, gender, and more — that occur around the production and distribution of material, ecological goods. Far from the shockingly persistent image of environmentalism as a principally "middle-class," "elite" (or in the bowdlerized cant of the know-nothing left, "PMC") concern, looking at these actually existing conflicts reveals a picture in which struggle is widespread, more frequent in the Global South and amongst the poor, North or South. In some cases, like with the international peasant movement La Vía Campesina, struggles are explicitly connected to a systemic ecological critique. In others, such critique is absent. But both can be described in the terminology first proposed by Joan Martinez-Alier and Ramachandra Guha: "environmentalism of the poor." However, others with even less likely bedfellows — as, for example, Murat Arsel discusses anti-coal coalitions in Turkey between peasants and former leftist intellectuals and officials — can have multiple motivations converging on what Arsel calls, tellingly, the "environmentalism of the malcontent."[13] The explosion of ecological distributional conflicts — 95% of all conflicts occurring since the crisis of the 1970s and 50% just since 2008[14] — and their occurrence precisely "along local and global commodity chains, from cradle to grave"[15] — tracks the acceleration and social metabolism of the extractive circuit.

One might describe these as among the initial "spontaneous" outbreaks of global exhaustion, with all the shortcomings and strengths that thinkers from Gramsci to Fanon ascribe to spontaneity. It is not only ecological distribution conflicts that are on the rise. Across the exhausted world, social unrest is increasing dramatically — only exacerbated by the pandemic, itself just another of the complex socioecological phenomena of climate change. An amplification of long-existing trends, current unrest is comparable not only to the social upheavals of the 1960s, but to those of the late nineteenth or early twentieth centuries. This is not just a casual inference, though the spread of social and political crises that ripple across the world in our moment can be easily observed.[16] Recent data from the Mass Mobilization Protest Database marks an approximately 58% increase in such events worldwide by 2019.[17]

Although the concentration of upheavals is greatest in middle-income countries, it is increasing everywhere, including in the Global North. As a team led by sociologist Şahan Karataşli observe, the current wave of social unrest is at least comparable if not *greater* than the mid-century breakdown of British imperial hegemony.[18] In fact, such extended periods of unrest are always associated with "periods of major *economic and political crisis* for global capitalism." Across studies, causal explanations are murky, although persistent themes are encapsulated in the Organization for Economic Cooperation and Development (OECD) "From Protest to Progress?" report: inequality, deteriorating conditions *despite* income, worsening labor conditions, and, of course, "the climate crisis." Adding current "megatrends" of globalization and technological acceleration, forced migration, and other similar conditions to those "key" factors only underlines "the existence of this generalized discontent" — and as we've seen, each of these trends is linked to the others and to worsening climate catastrophe.[19]

The OECD report notes that "rising stress levels and deteriorating mental health further attest to the difficulties people face today." This stress, this *exhaustion*, is also contagious. "Evidence of contagion of protests between countries and the emergence of global protest movements suggest that people in societies around the world are finding common cause." Here we begin to see the lineaments of the Exhausted; this is the foundation of a genuine political subject of left-wing climate politics. Perhaps ironically, these official aggregate analyses of global unrest are written in the anxious mode of risk analysis and crisis management. Data collection of this kind is often motivated towards the creation of models and strategies to predict and contain "systemic risk" or indeed strengthen systemic "resilience." Just as we saw with climate-related insurance in Chapter 1, this risk-management literature has led to burgeoning markets in insurance, provided by firms like Germany's Allianz, protecting businesses against property loss due to civil unrest as well as privatized security for corporations and the wealthy.[20] Even in nascent form, the Exhausted provoke the anxiety of the "crisis of crisis management."[21] At the same time, just as actually existing power will bet on long-term probabilistic climate outcomes that are not certain, they will also try to profit off the fear of their potential "gravediggers," or at least damagers.

There is a crucial difference between today's wave of unrest and previous ones. As Karataşli observes, today's wave lacks the organizational structure to "boost and spread the spontaneous and creative energy of the masses from below."[22] The decidedly non-radical authors of the OECD report note the same difference: declining unions, parties, and social institutions and associations more broadly. When Fanon observed "one can hold out for three days, three months at the most, using the masses' pent-up resentment"[23] before such spontaneity splinters, it was in an era of mass parties, politicization, and mobilization. In contrast, today we are only beginning to emerge from an era of mass depoliticization. While left-wing politics has always faced something of a time-bind, an under-examined aspect of neoliberalism is the way it colonizes the time for politics.[24] Extending and accelerating productive time as far as possible, well into supposed non-working hours, not only generates profitability but in the process specifically destroys the *time* for politics.[25] This makes the scale of contemporary social upheaval all the more striking and the imperative for organization all the more urgent. Part of the politics of exhaustion is stealing back time, securing space for ever greater intensification and politicization.

Andreas Malm — who describes himself as a kind of ecological Leninist — ends an affect-laden chapter, "Fighting Despair," with a perhaps surprising turn to Fanon. "Few processes produce as much despair as global heating. Imagine that someday the reservoirs of emotion that are built up around the world — in the Global South in particular — find their outlets. There has been a time for a Gandhian climate movement; perhaps there might come a time for a Fanonian one."[26] While Malm is emphasizing more the secondary (and for Fanon, passing) psychological dimension of Fanon's arguments about anti-colonial violence producing "dignity and self-respect," what is more vital is Fanon's understanding of affect, externalization, political *realism*,[27] and subjectivity.

There is no need to wait. The reservoirs of emotion are already finding their outlets. In mapping layers of disease, environmental conflict, and general social upheaval on top of one another, we can see these reservoirs and begin to outline the Exhausted as the mass political subject of left-wing climate realism. This is not complete, and, as with *all* social experience, it is not determinative. Nor is this *already* a subject or even a cohesive movement, as some might suggest. There may even be some groups and conflicts which fall away and others which are eventually drawn in. In laying this map over that of the extractive circuit, we begin to see mass *externalization* in proliferating if still disjointed conflicts. We begin to see what the Exhausted can be.

## Examining Exhaustion

Exhaustion should be understood not *only* in the sense of ecological exhaustion, of running past, beyond, or through planetary boundaries. Although in that ordinary environmental sense, the ecological conditions conducive for the mass flourishing of human life are indeed at the point of exhaustion.

Nor *only* in a sense of bodily enervation, of individual physical fatigue. Although in this sense, too, the labor regimes enabled by the extractive circuit have left many physically exhausted. But more than these, exhaustion is the experience, the sense, the *feeling* of how this global ecological niche is *spent*. We've already seen the social and ecological exhaustion of the extractive circuit. But this hardly captures the totality and particularity of exhaustion in this moment. Rather, exhaustion is an affective matrix — a "general affectivity," as Fanon once wrote, a particular "structure of feeling" — where these types of exhaustion are also bound up with the exhaustions of political forms; of and with whole ways of life; the exhaustion of living or resisting the 24/7 world; engineering, technical, and even aesthetic exhaustions. In ecological thought there is often a Malthusian impulse — intuitive yet fundamentally flawed — that materials will simply "run out"; that there is too little for too many; that resources will be exhausted.[28] But capitalism today, the world of the extractive circuit, is characterized less by this kind of running out — of fossil fuels, of "rare earth minerals" — than the *overabundance* of such resources and the social and ecological exhaustion that lies in the wake of their accelerating exploitation. As Marx already observed in the mid-nineteenth century, capitalist progress is "the art of not only robbing the worker, but robbing the soil."[29] Exhaustion traces the outline of a politics, of a broad agenda, and a struggle whose goal is the flourishing of a sustainable niche. Exhaustion proliferates; it is ubiquitous and yet it is specific.

The most prosaic way in which exhaustion has long been discussed is as a kind of pathology. In many epidemiological literatures, exhaustion is often treated as an extreme form of fatigue or, in more common parlance, "burnout." In turn, these are all classified as syndromes related to the inability to work, to "cynicism or negative feelings towards one's job," or "reduced professional efficiency" (WHO definition). There is a long history of this kind of analysis of exhaustion as essentially a labor management problem — on the factory floor, in the family, or on the battlefield. In this last case, "battle exhaustion" or "combat fatigue" is actually taken far more seriously than it is in many other arenas.[30]

This understanding of exhaustion in warfare generally traces back to the dawn of mechanized combat in the American Civil War and WWI. As with labor management approaches, this understanding of exhaustion was intrinsically modern. However, the battlefield situation did not provide the breathing room to experiment that the factory or the colonial plantation did. Workers or slaves were expendable; soldiers in active combat had to remain at the front at all costs. Exhaustion could account for mass casualties and quite literally turn the tide of a conflict. This led many militaries to develop sophisticated etiologies of exhaustion at a nexus of social, psychological, biological, and environmental factors. Although few turned to Fanon, this nexus is remarkably similar to Fanon's practical and political psychiatric theories and practice. Just as Freud, Benjamin, and Fanon would all come to view the different landscapes of modernity as battlefield-like, militaries independently came to see "battlefield exhaustion" as a particularly acute instance of what we now call PTSD. Interest from the American and Canadian militaries began during this early period but intensified by the time of the Vietnam War and the 1982 Israeli invasion of Lebanon, colonial wars in which exhaustion as an affective disorder proliferated. And in parallel with more contemporary writing on affect, such research often tried to understand the simultaneous individual and social natures of different kinds of "stress," "hopes," and "fears." Although these are expressed through varied individual experience and conditions, they are social in that they are contagious, shared *across* differences; "unit cohesion" depends on addressing both the individual and social levels.[31]

Probably the earliest pathologization of exhaustion was as "neurasthenia" or "nervous exhaustion" by the American physician George Miller Beard in the 1860s. Although ideas of exhaustion certainly predate the modern period, "before 1860 almost no medical or scientific studies of fatigue are recorded. By the turn of the century, the US Surgeon General's index listed more than one hundred studies of muscle fatigue as well as numerous studies of 'nervous exhaustion,' 'brain exhaustion,' and 'spinal exhaustion.'"[32]

The worker impervious to fatigue was a kind of utopian aspiration stretching from the dawn of "scientific" management to today's prescriptions to ease "burnout" with "compassionate" approaches that emphasize "recognition," "positivity," or "invite all the things that make us human to work."[33] That managerial utopianism — which, as Anson Rabinbach reminds us, mirrored the fantasy of the "endless productivity of nature"[34] — becomes yet another imperative to resilience, to internalization rather than externalization.

Workplace fatigue, the exhaustion of workers, as broadly conceived as possible, is a fundamental part of the equation. However, exhaustion is not simply synonymous with exploitation. In the extractive circuit, we already saw exhaustion in the "mental health plague" concentrated in lower- and middle-income countries but stretching to majorities across the world. At perhaps the broadest and fuzziest levels, we glimpse exhaustion today in International Labour Organization (ILO) reports and Gallup surveys describing 76% of the global workforce as suffering from "burnout," less than 33% describing themselves as "thriving," and a mere 21% expressing general engagement with work.[35] But beyond formal labor, we see a broader exhaustion in the vast "Global Emotion" surveys reporting 42% of the global population as anxious, 41% as stressed, over a third in physical pain, another third simply tired.[36] (Such statistics are overlapping and partially additive.) Next to a cheery photo of Gallup CEO Jon Clifton, the 2022 executive summary begins:

The world broke a lot of records in 2021. Corporate profits, venture capital funding, $CO_2$ emissions and the temperature of the oceans all reached record highs last year. But there is another record the world broke that hasn't yet made headlines — and it has to do with how everyone feels. As you'll read in this report, in 2021, negative emotions — the aggregate of the stress, sadness, anger, worry and physical pain that people feel every day — reached a new record in the history of Gallup's tracking.[37]

Such surveys — in their methodologies, in their interpretative gloss, and in their inherent limitations and assumptions — generally produce as rosy a portrait as possible. For example, many try to pin such findings on the COVID-19 pandemic despite data showing, as Clifton admits, the long-term nature of such trends.

As problematic as this kind of inquiry can be, Global Burden of Disease estimates highlight both the general epidemiological prevalence of disorders from depression to OCD across geographies.[38] These and related conditions are globally prevalent, as common in Malawi or Kenya as in Germany or France.[39] Rates of increase are remarkable in how they map almost exactly onto existing and projected global geographies of climate impacts.[40] Exhaustion — usefully for climate politics but confounding for some traditional researchers — seems to escape a precise medical definition, just as pathologies like depression are noted by medical researchers to defy strict definition and to exist on a broad spectrum.[41] As the editors of one interdisciplinary collection put it:

Our age, it seems, is the age of exhaustion. The prevalence of exhaustion — both as an individual experience and as a broader socio-cultural phenomenon — is manifest in the epidemic rise of burnout, depression, and chronic fatigue. It is equally present in a growing disenchantment with capitalism [...] in concerns about the psycho-social repercussions of ever-faster information and communication technologies [...] and in anxieties about ecological sustainability.[42]

This description is closer to the mark for understanding exhaustion as vital to the politics of left-wing climate realism. Exhaustion absolutely contains phenomena like burnout, depression, or fatigue, but also socioecological exhaustions — the reality of them and their *feeling*. The exhaustion of capitalism-as-we-know-it is conveyed in concepts like "zombie neoliberalism" that have permeated even mainstream business discourse. The exhaustions *in* existing social life and *with* that life; the exhaustions with, as Berlant suggests, "conventional good life fantasies."

As we've already seen, exhaustion appears throughout many canonical left literatures, but it's also unavoidable in contemporary analytic and theoretical research. "Exhaustion," argues Alberto Toscano is, "a prism through which to connect contemporary debates on the consequences of climate change to theorizations of the multiple crises of social reproduction."[43] In his 2015 book *The Burnout Society*, Byung-Chun Hal wrote that "every age has its signature afflictions [...] From a pathological standpoint, the incipient twenty-first century is determined neither by bacteria nor by viruses, but by neurons. Neurological illnesses such as depression, attention deficit hyperactivity disorder, borderline personality disorder, and burnout syndrome mark the landscape of pathology at the beginning of the twenty-first century."[44] Three years into the COVID-19 pandemic, we might revise this assertion: *past* a certain event horizon of anthropogenic climate change, it is the co-determining proliferation of such neurological illnesses in tandem with zoonotic viral infections and other socioecological phenomena that characterize such "afflictions."[45] Or, as Teresa Brennan puts it, "this is a world where inertia, exhaustion and the sense of running hard to stay in the same place mark everyday life. They are as much a mark of the present depression as environmental degradation."[46] Mark Fisher talked about this feeling in relation to aesthetic modernism: "while twentieth-century experimental culture was seized with recombinatorial delirium, which made it feel as if newness was infinitely available, the twenty-first century is oppressed by a crushing sense of finitude and exhaustion."[47] Bolivar Echeverria, writing from Mexico City more than a decade before Fisher: "at the end of this century — and millennium — under conditions in which capitalist modernity seems irrevocably fatigued, we perceive how illusory the political scene, so apparently realist, has been throughout the twentieth century."[48]

In a kind of strange harmony, the World Economic Forum's 2023 Risk Report highlights "rapidly accelerating risks clusters — drawn from the economic, environmental, societal, geopolitical and technological domains, respectively" and not only presents a comprehensive, literal mapping of what we might term "general exhaustion" — from individual disease (chronic conditions, mental health), to irreversible economic decline, exacerbated social crisis, and climate change — but emphasizes their fundamental interconnection at a macro-level, borrowing Adam Tooze's "polycrisis" concept. From this broad map of interconnected crises, they specifically map the intersection of human health and ecological degradation in particular.[49] They essentially draw, from the point-of-view of crisis management, the outline of the extractive circuit and all the nodes of exhaustion it produces. As the pre-pandemic trends accelerate, measures of global exhaustion become more volatile, more difficult to measure, but in all cases greater. Some global surveys find well over 80% of respondents reporting exhaustion (classified here as personal or non-work burnout[50]) and the WHO reports 25% increases in depression and anxiety since 2020.

As the Brazilian hydrogeologist Bárbara Zambelli wrote in an exasperated essay at the height of the pandemic:

> the capacity to produce and reproduce life is hierarchized, evidencing the gradient between living a full life and surviving. The ethical system we live in trivializes the exhaustion of the lives of some so the lives of others can, in fact, be produced and reproduced. The hierarchization of peoples, ecosystems,

and knowledges enable a certain ethical subject to prevail over the others. The pandemic is a snapshot of the system [...] The key approach to connect geosciences with the big interdisciplinary challenges faced in the world today is through social, economic, and political analysis.[51]

What she so acutely observes in the system of exhaustion is not only an ethic, requiring an ethical response, but that this connection, this synthesis, demands a *politics* of exhaustion.

As the joke goes, there's a German word for everything. *Zeitkrankheit* describes a disease particularly characteristic of the times. Exhaustion is our *Zeitkrankheit*. A literal translation of *Zeitkrankheit* would be closer to "time-sickness." Speed-up, acceleration, 24/7, always-on, lean, just-in-time; these are integral to the world of business-as-usual. People are sick of the times and sick from the times — "I am sick and tired of being sick and tired," as Fannie Lou Hamer once said. As with every other aspect of exhaustion, this is a genuinely transnational phenomenon: "a highly prevalent globalized health issue, present in all countries, that causes significant physical and psychological health problems." "Emotional exhaustion," when a subject feels "drained of emotional and physical resources," is one of the most common cross-cultural phenomena.[52] And, as expected, this is found in other extractive, sacrifice, and adjoining "zones" as well: migrants and refugees, native communities in Bolivia,[53] Indian women in domestic adversity.[54] These health trends are explicitly tied to the globalization already described in the extractive circuit.[55]

The exhausted world is not only sick from speed, it's *on* speed (and a lot more) just to get by. As one researcher discusses in terms of amphetamines and other stimulants:

the rapid rise in exhaustion related syndromes is not just a matter of the greater attention devoted to this topic but of an increase in manifestly experienced suffering [...] neuroenhancement is practiced in the hope of solving one of the key practical problems facing subjects under the conditions of contemporary capitalism.[56]

Again, this is not always correlated to labor questions: many prescriptions *increase* with unemployment, for example.[57] This is the medicalized response to exhaustion.[58] "Multinational companies ('big pharma') [...] already report massively expanding sales of their products in what they graspingly call 'pharmerging countries' in Asia and Latin America." There are also "local entrepreneurs ('little pharma') who trade on the generic leftovers of the branded blockbusters," a semi-formal market reflecting the manufacturing of pharmaceuticals in places like India and Brazil, as well as global distribution networks "off-the-books."[59] Amphetamines (and related stimulants) are the second most commonly used "illicit" drugs in the world (after marijuana).[60] One study of Brazilian truck drivers found as many as 73% using stimulants, but another highlighted the prevalence of amphetamines among British homemakers "increasing domestic efficiency, against the demands of caring for small children."[61]

Amphetamines aren't alone; a whole panoply of pharmaceutical products — from the opioids that are among the most prescribed medication in the United States, to SSRIs, benzodiazepines, and beyond — make up an affective maintenance and survival network for those at all the nodes along the extractive circuit. "One is justified in speaking of the commercial globalization of psychiatry through the medium of the pill, or 'psychopharmaceutical globalization.'"[62] The greatest increases in usage now are in low- and middle-income countries. In the intense debates concerning "global mental health," it is frequently noted how official numbers in the South are often *undercounted* since, in places like Brazil and India, much consumption is simply informal, while localized research shows the ready availability and use of a range of psychopharmaceuticals, including those (like antidepressants) with essentially no recreational value.[63]

The point here is *not* to ascribe normative value to the use or avoidance of medication, which can be necessary in many cases, even if their very design often reflects the limitations of capitalist development or their proliferation represents a kind of pharmacological colonialism searching for new markets.[64] (The recent COVID-19 vaccine apartheid has helped underscore the very real implications of restricting pharmaceuticals as artificially scarce commodities and pharmacological knowledge as IP.[65]) Even with a potential *politicization*, questions about such treatment don't vanish; they may be even more important. What we see here, as with epidemiological literatures, as with emotion surveys, as with theoretical inquiries, is just the beginning of understanding global exhaustion as a "structure of feeling" — as an affective matrix.

It is not only wealth and power that runs through the extractive circuit. *Feelings* flow through it as well; they pool around it. Such feelings and affects are not arbitrary or purely discursive. Nor are they politically determinate. They are informed by social position, but they do not constitute political subjectivity. What courses through the extractive circuit, alongside ever-accelerating extraction, expropriation, and exploitation is, for the vast majority of people, *exhaustion* in all its forms. Exhaustion *may* be the connective tissue between existing ecological and social upheavals across the world. Exhaustion *can* be the foundation for *externalizing* what are still too often individualized experiences of the relentlessness of the extractive circuit, for uniting and radicalizing. Exhaustion can be more than a "prism" through which to view debates regarding social and ecological reproduction; it can be a potent point-of-view and starting position for the politics of left-wing climate realism.

## Externalizing Exhaustion

The situation Fanon describes — the stretched class subject, the affective connective tissue — is *almost* isomorphic to our own. I don't turn to Fanon because he has *the* political theory for this moment. Just as it's all too common and tempting to misconstrue climate change as a universal experience for all people, it's equally common and erroneous to try and simply staple climate onto existing political theories and strategies without considering how much a shifting ecological niche structures a politics with its own dynamics, subjectivities, possibilities, conflicts, aspirations, and even temporalities. There are aspects of Fanon (from his existentialist flirtations to his Lacan-tinged misogyny[66]) — not to mention the different situation of the world today, ecologically and otherwise — which are hardly helpful for a left-wing climate realism.[67] But Fanon provides us with a particularly useful framework and jumping-off point for a politics attuned to the "urgency" and "immediacy" of a previously un- or under-theorized situation; for thinking affect, subjectification, spontaneity, and organization in "Manichean" conditions; for engaging the existing *fact* of an already violent situation, whether that violence is "direct" or "insidious."[68] We've already seen the contemporary world riven in two by fundamentally irreconcilable experiences, fundamentally different temporalities, even fundamentally different *climates*. "In the colonial world, the colonized's affectivity is kept on edge like a running sore flinching from a caustic agent."[69] This takes many forms; Fanon writes of the "North African Syndrome" which (like chronic fatigue today) confounded medical definition, manifesting principally as "lassitude, asthenia, weakness"[70] — i.e., dimensions of exhaustion.

An increasingly generalized colonial "affectivity" is ever more prevalent today. As we have seen, in business-as-usual, in the right-wing climate realism unfolding before our eyes — the traditional borders, boundaries, and hierarchies of Fanon's world are even more powerful. The Forever War is a permanent, omnipresent, colonizing terror and violence. Across the Global North, pockets of internally colonized territory and populations expand. The news of Empire's demise turns out to have been grossly premature. Rather, it has *metastasized*. It is increasingly the case that the "violence, fraud, oppression, and plunder" that described the particularity of the colonized world for Rosa Luxemburg "are displayed quite openly" in the imperial center. It is not yet the case that this happens, as she continues, "without any attempt to disguise them." But the "sermonizers, counselors, and 'confusion-mongers,'" who Fanon wrote of as crucially intervening in colonizing societies and largely absent in the colonies, are now working overtime. Socioecological violence — both in the form of genuine direct coercion and in the now-daily lived experience of climate change — are reaching populations in the metropole who have not experienced such conditions for over century, if ever at all. The "dominance without hegemony" that Ranajit Guha once identified as a peculiar characteristic of colonial and post-colonial societies is now increasingly, if unevenly, globalized.[71]

For Fanon, this affectivity is socially produced and yet far more generally shared than what strict class analysis might suggest. "This is why," as Fanon famously quips in *The Wretched of the Earth*, "a Marxist analysis should always be slightly stretched" in understanding the politics of the colonized world.[72] Although Fanon should be understood as a kind of heterodox Marxist thinker, his experience of the Algerian Revolution and observation of anti-colonial struggle in general demanded that class analysis be stretched. In anti-colonial struggles, people who would be classified in "orthodox" Marxist thought as peripheral or even corrosive to class struggle — i.e., peasants, *lumpenproletarians* (criminals and social cast-offs) — as well as those in "contradictory class locations" (like professionals and many others) were potential and even necessary participants in the politics of decolonization. At first glance Fanon's colonizer and colonized, settler and native seem just as neat an antagonistic pair as bourgeoisie and proletariat. But he understands that the native population is far from homogeneous; it is riven by distinctions of class, language, ethnicity, religion, and more. What connects these disparate groups is the general "affectivity" generated (albeit unevenly) through the shared condition of colonial domination, the omnipresent, ordinary daily violence of colonialism.

Like in contemporary climate politics, Fanon's theory neither abrogated nor abridged Marxist and other radical aspirations to emancipation. It is a *lateral* politics with a radicality all its own. Today, the sores "flinching from a caustic agent" are found across ever more boundaries. Exhaustion is in the Filipina migrant domestic worker; in the Californian digital laborer; in the gendered or racialized or orientalized objects of coercive repression; in the aspirationally employed and the precarious; and in those consigned to ever-increasing "surplus populations." Exhaustion may be felt as the experience of political conditions or the widespread impacts of ecological ones. Exhaustion can be experienced by South Asian peasants now facing unending monsoons and mudslides or by American farmers experiencing their first ever inland hurricanes or by communities discovering that the fires will never go out. Exhaustion is in the migrant camp and in mid-century models, not only of economic growth and labor practice, but of domestic and social norms, forms, and ideals. Exhaustion is capacious. It contains room for all of this and for novel, only recently catalogued climatic feelings: climate grief, climate mourning, climate rage.

What should be clear from the psychological, epidemiological, and sociological literatures is that exhaustion is being felt across each node of the extractive circuit. Within non-radical epidemiological literature, affect is noted as a way in which "social conditions 'get under the skin.'"[73] Contained implicitly therein is the idea that affect also spills over. Affect is not only a way in which the social flows through the individual but also, as Teresa Brennan observed, how affect courses beyond individuals. Brennan asks the rather ordinary question, "is there anyone who has not, at least once, walked in a room and 'felt the atmosphere'?"[74] As with Fanon's clinical observations, this is not just speculation; "the transmission of affect," as Brennan puts it, is observable everywhere from clinical situations to crowds. She notes, for example, the transmission of affect through recognizable, if not fully understood, mechanisms like hormones and pheromones, even while cautioning that these hardly give full expression to "capture a process that is social in origin but biological or physical in effect."[75] What Brennan observes, Fanon understood as the "atmosphere" in which "everyday life becomes impossible."[76] Just as so many different viewpoints converge analytically — though certainly not prescriptively — on capitalism as central to understanding climate change, we see similarly in so much of the normative and critical literature just how central and widespread an analysis of exhaustion is to understanding human life today.

Although Fanon is not an explicitly ecological thinker, his political thought is well-attuned to the divisions of this conjuncture, to the very real "direct violence" of colonialism and the "insidious violence" of the modern capitalist world:

> Between colonial violence and the insidious violence in which the modern world is steeped, there is a kind of complicit correlation, a homogeneity. The colonized have adapted to this atmosphere. For once they are in tune with their time. People are sometimes surprised that, instead of buying a dress for their wife, the colonized buy a transistor radio. They shouldn't be. The colonized are convinced their fate is in the balance. They live in a doomsday atmosphere and nothing must elude them [...] The colonized, underdeveloped man is today a political creature in the most global sense of the term.[77]

In the development of political consciousness, the colonized are ever more synchronized to their own time, "adapted to this atmosphere" bridging the immediacy of colonial violence and the "insidious violence" of capitalist modernity more broadly, but arrayed relentlessly, *politically* against it and all those who hew to it. Climate change is not "a doomsday" in the Christian eschatological sense of absolute existential destruction, but it does hold "fate in the balance" for this politicizing mass. It is frequently forgotten that Fanon *is* a Marxist, albeit one faced with an immediate project in the present, one not quite on the "orthodox" map. Decolonization is not only the obliteration of the colonizer's temporality and geographical order; it also aims at global reconfiguration.

The great political hope for Fanon is *externalization*. In his clinical work, Fanon viewed "affective disorders" at the nexus of environment, society, and biology. Unlike Félix Guattari — with whom Fanon shared many direct teachers and influences — Fanon didn't valorize serious mental illness *per se*. Rather, he diagnosed how the violence, "atmosphere," and degradation of colonized society interact with specific biological characteristics to produce pathologies. He mobilized novel clinical and therapeutic techniques, psychopharmacology, and social context towards the possibility of political externalization in tandem with individualized treatment that in many ways assists in adapting to existing domination. Even if the social conditions are, in the end, the principal cause — the "triggering factor" — of the disorder, Fanon's practical imperative required a dialectical approach. The pathologized individual (and the pathologized society) must come to understand their "internalization" such that it is even possible to choose "action [...] with respect to the real source of the conflict, i.e., social structure." Externalization — spontaneous or learned — is the beginning of both a political subjectivity and a political power. Far from his caricature as some kind of unyielding prophet of violence, Fanon simply saw violence as already a social *fact* within colonial societies. That there would be violent reaction is irrefutable. Even the most canonically "nonviolent" social movements — Gandhian *satyagraha* or the American Civil Rights Movement — occur alongside both spontaneous violent outbreaks and organized violence. It is *politics* — not violence exclusively — that is ultimately the outcome of Fanon's therapeutics.[78] It is easy to see the *fact* of socioecological violence and it brings us a step closer to *actual* climate politics.

Exhaustion is pre-political, inchoate, but far from the looseness so often assumed in affective language, exhaustion is more precise. It is socially produced, reflects material realities, and circulates among the vast majority along the extractive circuit. Exhaustion is not the way *everyone* — or just one tiny group — feels in relation to the current moment. Fanon makes many gestures to a possible new universalism in his writing, but his politics (and his clinical work) require emphasizing that "Manichean" separation. Even in the "compartmentalized" colonial world, in which the dividing lines and conditions should be plain to see, this Manicheanism must be emphasized — "it is necessary to keep the line of demarcation quite clear," as Hussein Bulhan writes — to connect general "affective disorder" and exhaustion with underlying socioeconomic realities and to emphasize that the conflict is fundamentally irreconcilable. Colonial conditions prime the pump, so to speak, but they do not automatically produce political subjectivity. This is of course, in part, the task of formal organization. Or, as Stuart Hall famously remarked, "politics does not reflect majorities, it constructs them."[79] One needs, as Fanon suggests,

> a well-defined methodology and above all, the recognition by the masses of an urgent timetable. One can hold out for three days, three months at the most, using the masses' pent-up resentment, but one does not win a national war, one does not rout the formidable machine of the enemy or transform the individual if one neglects to raise the consciousness of the men in combat.[80]

It is a similar dividing line we have been drawing between the world of business-as-usual — its anxiety of crisis management, its imperatives for resilience — and the world of exhaustion. The political project of organizing a flourishing, sustainable niche requires specific political actors. This is not a transcendental question, not one that unravels every philosophical knot, but an urgent social one in the everyday mundane.

As Kathi Weeks brilliantly observes, a particular project, in her case that of antiwork, cuts "across traditional class divisions" and requires a politics which "disrupt the functionalism of static class formations." These divisions are captured not only quantitatively but must be "grasped also in qualitative terms, as attitude, affect, feeling and symbolic exchange."[81] Affective attachments to the status quo run over and through pre-defined understandings of "women" or "class." As she argues earlier, "standpoint" theories which do so "carry the potential to enable and cultivate antagonistic subjects."[82]

Exhaustion is precisely such a standpoint, an organizing principle, a foundation, a lens, for the specific politics of a left-wing climate realism. The concept of the Exhausted does not — as with, say, "the multitude" — erase disparate differences or elevate any particular social dimension as the one prized characteristic. Whether heatwaves in South Asia, or hurricanes so prolific along the Atlantic coasts of the Caribbean and North America that they outstrip naming conventions — affective response to climate, how the social-cum-natural gets "under the skin" — is clearly proliferating.

These are not the *same* experience, nor does the politics of exhaustion need such similarity. Its power is in how radically different grievances can be connected with others; with their causal situation in the socioecological reality of the extractive circuit, in zero-sum conflict with the sated (but anxious) partisans of business-as-usual, in the collectively enticing promise of a flourishing, sustainable niche which not only presents its own quasi-utopia but ecologically *requires* relief from the relentless exhaustion of the extractive circuit. Davis's question again: "Who will build the ark?"

The question does not pose a quantitative difficulty. Part of exhaustion's precision is that what is socially and ecologically needed (within "a safe operating space" and with actually existing technical capacities) promises a richer, better, in many ways more abundant life, for roughly everyone who *might* be captured in the Exhausted. The *inexhaustible* technicolor splendor of their many possible realizations possible now, *necessary* now, is a qualitative question about *desire*, about who is most likely drawn to this different, sustainable, slower, even if rationally planned and centralized, life relieved from the unbearable relentlessness of the vicious cycle of exhaustion — rich, abundant, and flourishing, but in new and unfamiliar ways. Exhaustion guides us probabilistically to those who are already fighting and those who seem less *bound*, less *attached* to the world of business-as-usual.

Like Marx, Fanon understood political subjectivity as made possible by particular conditions within a mode of production. The extractive circuit is of course a mode of capitalism; a more developed, more *advanced*, specific arrangement of it, laid out in full socioecological expression. However, Fanon sees in the broader conditions of colonial society and its violences the possibility to forge "among the oppressed the consciousness of a shared condition and the habit of solidarity."[83] What connects the disparate social formations is that general "affectivity" engendered by the shared condition of colonial domination. The resulting spontaneous uprisings — like the increased frequency and intensity of contemporary social and ecological conflicts — do not constitute political consciousness in themselves. In Marxist terms, this is barely even recognizable as a "class in itself," let alone a "class for itself" (i.e., a self-conscious political agent). As some of the scholars of contemporary social upheaval note, while frequency and intensity are equivalent to revolutionary periods, this moment has produced little transformative outcome; there is little cohesion.[84] One of the questions that faces climate politics by dint of its "immediacy" and "urgency" is what mass politics looks like emerging from a largely *depoliticized* era as opposed to one of general mass movements and parties. For Fanon, it is only in the organization of the anti-colonial movement — in struggle with the colonizer, in the desires for decolonized life, in connecting momentary, often pre-political or amorphous affective responses — that even temporary "crystallization" of shared subjectivity is possible.

This formula is hardly unique; it is broadly similar to Marx's. However, the unexpected conditions and temporalities of the current socioecological conjuncture scramble the coordinates, directions, aspirations, and compositions of pre-existing political theories. Left-wing climate realism needs a genuine political subject. Not simply a movement of movements nor certainly anything resembling liberal moral individualism. The Exhausted are a stretched class, composed not uniformly in relation to production, but in disparate social blocs (using Gramsci's or Hall's languages) arrayed along the extractive circuit, in capitalism's long supply chains, and in those people and places consigned, as Ruth Wilson Gilmore writes, to "organized abandonment." This potential class subject shows every sign of *feeling* the irreconcilability and non-universality of the moment; social upheavals of such scale — especially in the depoliticized conditions of recent decades — are one sign of its political potential.

Exhaustion may seem intuitively a paradoxical condition from which to mount political struggle. But we should not view exhaustion through the lens of the management studies and related literatures that have dominated and defined it for over a century when thinking about the social or political capacities of the Exhausted. For Fanon, "lethargy" can be "galvanized."[85] "We are witness to the mobilization of a people who now have to work themselves to exhaustion."[86] Even the "violence" for which Fanon is so famous we are told "runs on empty."[87] "All those men and women who fluctuate between madness and suicide," he writes, "are restored to sanity, return to action and take their vital place in the great march of a nation on the move." This is hardly limited to the specific twentieth-century national liberation struggles. Cedric Robinson reminds us of slave revolts and other earlier anti-colonial insurrections, from the Haitian Revolution to slave revolts in "Jamaica, Suriname, and North America," and "Muslim revolts in Brazil" among populations being literally worked to death.[88] Mike Davis recounts "Millenarian Revolutions" in the contexts of extreme deprivation, from the Boxer Rebellion to Brazil, Colonial Asia to Africa.[89]

Marx — in the same breath as he marvels at the ingenuity of

the capitalist mode of production in general and of factory life in particular — laments "that factory work exhausts the nervous system to the uttermost, it docs away with the many-sided play of the muscles, and confiscates every atom of freedom, both in bodily and intellectual activity."[90] Exhaustion is everywhere in Marx's depiction of the world of labor — and the natural world as well. As Kohei Saito notes: "Since both labor power and nature are important for capital *only* as a 'bearer' of value, capital neglects the various aspects of these two fundamental factors of production, often leading to their exhaustion."[91] Marx recounts the historical fact of the ability of exhausted workers to win concessions, especially in the length and regularization of the working day. He also reminds us that exhaustion is not only fatigue. Similarly, what we see in rich Marxist histories like those of E.P. Thompson or in the Davis histories recounted earlier is not a theory of the singular proletarian subject, but rather the range of capacities that are possible even in a literally exhausted subject.

Right-wing climate realism is, in part, knit from managerial fears and anxieties as much as the pleasures of power, comfort, or supposed security of the powerful. Such affects are produced through one's place in the global caste system, but they are hardly "political" in and of themselves. Both the Exhausted and right-wing climate realism take true shape in a genuine political struggle. And just as Fanon had to stretch Marx to fit the colonial frame, we have to stretch Fanon, since of course the struggle for national liberation — fraught even if achieved, as Fanon and many others argued — does not map one-to-one on to left-wing climate realism. However, as Fanon wrote, "as long as colonialism remains in a state of anxiety, the national cause advances and becomes the cause of each and everyone." It is the job of a left-wing climate realism to keep business-as-usual in its own state of anxiety. Many on the left speak of the need to employ a diversity of tactics but balk at discussion of violence and even of property destruction. They cast violence as always a moral or strategic error, forgetting how much even militant *minorities* (as we'll return to) have been able to achieve using such tactics and strategies fitted to their precise time and place. No significant social change, certainly none at the scale of achieving a sustainable, flourishing human ecological niche, has ever occurred without some violent component. But it is not violence alone; there are also the institutions — not only parties but sports clubs, literatures, art movements, social occasions; the footholds in government; the class traitors in industry; the prefigurative communities joined (and enabled) by the force of realist power. All this nurtures the subjectivity of exhaustion as it spills over so many boundaries and borders into the lived reality of the Exhausted meeting the fact of socioecological violence with political force of its own, enough to keep the anxiety of crisis management in a heightened state.

The potential subjectivity of the Exhausted is not about the most abject or the most advanced class position. Exhaustion as an organizing principle is specific. It gives meaning to already existing experiences in, through, and around the extractive circuit. It gives form to "affective disorders" and actual passions. In the ecological conditions of this moment, it can connect radically disparate social locations, identities, and class strata. The possibility of the Exhausted is grounded in material conditions, connects ecological necessity with "good-life" desires long dated and disintegrating, and crystalizes into a genuine political consciousness through recognition and struggle against the partisans of business-as-usual and those who find safety in their shadow. These are people bound together through ordinary struggle but also by a pre-political feeling, an impulse, experienced differently, of the untenable and unbearable nature of their disparate situation. The "restoration" of all those "who fluctuate between madness and suicide" begins by *refusing* to be resilient, refusing to absorb maximal degradation with minimal support. Exhaustion here is the ground of a possibility — the possibility to finally let go, to externalize, to politicize, to return to a sanity that is not resilience, but rebellion. The current social and political uprisings and upheavals as we begin the second decade of the twenty-first century are, from the point of view of the resilience regime, "maladaptive" behavior. From the point of view of exhaustion, they are the beginning of political possibility.

To hold the line of a less than $\Delta 1.5$ °C world — a line already palpable in the ongoing catastrophe of our $\Delta 1.2$ °C world — requires, even in the filtered and reserved language of the IPCC, "rapid, far-reaching and unprecedented changes in all aspects of society." A flourishing sustainable niche will not, as we've seen, be "accomplished by the wave of a magic wand, a natural cataclysm, or a gentleman's agreement."[92] Decolonization marks the inceptions of a "new rhythm" of "urgent need." Marx notes that politics was always a fight for time; Benjamin famously suggests that perhaps what we need is "pulling the emergency brake." For Fanon, the colonized are caught between different *times*, experiencing the direct and "insidious violence" of the modern capitalist world, fighting for their own time. The politics of climate change is the urgent conflict "between two congenitally antagonistic forces." Ironically, the fight for time, for relief, for the minor paradise, must be conducted on the swiftest and most instrumental timeline.

## The Politics of Exhaustion

When Fanon was mapping his anti-colonial manifestos, when Gramsci was surreptitiously sketching subterranean strategies in Mussolini's prisons, it was a time of mass parties, a time of mass geopolitical powers, mass formations of the non-aligned states — mass *politicization* and material buttresses for the left in particular. Luxemburg's theory of the Mass Strike had the wind of the high-point of European labor organization and socialist mobilization at its back. Or to turn the clock back further to the nineteenth century, Harriet Tubman not only had the momentum of years of slave uprisings, but her underground railroad was a moving piece in one the largest civil wars in history. This held true across other geographies — when Rani of Jhansi took up the fight against the British, she did so as part of and propelled by the Sepoy Rebellion of 1857.[93]

While today, there are mass uprisings, mass movements, and social upheavals, this is *repoliticization* out of decades of neoliberal depoliticization. Even with new formations and increasing mass organizations, there is, in the North, little approaching the kinds of organization, the kinds of infrastructure needed to extend "momentary passions" into the intense conflict already at hand in the politics of climate change. Organizations from La Vía Campesina to Extinction Rebellion, ecological distribution conflicts burning across the South to nascent eco-socialist formations in the North, remain disjointed. However, those 3,000+ conflicts, these organizations, and recent social upheavals more broadly are a beginning, a significant break with the quiescence of depoliticization, with the once-hegemonic Thatcherite mantra that "there is no alternative." The intellectuals of business-as-usual, though, already panic at these stirrings, fueling a whole luxury publishing niche for the consumption of concerned capitalists, anxiety-ridden crisis managers, and closely bound partisans — even while there is yet little infrastructure to support the transnational aspirations of radical local projects and movements like Cooperation Jackson in Mississippi. The left is in power in precious few places with differing powers, constraints, challenges, and contradictions — Cuba, the "Pink Tide" governments in Latin America, Vietnam, a few smaller strips and states in tiny dots across the Global South. Even if one includes China (controversial for some) in such a list — the world's largest carbon emitter but also the global leader in green technologies, South-South transfer, and more[94] — the politics of exhaustion, of left-wing climate realism, faces a *political climate* not seen since the nineteenth century.

In the face of such conditions, there is the desperation-fueled Climate Lysenkoism and various other (anti-) "realist" permutations trying to thread the need for mitigation and adaptation with the status quo. Alongside these are well-intentioned plans, policies, visions — many of extremely salutary contribution but lacking serious contention with power.

On ecological and other left grounds, renewed calls and debates about parties and party-like structures abound. Expanding upon Samir Amin and countless other proposals for new internationals and new parties, Karataşli argues, much like Brazilian political theorist Rodrigo Nunes for hybrid party forms. Networked and centralized structure, horizontal and vertical, direct and formal, local and mass; questions debated *ad infinitum* have real historical and theoretical grounding in such cases. In the Anglophone world, Jodi Dean makes perhaps the firmest case for a new party precisely in terms of extending, cultivating, developing, and organizing quickly dissipating moments of affective efflorescence.[95] However, Gramsci (and contemporary Gramscians) complicate these pictures by observing how a party isn't always the formal political structure imagined even in Gramsci's time: a "newspaper" or a "review" or by extension seemingly apolitical organizations of recreation (a sports league) or of primarily technical focus (a mechanics', nurses', or engineers' club) can all function *as* parties.[96] The landscape of today's left is largely bereft not only of the ambitions of new internationals but in many places of the counter-hegemonic institutions — the spaces from dance halls to clinics to sites of "maroonage" — that are so pivotal towards mass or "stretched" class feeling, from Thompson and Davis to James and Robinson.

There is no single overarching blueprint for these kinds of questions, certainly not across all geographies, and even in relatively localized forms. That, as Gilmore observes, is determined in the course of struggle and the indeterminacy in which practice unfolds.[97] This is the very basis of praxis. Instead of another just-so story or an imaginary blueprint, we can chart the barriers, historical antecedents, and geographical comparisons a left-wing climate realism faces, as well as some of the necessary strategies and tactics a politics of exhaustion will have to countenance, consider, and employ.

Recall the modest attempts to quantify the costs of a less-than-$\Delta 1.5$ or even a $\Delta 2$ world: compounding with non-GHG boundaries, even by a relatively moderate estimate, these amount to approximately $80 trillion over the next 20 years, with massive write-offs of stranded assets and lost profits in the immediate term, of roughly $4–16 trillion. There are not just the costs of a green transition; there is, by necessity, wealth destruction (in the overdeveloped world, this means additionally not just GDP reduction but destruction *and* redistribution of accumulated capital). Like climate politics itself, even this basic reality presents a zero-sum (or "negative-sum") game.[98] Even capitalist commentators like the *Financial Times'* Martin Wolf recognize this as a reality: "If, as now, in the UK and other high-income countries, the economy stagnates, politics becomes fraught, since one group cannot have more without others having less." And as he observes, there is no road to renewed growth, desired or not.[99] Such realities are not *per se* incompatible with ideas like a Green New Deal, but they do place an imagined model of *politics* based on Rooseveltian or European welfare compromises outside the realm of possibility. At the same time, to borrow Malm's time bind again, the idea of not only *achieving* but actually *building* a full socialist or communist society has not only proved historically difficult, but the idea of doing so within the temporal constraints of mass climate mitigation and adaptation is equally impossible. When acknowledged, this intra-left impasse drives a renewed, largely fruitless rehash of century-old debates about reform and revolution.

A left-wing climate realism does not depend on a fully classless society, the abolition of all private property, and the establishment of post-monetary markets, let alone the negation of the "value-form" or the withering of states. In fact, not only is the total revolution quixotic, it misunderstands the unanticipated, unexpected, and lateral nature of emancipatory climate politics. These features of a fully socialist society would actually be *impediments*, slowing down necessary socioecological transformation in both technical and political terms. The relief and reversal of exhaustion has myriad appeals to actual majorities generated through the extractive circuit. For many emerging and vital constituencies of the Exhausted — from peasants and smallholders to small market actors and even *some* elements of capital (there is *some* profit to be made after all), among others — the project of full socialism put in these terms (even if possible in the urgency of climate) is not necessarily desirable and has considerably less mass purchase than exhaustion. I don't necessarily mean the word itself but the logics and projects it entails; the world it can build. This is not a discursive game or a trick or "bribe": all the coordinates of exhaustion and its relief are structured and primed by the conditions of this conjuncture.

While disaster communist, communization, and other similar projects will almost certainly happen, and are practically and strategically useful, without forces of mass power they will be rendered irrelevant in the face of the mobilized forces of right-wing climate realism. Goldman Sachs might not care if you raise chickens, as Dean famously quips, but it *does* care about maximizing shareholder value. Recent statements and notes from the International Monetary Fund (IMF) have all but abandoned mitigation and its drastic — catastrophic for them — costs in favor of a largely exclusive adaptation strategy in which climate change is understood as fundamentally just another in a long line of "societal challenges" to be met with "technological diffusion, innovation, and behavioral change." Unlike mitigation, which is plagued by difficult "coordination problems," "individuals and firms have strong incentives to adapt because many adaptation benefits tend to be local and private."[100] As the climate finance advisor and journalist Kate MacKenzie put it in macabre humor: "You could come away from this thinking that adaptation is well understood and straightforward; the main challenge is simply making sure you don't spend too much on it."[101] More darkly, this is yet another turn in favor of modes of right-wing climate realism, of ecological fortress, theoretically sheltering zones of wealthy safe havens behind risk-shifting sea walls in which adoption of knowledge and technologies from "warmer" geographies are grafted, a perverse inversion of the minor paradise in service of a continually exhausted future.

However, there is a gulf that lies between reformist models based on the "win-win" model of the mid-century and full-scale revolution. A string of analyses from nearly a decade ago argued that the nearest historical model and comparison for the wealth destruction, redistribution, and costs for mass climate mitigation and adaptation is slave abolition in the United States; in other words, "civil war."[102] Slave abolition, as measured by Piketty, was the largest destruction and transfer of wealth in human history.[103] Its costs, when measured in comparative economic historical terms, are eerily reflective of that that $4–16 trillion near-term climate transition price tag. This is an illuminating analogy for the scope, stakes, urgency, and intensity of climate politics.

It is simply a fact that this much power, this much wealth, as Fanon began his political theory, will not give way without a movement on all fronts, without force, without violence. This does not mean *only* or even *principally* violence, but in that gulf between the social democratic fantasy of smooth electoral and legalized strike action and the full-scale armed "war-of-maneuver" of direct revolution, something like "civil war" surfaces as a realistic political model. Even understanding that there cannot be a one-size-fits-all model of left-wing climate realism, a set of strategies — differing based on the specific conditions of each geography — does emerge in the particularities of this unexpected conjuncture. Civil war is the option that lies between two impossibilities; its realistic radical possibility (assiduously avoided and downplayed in the ideological fictions dominant in so many academic fields and activist playbooks, or discussed only in pursuit of avoidance and suppression) only matches its *already extant* radical reality.

Talk of civil war conjures images — not without good reason — of full-scale total war, of the American Civil War or, as with Fanon, anti-colonial civil wars like in Algeria — protracted conflicts of mass mobilized and militarized societies. But thankfully this does not define the concept. As Gramsci described Gandhi and the Indian anti-colonial movement in the 1920s, that was "a war of position, which at certain moments becomes a war of movement and at other times underground warfare."[104] It was, in other words, civil war. "Boycotts are a form of war of position, strikes of war of movement, the secret preparation of weapons and combat troops belongs to underground warfare. A kind of commando tactics is also to be found, but it can only be utilised with great circumspection." In Gramsci's terminology, the war-of-maneuver (direct confrontation, i.e., revolutionary insurrection) is juxtaposed with the war-of-position, the more careful preliminary political and cultural strategy through "footholds" and "trenches" across civil societies and states. Here we have something in between "underground warfare" and "commando tactics" only to be employed with the greatest of caution. While the popular image of Indian independence is of non-violent success through Gandhi's *satyagraha*, Gramsci is actually understating the case. Before and after Gandhi returned from South Africa, there were both spontaneous uprisings and planned attacks while Gandhi and the Indian National Congress proceeded in fits and spurts with formal politics and Gandhi's famous non-violent actions and prefigurative communities. As all serious scholarship on possibly the world's best-known case of mass "non-violent" social transformation attests, it's success rested in no small part on violence.[105]

Although Fanon's direct model was the *Front de libération nationale* (FLN) and a full-blown anti-colonial civil war, he recognized that even modest, "symbolic violence" is pivotal even in cases where it is far from the determinative factor. Ironically, Fanon argues this not in his *realist* prescriptions but in the much-overplayed moments of "self-respect" and recognition.[106] Actual violence, as Fanon so carefully traced in the book's concluding psychological case studies, is corrosive and debilitating to all. Far from the kind of celebration of combat as the quintessential human experience you find in thinkers like Ernst Jünger, Fanon's civil war is a tragic affair, imposed only out of necessity. These are the modes of "civil war" we should be thinking of in terms of left-wing climate realism; adapted to today, not carbon-copied from the past. "Civil war" becomes the political model beyond the empty liberal admonition to "vote harder" and the quixotic thought that the world revolution will come before ocean acidification reaches levels not seen for 26,000 years.

Talk of civil war has also found a recent resurgence in the North, in the United States in particular. The anxiety of the crisis managers of business-as-usual, already jittery from the slow return of mass demonstrations, mobilizations, and nascent repoliticization, goes into overdrive along the "liberal" edge of right-wing climate realism, where they see the specter of "populism" (read: *any* politics outside the ever-narrowing bounds of ever more tenuous hegemonic "consensus"). In this panic, scholars like Barbara Walter ring alarms about a descent into new civil war, which they understand as between Republicans and Democrats, red states vs. blue states. Walter correctly theorizes that a civil war in the US today would not look like the Civil War of the nineteenth century, that due to how much conditions have changed, it is more likely to take the form of insurgencies and counterinsurgencies.

In critical response, scholars like Nikhil Pal Singh observe that by Walter's definitions, based on the globally exceptional scale of US police violence alone, "in the zone of precarity where many people dwell [...] civil war is ongoing."[107] Singh correctly notes that the framing of civil war talk by Walter and others is fearful catastrophizing which turns away from the real disintegration, degradation, and disillusionment of American society and instead reduces these to a lament for "waning confidence" in American Empire. Civil War between Democrats and Republicans is absurd, a hobgoblin in the psyche of crisis managers everywhere. Singh dismisses this as a distraction from the "daunting fight to salvage a democracy of common purpose and shared prosperity." But he does observe the possibility of an already existing "civil war" waged from *within* the national security apparatus. And, by way of comments from Newt

Gingrich, he records that the right waged a successful civil war through the uniquely minoritarian institutions of the American liberal republic.[108]

While, again, politics of left and right face different challenges, it is worth returning to what even a militant minority (let alone a burgeoning majority) can achieve. Singh mentions, for example, the successful rollback of abortion without acknowledging the plain reality that it was the coordination of "respectable" (and plausibly distanced) juridical organization and formal electoral politics; strategies across media and communications; gaining footholds in key state institutions; taking advantage of cultural sites — like churches — to cultivate affective intensity; and, almost universally ignored, a successful campaign of political violence in the plainest terms: intimidation, harassment, assault, arson, bombing, and assassinations. This campaign of violence was much in the model of insurgency; it created (as military occupiers like to say) "facts on the ground," in this case the literal destruction of enemy positions and deterrence of enemy personnel (doctors, nurses, clinic workers, and volunteers). Through these and other tactics it created *de facto* regions without access to medical abortion, formalized through the more "proper" and "moderate" factions after the fact into *de jure* restriction.

Raymond Geuss is fairly open about civil war as an operative and useful political category and model. He identifies just this kind of unraveling with Hobbes' definition of civil war.[109] Geuss takes as foundational Marx's claim that all class society is a "more or less veiled civil war."[110] Geuss has no sympathy for the waning confidence in the "rules-based international order" (which he pillories as the "sheltered internal space of *homo liberalis*") or the "nostalgic breeze of late liberalism" that laments its weakening. (Singh too argues that this is already waning.[111]) Political realism for Geuss is, as we've seen, not the suffocating ideological "realism" of Fisher's "capitalist realism" nor the normative "realism" of the officially sanctioned and bounded terms of "proper" politics. From the Greeks to now, civil war is a distinct and important model of politics. The question of using that model or violence is posed, Geuss argues in discussion of Fanon, in the "context" of and "comparison with" what violence is ongoing. Civil war and violence are historical facts; even if one is morally convinced by nonviolence, Geuss argues in a particularly illustrative example, "the goal of preserving the viability of our planet should reasonably take priority over the avoidance of violence."[112] At the end of an essay on Marx and morality, Geuss paraphrases Brecht: "I have nothing to say to anyone who, in view of the poisoning of our earth, can think of nothing but the question of gross national product."[113] Even if contemporary liberal commentary on civil war is an absurd distortion, it should not blind us to it as a mode of politics.

## The Politics of Exhaustion: On Violence

As the US-supported, neo-fascist coup unfolded against the *Movimiento al Socialismo* (MAS) government in Bolivia, protestors took to the streets, chanting "here we go, civil war."[114] And this is precisely the category which best characterizes the period between the 2019 overthrow and the eventual 2020 capitulation, election, and restoration. It gives us one model of what "civil war" means and can look like today. This episode, and MAS more broadly, are instructive. MAS is less a classic, formal party than a sometimes tightly, sometimes loosely bound set of social movements, organizations, peasant and trade unions, indigenous radicals, and Marxists. MAS illustrates how a formation like this can radically transform a society, but also how in isolation and subordination in the American "rules-based international order," parties like this can be riven, particularly along socioecological fractures. MAS demonstrates how important it is to carve out physical space for left-wing thought and politics through national self-determination (as my colleague Max Ajl notes[115]). It's this that provided the material conditions to convene the 2010 World People's Conference on Climate Change in Cochabamba, Bolivia, with participation from over 100 countries. The Cochabamba

Agreement is probably the most radical official transnational climate politics document ever produced. It shows that most Global South states are committed to what I would call a form of the politics of exhaustion, but lacking a path beyond the layered caste system of the global economy, and without Northern partners, such an agreement is not possible to actually carry out.

MAS is, as Santiago Anria notes, among a set of "movement-based parties" which emerge from organic social movements, in contrast to the supposed "norm" of parties (like modern US Republicans and Democrats) organized by political apparatchiks and legislators as electoral vehicles.[116] Anria provides illustrative examples across a wide political spectrum, although almost all are counter-hegemonic in some way, like the African National Congress in its anti-apartheid formation; the Socialist Congress Party of Kerala (which later became the Communist Party of India (Marxist) (CPI(M))); the Muslim Brotherhood; and, both germane and ironic, the historical case of the Republican Party "emerging from the abolitionist movement in the context of the American Civil War."[117] Most of these parties share experiences with civil war and violent conflict but also — just as importantly — with formal politics; fostering institutions, sites, and organizations for care and recreation; and the full repertoire of labor strikes (extending far beyond the shopfloor). In other words, they vitally and strategically nurture forms of Bosworth's "affective infrastructure" or Gilmore's "infrastructures of feeling."

In the 2019–2020 Bolivian civil war, even the organizations and constituencies whose ties with MAS had frayed in the vice-grip described above mobilized with differing degrees of coordination. Estimates of casualties are difficult to find and range from dozens of deaths to more than a thousand, with considerably larger numbers injured.[118] By pretty much any standard (and particularly the prevailing skew favoring nonviolence in movement studies and climate movements in particular), violence was omnipresent. Most deaths and injuries were at the hands of coup forces, but pro-MAS mobilizations (alongside "peaceful" protests) took part in arson, blockades, infrastructure destruction, and direct killing with firearms (both defensive and offensive).[119] The neo-fascist coup regime (which, in a familiar tale, claimed legitimacy by way of power-erasing gestures in the garbled language of constitutional law and friendly media spin) enjoyed support from the United States and other powers, and ruthlessly suppressed opposition, all but announcing in its early days a campaign of genocide against the informal and indigenous populations. Still, through the long mobilization and particularly the violent blockades of key political spaces and upper-class neighborhoods, it was eventually forced to concede.

In a fit of the absurd, in its final days, the Áñez regime submitted the MAS blockades to the International Criminal Court (ICC) as "crimes against humanity," specifically arguing that the blockades cut off supply chains needed for healthcare in the burgeoning COVID-19 pandemic.[120] (That the healthcare system's large-scale development and public provision was one of the signal achievements of the MAS government was apparently lost on them.) The coup's genocidal intentions (for which Áñez and her co-conspirators currently face local trial) and large-scale massacres against protestors (which only intensified counter-mobilization and support) have never faced serious international scrutiny in the "rules-based international order." The ICC took the charges against MAS blockaders seriously, even if ultimately dismissing them largely on grounds of intent. It is an undeniable fact, though, that "innocent" people died in the struggle, alongside state security and irregular "combatants." Much in line with the full scope of Fanon's political theory, blockade organizers apologized for harms, however unavoidable, to the Bolivian people.[121] This is the inexorable truth of a real politics with urgent, radical stakes.

The MAS case helps answer two of the great riddles of the conjuncture. First, what is the "class capacity" of this "new wretched of the Earth"? What is its power? Here among other cases, we see the answer. Informal workers, from peasants and indigenous activists to the urban precariat — the majority of those engaged by MAS — wield a pivotal power far greater than many traditional theories in this moment: the power to fight, to blockade, to cut off enemy strongholds, to take and hold space. Second, as Edwin Ackerman notes, MAS demonstrates how neoliberalism — which has proved so successful at repressing and dismantling traditional labor parties and institutions like unions — is surprisingly conducive to the rise of new models and modes of mass parties and mass politics.[122]

> It was the "success" of neoliberal societal dislocation — its effectiveness in fragmenting existing social blocs — that produced the conditions for party articulation and organization [...] it is in the context of erosion — of dispossession of the means of production — where the very possibility of articulable interests emerges.[123]

These interests, these feelings, are global but still largely disconnected. "Who will blow up the pipeline?"[124] many astute critics and commentators ask in response to Malm's polemic, just as we asked earlier who will build the ark of a sustainable niche. But the extractive circuit *has* primed the potential of the Exhausted. The tasks of organization are no less or more difficult, but they require a radical reorientation towards the likely and unlikely, abandoning the futile hopes of universality and transcending powerless, dated, or dogmatic politics.

Among Southern states, Bolivia is noted to have a particularly "thick" and developed civil society, and civil conflict is almost always simmering and occasionally explosive — much like in the United States. The extractive circuit has instrumentalized states, even in the North.[125] France, the nation-state *par excellence*, once the key example of the disjuncture between metropole and periphery, of the mid-century failure of the European working class to join with their colonized brethren, today sees authoritarian liberalism (or its counterpart, open fascism, and the burgeoning eco-fascism of the National Rally) raining the power of the state on Muslim minorities, unions, environmental activists, and the population at large.

The extractive circuit chews holes through social and state fabrics, which can fill anew with counter-hegemonic organization; ironically, it makes the politics of exhaustion possible. Exhaustion's externalization in everything from uprisings and riots, mass protests, and new models of strikes (many quite similar to pre-legalization strikes) demonstrates an affective inversion that Fisher writes about not long after his piece on the privatization of stress. He encounters several of his students — formerly disaffected, desultory — in a kind of spontaneous formation during the 2011 London riots that took place under then Prime Minister David Cameron. "They show no surprise, no self-pity or hyperbolic self-dramatization, just a resolute sense of what needs to be done and a delight in doing it. *I enjoyed it, looking forward to next week.*"[126] Atomization, systemic "total integration," and widespread depoliticization are all too real, but part of the solution to "bowling alone" is found in these modes of exhilaration, although not in these alone. But for the moment, see the affective register: anger and joy, formerly apathetic and disillusioned students now find delight in mobilization.

The politics of exhaustion are not simply transposable or imposable from one site to another. The exhilaration of a London riot lacks the political and ideological formations — the political education so downplayed in revivalist thinking but so important in nearly every historical case — of, say, Germany's *Ende Gelände*, which in turn is embryonic in comparison with what we see in MAS, no matter its tensions. None of these are yet a full realization of the politics of exhaustion. The new conditions which centripetally draw the components of MAS and its allies together and also centrifugally pull them apart are already socioecological in the broadest sense possible. The constraints on MAS's ambitious economic and ecological goals are not biophysical or technological. They are the reality of today's unipolar, US-centered, "rules-based international order."

Even if we conceive of some kind of global civil war, that does not mean it would play out the same everywhere. One clear imperative is that a left-wing climate realism must build lines of communication and transnational organization — however loose, however tenuous at first — that can not only coordinate collective action but can translate into altered or new structures of international governance to replace, supplant, or take over existing institutions of the "rules-based international order." In other words, fulfilling, even if altered in the unexpected socioecological project, what Adom Getachew has called the transnational worldmaking aspirations of anti-colonial nationalisms, in proposals like the New International Economic Order or, as Ajl highlights, the more radical visions of the Bandung Conference. This work is neither as simple as calls for a new International might appear, nor as difficult as some armchair radicals suggest.

Bonds of trust — of solidarity — will at first likely be transactional or, perhaps, discursive. Solidarity, as Gilmore suggests, is not built through "sentimental care for the 'needy'" — the general framework of *official* climate justice theory and practice — but "through constant interaction."[127] This can be as straightforward as establishing *contact* from one place to another, between elements that see themselves in the Exhausted (whether by that name or any other). And yet it forces issues not only of trust (particularly in the cases of the US, UK, and EU) but the challenges of a world of networked communications that are enemy terrain. But just as neoliberalism and the extractive circuit writ large have rendered new mass politics and new party forms possible, they have created the opportunity to piggyback, circumvent, and, as McKenzie Wark argues, "hack" their infrastructure for counter-hegemonic organizing.

## The Politics of Exhaustion: Internationalism

For the Southern reader, the politics of exhaustion might seem quite familiar. The minor paradise of the sustainable niche does not require utopian perfection; in all its earthly splendor, it is imperfect. The intensity and urgency; the temporal socioecological pressures and the "ugly" nature of even a radical realism; self-determination and sovereignty, a globally *rare* condition (dirty words for some in the North but familiar and desired horizons for subalterns the world over) — all mean accepting myriad forms of social and political organization which many (myself included) might find dissatisfying or unsavory. But that is a matter — internationally — for sub-political discourse, for ethical, not coercive, suasion. What is non-negotiable is the socioecological agenda of the sustainable niche in its rough contours and resplendent expressions. The Davos set might fret about the rise of the South as the preeminent ecological threat. But it is as ever the North, and Americans in particular, who need to begin the trust-building.

The South has every *right* to develop hopefully through greener alternative paths and modes of development outside of imposed restrictions and models. But questions of "right" miss the point. More importantly, it can and likely will do both. This is a given fact, not an ethical dilemma. While green transitions in petro-states like Ghana or Iran are certainly possible,[128] part of any sustainable transition must also see such states (extending for example to Angola, Nigeria, Equatorial Guinea, and so on[129]) as the few remaining privileged exporters of fossil fuels at extreme costs within the overall carbon budget, just as with the DRC and cobalt.

It is not that the South is either entirely passive or immaterial. It is that it falls on the North, and the United States (or elements from within to push the United States), to take that first step of solidarity, to prove willing to let go of wealth and power accrued through literal *centuries* of catastrophic destruction.

This also helps explain the seemingly ambiguous climate politics of China. On the one hand, it is the world's largest carbon emitter[130] and on the other, a leader in the development, production, and spread of green technologies, particularly in South-South relationships, loosening destructive international

IP regimes and other policies consistently demanded by most Southern states. These are driven not by some kind of despotic Oriental duplicity but by the constraints of global political economy, recent US national policy, the development needs of its own population, and respect for the internal affairs of other states. While many Western commentators portray this as solely driven by factors like the pandemic and development, the key shift is US policy. Beginning with the Trump administration and *intensifying* with the Biden administration, the United States — ever militant in enforcing that safe space for capital, the "rules-based international order" — switched in this period to a posture of a "new Cold War."[131]

As international analysts reiterate, and as is evident from the relatively plain language of Chinese policy, the PRC is more than willing to play ball on genuine climate action, towards global "ecological civilization" on Southern terms.[132] Even under current conditions China's per capita carbon emissions have been leveling since 2013.[133] But it is not willing to go it alone or at the cost of remaining — even with perfect redistribution — a middle-income nation, leaving its largely poor population below levels of sustainable flourishing. None of this makes China *nice* or *never racist* in moralistic terms, but it is not a neocolonial or imperialist state; that is a pure myth — or telling projection — of Western origin.[134] As both mainstream and heterodox American scholars have noted, whether for ideological or simply pragmatic reasons, the only path open for Chinese aspirations is essentially the *opposite* of colonialism or neo-colonialism; the more peripheral states develop, the less power the imperial core has — outside of military means — to impose its will through existing international institutions.[135] And China would like that development to be as green as possible. As an interdisciplinary team recently wrote in *Science*, the US national security stance will "slow technology deployment and the global low-carbon transition."[136] Hysterical American claims about Chinese "debt trap" diplomacy and the absurdly ironic accusation of neocolonialism have been shown time and again to be empirically false and logically insane. China is not only leading in development funding, it is doing so on terms which specifically work against the logic of Western colonialism and often at a loss, willing to forgive debt and extend interest-free loans to ensure local development capacity rather than permanent dependency.

So it is not the 100+ signatories to the Cochabamba People's Agreement, nor China, the world's technical largest carbon emitter, whose climate politics pose the greatest international risks. On the other hand, there are the wildcards of neo-fascist India and the greatest "rogue" petro-state of them all, the United States. The civil war of the Exhausted is global, its uphill battle needed everywhere, but nowhere more than in the North, and the United States in particular. To carry out the politics of exhaustion in the United States is not to fashion a new benevolent and "indispensable" America. For most *Americans* and for the world, the value lies in proving the *dispensability* of American empire. Even radical activists and thinkers who understand that material throughput use must decrease in the overdeveloped world to create space for low- and middle-income states to grow, who profess opposition to climate colonialism, are often still under the fantasy that the waning "rules-based international order" is anything other than a global Behemoth for capital, protected by American military might, accelerating the extractive circuit, keeping the South underdeveloped (and much of the North desperate, precarious, exhausted).

In this light, the most significant climate document of this moment is the 2022 American National Security Statement. A first step to international solidarity, toward crystalizing the Exhausted, requires not only creating that economic and ecological space, but *fighting* for the disintegration of even a semi-sovereign American-centered empire, which holds back the sovereignty and self-determination of the South, which enforces the global caste system and color line. There is a parallel between longstanding fears of the dangerous appetites of the darker peoples of the Earth and contradictory fantasies about violent struggle. Either violence belongs "over there" (as my colleagues Suzanne Schneider and Patrick Blanchfield have discussed) in the designated non-Western conflict zones among the barbaric natives *or*, in an updated, self-flagellating iteration of the "Noble Savage" myth in which the colored body is somehow magically, ontologically "pure," these same zones should be free of even strategic considerations of conflict and violence.

A better place to start, as Fanon would suggest, is acknowledging the already existing facts of violence and civil war. Perhaps one of the starkest representations of how anti-colonial politics are isomorphic to the twenty-first-century socioecological conjuncture is Eyal Weizman's mapping of "the conflict shoreline."[137] Most Western drone strikes occur across Africa and Asia precisely along the longest aridity line — the conflict shoreline — demarcating a toxic admixture of meteorological, agricultural, and political considerations. This too is a form of right-wing climate realism; it is climate apartheid being enforced in real-time. Right now. These kinds of warzones are not only more likely if business-as-usual continues; they are an actual *front* in the quite intense and immediate violence of climate change — direct and "insidious"; social, ecological, and "structural."

In this light, the civil war in Bolivia — long simmering, in some ways like the US — should be understood as a socioecological conflict. Not because of some supposed plot to control Bolivia's lithium, but because, as we've seen, the conflict — like so many others — is waged at the forced fault lines of socioecological constraints, on today's unipolar landscape, within capital's utopia, the extractive circuit. It is not counted on the sprawling maps of ecological distribution conflicts (although Bolivia's possibly more intense Gas War is), but it fits precisely within Martinez-Alier and Guha's definitions of "environmentalism of the poor." The logic (if not the stunning visual record) of Weizman's conflict shoreline transposes onto this case and many others. The full range of the Bolivian struggle against reaction — from official politics to peaceful protest to blockades to bombings and assassinations — should be understood as *already* climate politics.

Malm's polemical provocation in *How to Blow Up a Pipeline* is one of a handful of major Anglophone texts to broach the question of violent tactics as part of Global North climate struggle. (The book, controversial in academic climate circles, is now a motion picture quite well received in more popular venues[138] and immediately cited by the FBI and "23 separate federal and state entities — a veritable alphabet soup of angst"[139] as an immediate threat to fossil fuel infrastructure.) Despite the remarkably restrained arguments about engaging in property destruction and sabotage (classic tools of the labor movement), the pamphlet received near-universal academic pushback. The book is far from comprehensive, but its critique of the hegemonic Northern environmental mantra is compelling and, if anything, *too moderate.*[140] Tactics along the lines of what Malm proposes are far more common in existing ecological conflicts than is often stated.[141]

To cite some exemplary cases, in 2006, La Vía Campesina (probably the largest environmental justice organization in the world, representing some 200 million peasants, indigenous peoples, and migrant and rural workers[142]) organized some 3,000 members, including 1,000 from the Gramscian-Marxist Brazilian Landless Workers' Movement (MST), to occupy the nurseries and labs of Aracruz Celulose (a Brazilian-based multinational that is today part of Suzano Paper and Pulp). They protested the eviction of indigenous people in Espírito Santo and more broadly for "defense of family agriculture, food sovereignty, crop diversity, especially keeping native seeds, the opposition to monoculture and the expansion of international capital companies."[143] The women occupying the corporate facilities destroyed over five million eucalyptus seedlings as well as one of the laboratories, property destruction valued at millions of US dollars.[144] However, this is not reported as "violence" or "property destruction" or "sabotage." The EJAtlas, which catalogues ecological conflict, characterizes it vaguely as "direct action."[145] Similarly, the successful shutdown of an Algerian fracking operation is described as "fierce protests,"[146] or elsewhere as "violent confrontations between police and protestors," closer to the truth, but still couched in the evasive language of media partiality ("clashes between X and Y"). In actuality, the protests were mass mobilizations – on explicitly ecological and social grounds — that set fire to a local government building and a police transport (injuring but not killing 40 police officers) which were then met with live fire from security services.[147] Eventually the project was forced to be abandoned, a victory whose academic celebration essentially obfuscated the efficacy here of not only property destruction but direct interpersonal violence.

This is not a critique of the EJAtlas, which is an invaluable project and tool,[148] but rather emphasizes how the poor, indigenous, peasant, working class, and middle-class intelligentsia of environmental movements in the South (although there are cases in the US as well) are frequently utilizing the tactics Malm suggests, or even more radical ones. And at extraordinary peril. In the aptly titled study "Not Victims, But Fighters: A Global Overview on Women's Leadership in Anti-Mining Struggles," Francisco Venes, Stefania Barca, and Grettel Navas observe that in about 30% of women-led anti-mining struggles, direct action — from Bolivia-style blockades to sabotage — was not only more prevalent than is usually supposed, but was in fact the most common tactic.[149] Their work challenges pre-conceived, romanticized academic notions about gender, indigeneity, race, and class. Importantly, the study also highlights affective dimensions — "feelings of sadness, anguish, shock, powerlessness, emotional stress and lack of self-esteem" — as well as complementary practices and tactics: "direct action; organizing public events; territorial oversight; consciousness raising; legal procedures; advocacy and campaigning; creating socio-political spaces and fostering community livelihood; and promoting care work."

These disparate groups — one a massive achievement of international organizing from which any new meta-parties or transnational networks can learn, the other largely *ad hoc* and spontaneous, although militant and clear-eyed — share what I have been calling the "affective matrix of exhaustion," produced in and around these nodes of the extractive circuit. And they choose externalization, at great risk, as opposed to resilience, apathy, fatalism or (however unlikely for these groups) the flight to the safety of the armed lifeboat.

Violence is not the only or even the principal element in the strategic repertoire of civil war — which looks different today across both histories and geographies — but its erasure from the existing landscape and from theoretical imagination has left a gaping hole in the way even the left talks about climate politics. It is easier to imagine the end of capitalism, apparently, than to imagine the strategic use of violence in achieving mass sustainable flourishing. While there have been many critical and engaging responses to Malm's polemic, it is surprising that seemingly none have asked whether this modest call to arms goes far enough, emphasizes a different logic of targets, suggests how violence might be utilized across the terrain, or asks whether Malm is *too* reticent, *too* reserved, *too* reluctant to probe the blurry lines between property destruction, sabotage, and violence against people. A "civil war" today would not be the full-scale mass military conflict of yesteryear but something closer to these multifaceted insurgencies already haunting the "crisis of crisis management" today in reports, risk assessments, and surveys. Malm provides many of the initial pieces needed for consideration, but there is much left to do to understand the whole board, at local and transnational levels, even if the rules of the "game" are continually shifting, probabilistic at best.

Strikers in Bolivia didn't bring down the neo-fascist (indeed eco-fascist) government by some abstract notion of denying surplus value at the point of production, they did so by escalating force and escalating mobilization at the centers of wealth and power themselves. Thinking locally in the United States, we might begin with targets of a "class-based punitive ecology,"[150] to borrow a phrase from Cedric Durand: private jets, superyachts, sprawling estates. We've already seen the broad-based agreement about concentrated wealth and its climatic powers in both consumption and production. A recent study sees ultra-high-end consumption *alone* (*excluding* more determinate influences of investment) — the top 0.7% of global wealth (i.e., those with a net worth well over $1 million, above the wealth threshold of the US top 5%) — eating 72% of the remaining carbon budget to remain within Δ1.5 in the next 20–30 years. Furthermore, total carbon emissions from this 0.7%, assuming even modest growth in numbers, exceed all annual thresholds by 2037.[151] But more important than the direct ecological benefit of taking out such targets is their *political* significance.

One of the shortcomings of left-wing *populism* (as Bosworth observes) is how easily a vague anti-elitism can slip into right-wing chauvinist nationalisms. In concert with other aspects of a left-wing climate *realism*, this kind of "soft target" helps to *begin* or *build up* ever more widespread recognition of the actual enemy. (Note that this is a strategic, not moral or ethical, question.) This destruction (there are only 5,245 superyachts in existence) points at the representative symbols of wealth and power rather than, say, foreign states or minorities. It can also begin to isolate and sow fear among the partisans of right-wing climate realism. One might (depending on specific conditions) soon add the destruction and sabotage of *new* fossil fuel infrastructure. It will only be under tremendous pressure, across a wide set of "trenches" and "footholds," across societies and states, that actually dismantling existing fossil fuel infrastructure, stopping its operations, enforcing energetic and material boundaries, and all the finer points of necessary climate mitigation and adaptation will occur. This means far more than just strategic violence. But we might look at even these cautious examples as the tactical equivalent of progressive taxation: incurring costs first to the best off. Only once a transition is occurring, or after large-scale subjective crystallization (or at least mass sympathy), would sabotaging operational fossil fuel infrastructure make any sense. And that might be a strategic tactic if sufficient, credible threat is already achieving swift drawdown.

## The Politics of Exhaustion: Structure, Affect, and Conflict

One of the most compelling and consistent arguments from the mainstream psychological literature on politics, risk, affect, and climate change is that the more a particular catastrophe is associated with a specific (i.e., non-universal) human cause, the more likely people are to react with anger and action.[152] Ironically, much of this research is related to or derived from the psychosocial resilience literature we began with and is found with most frequency in studies on crisis management. Disasters attributed to natural causes, or "Acts of God," are often met with feelings of powerlessness, despair, or quiescence.

However, if the disaster is human-caused, there is an increase in prolonged "negative affects" like anger which are difficult (read politically destabilizing) to control; even further, if causal agency can be isolated, there runs the "danger" (read political potential) of polarization (drawing the political line) and increased proclivity to collective action.[153] There is a continuum from "Acts of God" to "accidents" (which result in mild fault-finding and response) to "events that are deliberately perpetrated," which at a societal level can lead to "war and civil strife."[154] Many scholars in these areas see this kind of response as fundamentally flawed or biased. The same events are perceived as being characteristically worse if understood in a framework of human causality as opposed to natural causality. "The death of an identical number of birds due to an oil spill will not cause less suffering when caused by a natural event compared with a human-made event"[155] is evidence of "bias." "Estimating the suffering of someone who dies due to a hurricane to be higher when that hurricane is perceived to have been induced by human-caused climate change than when the hurricane is presumed to have been induced by natural climatic changes indicates a fallacy."[156] Across this research there is a tendency to ignore and obfuscate social conditions and to posit transhistorical theories of human cognition. But even understanding these basic, methodological flaws, this negatively valenced so-called "bias" is in fact more accurate political perception.

While crisis management literatures tend to naturalize *certain* forms of affective bonding — for example, "rally-round-the-flag" patriotism — they avoid most assignment of specific causal agency and responsibility as this (psychologically) inhibits resilience and sows mistrust in institutions. The social psychologist George Bonanno, for example, (widely cited in the gray resilience literatures), while broadly endorsing Antonovsky's "sense of coherence," underlines the importance of what we calls "national resilience," "patriotism, optimism," social integration, and trust in political and public institutions," particularly when "the nation" is under threat of human attack.[157] At the same time, "human-induced disaster" outside of such a frame, in which responsibility or cause are identified, are often "*toxic* or *corrosive*," resulting in "polarization and antagonism."[158] In this view, the collective action impelled by such situations prolongs traumatic effects. Similarly, "generous aid" (i.e., material aid) can potentially "undermine the struggling communities' cohesiveness and sense of collective efficacy"; instead, "informational support" from the government and media "after human-caused disasters that are typically characterized by collective confusion and uncertainty" is vital.[159] This "support" must be carefully coordinated to promote consensus and avoid polarization" (i.e., state propaganda).[160] Above all, a *return to normal* is imperative. Bonanno's "informational support" and "psychological educational factsheets" are postcards to Gilmore's spaces of "organized abandonment," sent alongside police officers armed with his advice to cultivate "hardiness," "self-enhancement," "repressive coping," and "positive emotion and laughter."[161]

For a left-wing climate realism, political strategy must be grounded in the specific antagonism towards the actual causal agents of structural exhaustion, i.e., the specific configuration of capital and its partisans, of right-wing climate realism, in this moment. A host of research supports the intuitive "common enemy" theory, but it is argued to be limited in the case of climate change because causality is ascribed to humanity in general or to abstractions such as "climate change" or "global warming."[162]

A resulting perception of unjust group-level disadvantage leads to group-based anger, which in turn motivates individuals to take collective action [...] emotions in the face of group-based environmental destruction may be different from those experienced in the face of group-based disadvantage: After all, each of us contributes to the problem of climate change more or less via our carbon footprint. It is therefore much more difficult to identify a well-defined "outgroup" to be angry at in this context [...] everyone is more or less responsible for the problem of climate change.[163]

Violent acts against "soft targets" of a "punitive class-based political ecology" are part of a process of identification but insufficient for the needed politics of exhaustion. Rather, it is across a concert of spaces and institutions that both political education — as the dialogic connection of experience, affect, structure, and socioecological cause — and affective reverberation can occur. There is remarkable agreement across discipline and disposition about the escalating radicalization and intensification of collective subjectivity (the dynamics of Elias Canetti's *Crowds and Power*) as struggle proceeds.[164] It is only in organization that such "incandescent" passions can be extended, connected, and strategically oriented.

Part of what Fanon understood about violence is that it can demonstrate the vulnerability of a seemingly indomitable enemy. (And as Du Bois argued, much in line with Fanon, a process like this works in horrific reverse — the expendable or dominated subject is suddenly taken seriously when they prove capable of inflicting violence.[165]) This observation has essentially been reverse-engineered or reconstructed in many of the studies discussed here. There is no reason to restrict these kinds of strategic considerations to sabotage and property destruction. People will die in even the most restrained violence: "combatants," police, guards, private security, but also "collateral damage." Theorists and activists who refuse to countenance this do not understand the history of large-scale social transformation. Nor do they acknowledge that *all* politics, even those of the mildest variety, put the prospect of direct coercion, force, violence, and death on the table. All politics — in fact this is the very foundation of strategic *nonviolence* — involves risk, including deadly risk to self and others. Assassinations, attacks, making life unbearable for the enemy — while capital has developed so many systems to thwart traditional challenges, it still needs space to operate and for security (of investment and of its agents and partisans).

The logic of decolonization — whether indigenous,[166] "the internal colony,"[167] or the growing colonial relation — begins here. Through the escalation and radicalization of struggle you have the possibility of Bolivian blockades, civil war or strife, or in transposed Fanonian terms, an ecological Điện Biên Phủ.[168] The achievement of one or even many of these — of spaces denied to business-as-usual — may have only limited ecological value, but its political value is inestimable. This value will vary from place to place, but just as today's depoliticized landscape offers considerably less support and institutional organization than that of the nineteenth and twentieth centuries, the extractive circuit has rendered states (even in the metropole) more directly coercive, more brittle, and yet less capable of "repelling" new footholds and trenches.

Of course, this is not only about violence. There is also the necessity of formal, official electoral politics (to take state power or to have pressure points within states), the lever of labor and strikes (understanding the nature of political strikes and the blurry historical lines between "lawful," peaceful, and violent strikes). There are the institutions of counter-hegemonic organization, education, and care. In a philosophical examination, Maggie FitzGerald argues that an ethics or politics of care does not preclude or logically dissolve Fanon's political theory, including his understandings of violence.[169] I would add that this is not only *not* a theoretical problem, it *is* a political necessity. Spaces and practices of care are part of the transactional beginnings or, again, as Gilmore puts it, "the constant interaction" of solidarity. FitzGerald astutely argues that feminist and queer ethics of care are particularly compatible with Fanonian politics. Moreover, Fanon's own writings and psychiatric practice — which do not celebrate some kind of triumphant, swaggering warrior made "whole" by violence, but rather emphasize how inflicting violence, bombings, attacks, deaths leave so many anti-colonial fighters fundamentally broken[170] — alongside his serious lacunae, demand them. As Sujaya Dhanvantari meticulously reconstructs, there is a mutual influence between Simone de Beauvoir's feminist arguments about justified political violence and Fanon's.[171] And yet gendered and racialized lacunae blind us to the practical implications of political militancy and ethical care work as complementary and reinforcing.

Historical examples of this complementary and reinforcing concert are not hard to find. We might start with the Black Panthers' free clinics and breakfast programs. Or the freedom schools and community health centers organized as part of the loose broad coalition — from the Panthers to the Student Nonviolent Coordinating Committee — of the American Civil Rights movement. Or we could return to the Keralan case, where those same sites of radical literary production work alongside village-level healthcare; both are *praxis* in motion, and complement the CPI(M)'s militancy. In the Indian Independence struggle — long before Gandhi arrived on the

scene, swapping his finely tailored barrister suits for loincloths —
the *Swadeshi* movement established spaces for local economic
activity and anti-British boycotts (i.e., specifically targeting the
political economy of colonialism) and spaces for education,
cultural production, health, recreation, and more. These were
sites of radical organization and often violent struggle. We
might also think of the centrality of "palaces of the people" to
labor, social democratic, and revolutionary movements across
Latin American and Europe, which housed "special buildings
for working class meetings, education, culture, and welfare,"
including but not limited to "libraries [...] cinemas, gyms,
and recreational space."[172] Varieties of formal and informal
community care, support, and safety can be found in political
movements across the world, often from sites that are today
zones of "organized abandonment."

To use an example outside of mainstream ecological
thinking, the relative success of the Muslim Brotherhood
(*Ikhwan al-Muslimin*) is due in large part to their presence
across North Africa and the Levant at this micro-political level:
the *Ikhwan* is the neighborhood clinic, repair shop, school, soup
kitchen — the center for care in the increasing absence of state
services or where they never existed at all. The *Ikhwan* began
largely politically disengaged, seeing itself as a movement for
Islamic renewal through this kind of community organization.
Part of their success is in understanding both an intrinsic and
social value to this work: many are initially drawn to the *Ikhwan*
not because of any or particularly strong Islamist beliefs, but
because they come to associate the *Ikhwan* and its ideology with
this care work. The *Ikhwan* is there when you need childcare,
medicine, shelter, or a tune-up. Despite their largely top-down
structure (as opposed to more "organic" bottom-up cases
or hybrid cases like MAS or CPI(M)), there is no trickery or
bribe here. Such activities are part of the *Ikhwan*'s foundational
ideals and practices as well as being politically important. Their
cautious, organized militancy and their ethics of care are two
sides of the same coin. Islamist movements are notoriously

difficult to categorize on two- or even four-dimensional planes of political analysis. One might find the theological politics of the *Ikhwan* discomfiting (I do), but as Dean observes in the Egyptian context, its organizational structure proved uniquely capable of capturing the wave of social upheaval.[173]

Both Nat Turner and John Brown claimed a direct calling from God for their early abolitionist insurrections and insurgencies. And as Angela Davis characterized them, "Nat Turner and John Brown were political prisoners in their time. The acts for which they were charged and subsequently hanged were the practical extensions of their profound commitment to the abolition of slavery." They were not lunatics, criminals, or adventurers, but representatives of the "persistent challenging-legally or extra-legally of fundamental social wrongs fostered and reinforced by the state [...] with the ultimate aim of transforming these laws and this society into an order harmonious with the material and spiritual needs and interests of the vast majority."[174]

These are just a few examples that might provide models for the politics of exhaustion, for a left-wing climate realism. The Indian Independence movement, the African National Congress, Sinn Féin and the Irish Republican Army, and indeed Fanon's FLN all demonstrate the plausible deniability (and the popular attraction) of radical, violent flanks and formations within the "respectable" factions of labor and community organizing. Lost in the false dichotomy between violent struggle and care is how contemporary socioecological organizing that borrows from these models can provide multiple modes of support and engagement, and radically alter social norms. We cannot simply copy historical forms — even these more recent ones — onto the present. But we can learn from them.

Another example from the EJAtlas shows both radical potential and the existing limitations. The Coal River Mountain Watch (CRMW) was started and led by local working class women fed up with the clear health and environmental impact of the mountaintop removal projects being pursued, radical ecological activists (including members of Earth First! who had settled in the area some two decades prior), and other community members from the relatively small West Virginian area. Initially, standard and predictable divisions emerged: miners were hostile to the organization on economic grounds. However, as anti-union measures from both the state and the Massey mining firm increased — what Jeff Feng calls "affective attachments" to the industry — the firm, and coal itself, waned. At the same time, radical environmental action, including blockades and sabotage — characterized as "terrorism" by the firm — provoked extraordinary crackdown. Against the prevailing nonviolent wisdom, the militant actions garnered increased sympathy, owing in part to the community's memory of violent repression and union-busting dating back to the 1920s. Within seven years from the first actions, many miners joined the organization.

Leveraging the "common enemy" effect, "bridging" became possible, focusing on shared negative affect towards the firm and positive affect to the local environment. Feng even records instances of "solastalgia" (a feeling of powerless sorrow in relation to immediate natural environment degradation) among miners.[175] CRMW had a string of victories — stopping or delaying a number of projects and permits.[176] It also established parallel projects — a local farm, a cemetery, and a water education school for children. These sites — connecting care with militancy — were vital in providing space for more moderate supporters as well as reversing gender norms in a highly gendered movement. The promotion of new economies centered on care and social reproduction is common in green discourse. But often, as my colleague Alyssa Battistoni has argued, these can reinforce a gendered division of labor. However, these spaces proved more welcoming to ambivalent *men*, who would tend the gardens, teach children, or mind the cemetery.[177] Thus in socioecological formation, even while gender identity was omnipresent, gender roles began to become denaturalized. CRMW support was hardly universal; there was still much antipathy, but the social base had fractured, even at the family level. Still, the seemingly perfect picture of grassroots environmentalism attracted outsized national attention — bringing extensive media coverage, celebrity endorsements, and the involvement of "Big Green" NGOs (instead of dismissing all movement organizations, it's probably useful to distinguish the Sierra Club from an organization like CRMW).

However, by 2015 this local model with so much potential began to recede. That year, many of the coal firms declared bankruptcy and the media declared the situation over (the bankruptcy claims merely offloaded liabilities, including hard-won pollution lawsuit damages, regulatory fines, and union pensions).[178] The national NGOs departed and attention moved to fracking, which was tearing up communities not terribly far away. However, Feng also points to a fundamental organizational limitation and error: in many minds, "bridging" aimed to overcome "difference," and emphasized "sameness" and "*depoliticizing* [...] through assimilation."[179] This political error is common to liberals and "class abstractionists" alike. The logic of the politics of exhaustion, even of the idea of a crystalized "stretched" subject of the Exhausted, and any of its organizational structures — call them parties if you want — does precisely the opposite. It captures, works with, organizes, is in constant interaction with structurally produced inchoate affect, expressed in disparate social upheaval or still dormant.

The politics of exhaustion sees amplification, radicalization, escalation, politicization through the recognition of those affects in wildly *different* social identities and positions, through the recognition of common enmity *across* difference, through a shared horizon of the minor paradise of the sustainable niche, by solidifying positive affect in shared struggle. Its components come together not in the pieties of liberal individualism, but in a civil war which has already begun, a war whose precise divisions are likely to cut through existing bonds and find *power* in seeing shared exhaustion, shared desire, shared enmity *in* difference.

# The Politics of Exhaustion: How Are You Feeling?

The "the more or less veiled civil war, raging within existing society" is already here; it has been for a while; we're just behind the times.

Different prized constituencies might end up among the Exhausted — and some may be divided along lines that become more indelible the more left-wing climate realism advances, the more the Exhausted moves from potential to reality. I suspect that far more than fault lines within false unities and fabulated universalisms — whether these are the essentializing mystification of a particular indigenous group or the purely theoretical abstraction of working class unity — it is some of those who profess greatest *environmental* commitment who may find the necessities of real climate politics unbearable. (These hesitaters will range from the socially detached "deep ecologists" to the romantic conservationists to the Climate Lysenkoists.) They will either find their place reluctantly in one of the nonconfrontational institutions or wings of the Exhausted, or they will take their "environmentalism of the rich" to the decks of the armed lifeboat. Civil wars are great sorting devices.

There is room in the politics of exhaustion for the migrant care worker and the theoretically middle-class desk-jockey, the informal surplus populations and the "traditional" worker, peasants and smallholders, the Brazilian MST and the newly radicalizing Teamsters. Fossil capital is not just Shell but JPMorgan Chase, Mitsubishi, H&M, and many in between and far beyond. Extractive capital is not just Occidental; it's Apple, Microsoft, Tesla, Amazon, Alphabet, Sony; it's Partners Group (private equity), UBS (banking and finance), and Swiss Re (insurance).[180] But there are even allies to be found in those segments of capital which see possibility — and profits — in green transitions. Firms — from some sectors of energy to all manner of smaller firms in care work to construction to "content creators" — who see new openings and potentials even within a zero- or negative-sum market. This is not the first energy transition in history, but it is the first which must net *less* energy use on the other side.[181]

It is not only among the potential Exhausted that there is the near inevitability of rending and separation. Within capital there can be splintering in some sectors — it's hardly difficult to find capitalists willing to take advantage of short-term opportunity *outside* of capital's aggregate interests. After all, as Marx wrote, "in every stock-jobbing swindle everyone knows that sometime or other the crash must come, but everyone hopes it might fall on the head of his neighbor, after he himself has caught the shower of gold... *Après moi le déluge!* is the watchword of *every* capitalist."[182] In the extended socioecological present, the politics of exhaustion must seek to underline existing divisions within capital, to hijack the "coercive force" which confronts "the individual capitalist" and lead them away from the Rex Position and into taking advantage of the political opening.

This is *one* of the roles of formal politics in left-wing climate realism. Even before or without taking full state power, having formal officials within governments presents a strategic advantage in the forward motion of the greater movement; it can also be a monkey wrench, disrupting the political organization of capital. There is no left-wing climate realism without formal politics, without labor organization in and beyond familiar modes, without alternative institutions, and without violence. The civil war model of today doesn't mean pitched mass battles or even the same arrangement of forces in every geography. It means recognizing the *depth* of the real conflict and the tactical *breadth* necessary to achieve a sustainable world.

Moreover, civil war isn't just the height of conflict; it's the stirrings before and the radical reconstruction during and after the fighting. Algerians shutting down Tunisian fracking operations, MST delaying Brazilian monoculture and deforestation, American eco-socialist coalitions getting New York to commit to building out public renewables — these are all tremendous achievements, but they are still the stirrings.

We have yet to mount the challenge to *stop* fossil capital, to reverse the extractive circuit, to seize the opportunity found in crisis. And crises are not in short supply. They proliferate along the "hockey stick" of atmospheric carbon, the line of species extinction,[183] or the unprecedented, ever-increasing, intensifying marine heatwaves rippling like wildfires across the surface and the depths of the ocean.[184] Here we either find ourselves right back at the beginning of the feedback loop of the extractive circuit — of the Philippines, coral bleaching, migrant labor, social instability[185] — or we embrace the politics of exhaustion. So many frame COVID-19 as a "rehearsal" or "trial run" for climate change. The ongoing pandemic *is* climate change. We are past the event horizon now, even for some of the most distant from "the front." Early pandemic responses showed a tantalizing possibility, an almost tragicomic hope, in which the logics of the extractive circuit would be reversed by the sheer "revenge of nature." These gave way quickly in many places to the sacrifice of expendable "essential" workers, to ever more spaces of social abandonment, and to an astonishing accumulation of concentrated wealth. Hell is not something that awaits us; it's this life, here and now.

Mike Davis's last interview focused on this double meaning of "catastrophe." Catastrophe as disease, as the pandemic, as current trajectories toward a Δ3 world "which is almost unimaginable." But also in the notion that catastrophe, "in resonance with Walter Benjamin, is the belief in the sudden appearance of opportunities to take leaps into an almost utopian future."[186] What the pandemic showed was just how little the left had developed the capacity to turn disaster into rupture. That things go on like this is the catastrophe. "Civil War? Some analogy is inevitable and should not be easily dismissed," Davis wrote in reviewing the 2020 American election.[187] His final essay — a whirlwind catalogue of social, political, economic, and ecological catastrophe — ends with a call, perhaps surprising for some, to "pay homage at the hero graves of Aleksandr Ilyich Ulyanov, Alexander Berkman, and the incomparable Sholem

Schwarzbard"[188] — each a communist or anarchist "terrorist" associated with violent action. Davis's lacuna was not violence. It was rather in overlooking, as Laleh Khalili argues, how

> in the mega-cities of the Global South, the capitalists live lives not too dissimilar from those of the bourgeoisie in New York or London. And it might be worth remembering that the lives of working classes — especially if racialised — in littoral and hinterland cities in the US are as cheap if not cheaper than lives in the mega-slums of Mumbai, Lagos, and Rio de Janeiro.[189]

These are the interwoven worlds of capital, climate, communication, "political mobilization [...] and dense and sustaining social relations." Khalili views violence (in counter-hegemonic politics) not as a moral question, but a fact to be understood, particularly within radically new contexts.[190] Different conditions, times, geographies, different emphases.

There is a kind of nervous tic in the "bloody, pitiless atmosphere" in which people are "witnessing a veritable apocalypse."[191] When Fanon was writing this, he was drawing attention to the *limits* and *self-destructive* qualities of violent politics. Yet even while we see upheaval across the world, violent and otherwise, the tic today in radical politics seems an almost automatic adherence to liberal norms, not only concerning violence but even negative affect. A chorus of familiar objections are raised at even the most tepid suggestion toward violence: *They will call us terrorists! There will be reprisals!* Yes, of course, and yet that doesn't stop any of the movements already in motion. Or — freezing Gramsci in time — *violence is not acceptable in the North! People are disillusioned by violence and radicalism!* As I write, protestors are burning police and government buildings in Paris. The George Floyd uprisings in the United States similarly saw rioting and the burning of police stations. The idea of violence is countenanced (at arm's length) as perhaps appropriate *there* — in Algeria, with the two years of escalating anti-fracking resistance, or for the MST in Brazil — but not *here*.[192]

*Repression will be brutal and disproportionate!* The ugly reality is that this is precisely what one wants. That response sparks escalating waves of mourning, protest, and further militancy. That is precisely what has happened in countless moments and projects of radical social transformation.

Franz Neumann — an author who took very seriously the intersection of affect and politics — called the desire for a "riskless" politics self-defeating and apolitical, amounting in fact to an anti-politics. It wants "to achieve everything, but risk nothing." It does not know "or does not want to know that the struggle for political power [...] is the agent of historical progress."[193] It is itself a *desire,* perhaps the cruelest optimism of them all — one Berlant acknowledged but also held onto. There is no riskless politics of mass radical social transformation. The amount of power represented in those trillions of dollars of assets which must be essentially destroyed, trillions more shaved off in market capitalizations, the amount of power it would take to reduce capitalist market relations to a controlled circular competition within the new socioecological *chumra,* the necessarily decommodification of vast swathes of economic life, is not possible through a "gentleman's agreement" or "wave of a magic wand." Achieving that new paradigm of risk — predicated on inverting this exhausted world — requires taking maximal *political* risk.

Fanon begins from the fact of violence. For a left-wing climate realism, that is the fact of the violence — slow and fast — of the extractive circuit, of the violence inherent in right-wing climate realism. If you're scared about non-strategic violence, organize its fact into part of a broader strategy. If you're scared about the existing landscape of violence, begin to tilt the scales.

While ideology is clearly rending across the world, the Northern left in particular seems in thrall to a kind of cheery magical thinking, a prosperity gospel or positive psychology for a left scared of "our own capacity for violence" or of "the uncontrollable force of the people," as Dean suggests.[194] Or is it that we associate negative affect intrinsically with the right? According to Ngai, "the fact that the political right has more visibly and unhesitatingly instrumentalized its disgust throughout history does not mean, however, that the left lacks or should suppress its own."[195] If there really is some psychology-based disposition based on the vague social position of so many left intellectuals in the academy or media, it's a cloying attachment to liberal respectability.

There are serious limitations in comparing explicitly right-wing movements with left-wing ones, as we have already seen in the case of left-populism: in terms of ideology, structure, and material conditions. One could add even more — stretching from questions of media to states. Additionally, as Gramsci observed, one of the most basic differences between left and right political formations is that the constituency of the left simply has to work every day in a way that the those with "ample financial resources" and who are not "tied down with fixed work" do not.[196] Still, the far-right today contains lessons — many learned from the Marxist left — beyond purposeful ruthlessness. The far-right is simply better at affective politics, at helping inchoate affective exhaustion find shape, directing and recapturing it for their specific political projects.

The philosopher Myisha Cherry — in her *The Case for Rage* — observes the ways in which neo-Nazi groups create space, particularly for young white men who "felt marginalized, angry, and broken."[197] They cultivate a vague "rogue rage" or a "wipe rage" — one focused particularly on racial elimination. But this example hardly means rage or anger are to be avoided. Rage has political utility. Indeed, for Cherry, who synthesizes elements of realism with moral deontology, it can be an imperative.[198] Even from this point of view, she endorses the range of violence proposed by Candice Delmas's "uncivil disobedience," which stretches from sabotage (including specifically eco-sabotage) to guerilla struggle and interpersonal violence in response to what I have been calling, following Fanon, "the fact of violence." But this is not meant to emphasize violence yet again. Pointed the right way, rage is, as Cherry proposes, "what we need desperately. We should cultivate it, guard it, and use it in anti-racist struggle." She reframes the debate about potential alienation, not only through questions of filiation and strategy, but also in noting that those alienated are likely not partisans of the conflict. In other words, it helps draw the battle lines. It moves people to necessary risk: "those who are angry — compared to those who are fearful — are more prone to make risk-seeking choices, and these choices are influenced by people's beliefs about themselves." *Risk-seeking* is a political necessity. Cherry dubs knee-jerk counteractions to these arguments — which are, again, quite moderate, quite reserved — "the anger police." And cautions that "the anger police may be a policing wolf in solidaristic clothing."

When we think about left-wing climate realism as working across new party forms, loosely or tightly coordinated, requiring the sites and organizations needed for war of position, war of maneuver, and "subterranean" guerilla war, these are also spaces and institutions for the specification and intensification of structurally produced affect. Their necessity is complementary across instrumental and affective planes. For the left, this cannot be achieved by "bribes" or "tricks," which don't sharpen disgust or rage, but render them still slippery. Affective resonance is not the imperative of relentless positivity and optimism. As Khalili observes, enforced "happiness" in the mode we've encountered it in literatures of crisis management and counterinsurgency "reflect[s] the hegemonic meaning of happiness in neoliberalism," an imperative "around which so much economic and psychological theory is built."[199]

Here we see convergence with James's arguments about political education as fundamentally grounded in the material conditions and trust that everyone can handle the truth, no matter how dire. The Exhausted do not require some kind of shared epistemic or metaphysical framework or even a homogenous identification — only *praxeological* crystallization. Across all those sites, footholds, and trenches, political education and struggle connect feelings of exhaustion with their proximate and ultimate socioecological causes, with the recognition of the enemy already working, with the vision of relief from the relentless acceleration of the extractive circuit. Not as a catechism to be drilled, but as a story — all too true — to which each can contribute and communicate their particular experience. The politics of exhaustion require the space for inchoate affect, for "momentary passions" to become coordinated political action.

In many ways, Cherry's case and politics are moderate, and yet in this she demonstrates just how much the left has ceded the affective field. As it's not just anger or rage in question. As early as the interwar period, Benjamin castigated the German SPD for its faith in mechanistic progress, which made "the working class" forget its "hatred." China Miéville surveys a world which

> thrives [...] on sadism, despair and disempowerment. Alongside which are thrown up species of authoritarian notional 'happiness', an obligatory drab 'enjoyment' of life, a ruthless insistence on cheerfulness, such as Barbara Ehrenreich describes in her book *Smile or Die*. Such mandatory positivity is not the opposite, but the co-constitutive other, of such miseries. This is the bullying of what Lauren Berlant calls 'cruel optimism', including on the Left: no judicious earned optimism but a browbeating insistence on the necessity of positive thinking, at the cost not only of emotional autonomy but the inevitable crash when the world fails to live up to such strictures.[200]

Miéville quotes Sophie Lewis:

> Hate is almost never talked about as appropriate, healthy, or necessary in liberal-democratic society. For conservatives, liberals, and socialists alike, hate itself is the thing to reject, uproot, defeat, and cast out of the soul. Yet anti-hate ideology doesn't seem to involve targeting its root causes and points

of production, nor does it address the inevitability of or the demand — the need — for hate in a class society.[201]

Hate is not simply an affective disposition; it can lend analytic clarity.

Perhaps it is fitting that some of the best works to imagine a left-wing climate politics are fictional, operating just outside the "ruthless insistence on cheerfulness" and the last bastions of ideological "realism" and its reality-denying political logics. After all, it is in fiction where the first examinations of a "structure of feeling" were discussed before they would become part of Thompson's class analysis, Bosworth's "affective infrastructures," or Gilmore's "infrastructures of feeling."

Kim Stanley Robinson's *Ministry for the Future* provides one portrait of what global socioecological civil war might look like in the present moment. Robinson's cli-fi tale imagines the seemingly unlikely alliance of barely funded and supported international bureaucrats, semi-connected black ops operatives, organic eco-terrorists, seemingly disparate and only loosely united mass sympathizers, and even central bankers (including, wisely chosen, that of the People's Republic of China). The weaknesses of Robinson's novel lie almost entirely in its modes of technological optimism. (Robinson himself has walked back some of the most egregious examples, such as his blockchain-powered "carbon coin".) Its strength as a *political* model has gone largely overlooked. Robinson does partake a bit too much in the logic of political violence coming from "over there," while fun financial experiments are "over here." Some of its immediate events — like the quick fall of Hindutva fascism in response to a protracted crisis of deadly heat in South Asia — seem perhaps too simple or implausible, although I guess I, too, can hope.

But Robinson's fiction contains more political reality than most radical theories. It thinks in a timescale of the immediate present, it contends with existing institutions and powers. It is highly unlikely that international institutions or major powers like China, the United States, India, or Europe will simply disappear, or that their different state forms will be entirely demolished in the timescale of a left-wing climate realism. In this Robinson produces unlikely echoes of the late Samir Amin, who posited just such a radical transition and transformation of even the most loathed international institutions towards a new emancipatory socioecological world. And not only does Robinson's novel concord with historical precedents in its "diversity of tactics" situated within concrete situations, it imagines a politics coming not out of the sweeping mass movements of the nineteenth and twentieth centuries, but rather an unprecedented (and unanticipated) emancipatory politics emerging from a long period of widespread *depoliticization*. We see structurally produced affect at local, national, and transnational levels as well as a loosely coordinated set of struggles across terrains of formal politics, finance, organization, terrorism. We can think of this not as *the* model for such a politics, but as *one* way to conceive the actuality, the real possibility of left-wing climate realism at the macro-level.

On the micro-level, though, Daniel Goldhaber, Jordan Sjol, and Ariela Barer's improbable film adaptation of *How to Blow Up a Pipeline* depicts not only the titular act of sabotage but — more importantly — the mechanics in miniature of a crystallizing formation, in this case a terrorist cell, bound together by the logics of socioecological exhaustion. There are different reasons for the eight members to be there. They hardly share the same experience. The group's quasi-leader, Xochitl, is motivated by ideological commitment but also personal grief and a sense of practical "urgency." There are a pair of classic black bloc anarchists, pulled from spontaneous outburst into organization. There is Xochitl's partner, Theo, seeking some measure of meaning as she dies of pollution-induced cancer. There is Shawn, a Black college student, who is ideologically inclined but is drawn to the cell as much from the experience of activist inefficacy and a feeling of powerlessness. There is Dwayne, the Texas smallholder, who is motivated by a mixture of local environmental understanding and *property* attachments, contradictions straight out of Bosworth's *Pipeline Populism*. Dwayne is the quintessential representative of how climate politics can sharply reconfigure existing political lines, even as it inscribes more powerful ones. But Dwayne could also be a smallholding farmer or a peasant basically anywhere on Earth. The actual knowledge — a running joke in the movie is characters reading Malm's book in the background of scenes, with Shawn dryly remarking at one point, "You know, it doesn't actually tell you how" — is provided by Michael, a native American who bristles at the encroachment of fossil fuel infrastructure onto indigenous land, at the ecological and social devastation it brings, but also at the essentialist notion of the romantic, peaceful Indian. Michael starts the process of teaching himself how to make an IED and catalogues it — abstractly — on TikTok. The others seek him out. If Xochitl represents urgency, Michael represents intensity.

The formal political logic of the film is as insane as Robinson's carbon coin. They're going to drive up the cost of oil to force market transitions. But that is not its political salience. As observed by not particularly radical critics, the film shows a collection of "the permanent underclass,"[202] "a true coalition of people working together to make a real difference against the forces wreaking havoc on the environment for their own profit."[203] "What binds the characters together is how they have all exhausted going about the fight in the 'right way.' They have protested and lobbied only to be ignored as the wheels of death keep spinning. Any supposed legal remedies have been rendered useless as profit gets put before all else. They have endured devastating losses of loved ones from sudden heat waves and are themselves suffering from the impacts of pollution taking place right next door."[204] "Crucially, *How to Blow Up a Pipeline* shows that environmental activism isn't just for academics. This is clear from the makeup of the group — most come from working class backgrounds — but it also comes up directly in the characters' interactions with each other and the world."[205] This is the affective matrix of exhaustion. This is the sharpening of externalization against the enemy and the increasing crystallization *in* difference.

In both Robinson's novel and Goldhaber, Sjol, and Barer's film, we also see the lateral nature of both climate politics and its project, lying somewhere along or rather *apart* from the spectrum of reform and revolution. The novel sees forces of coercive power and alternative institutions hemming in and changing capitalism-as-we-know-it; the result is neither its full abolition nor just an impossible "green" version. In both, we see how classic Marxist languages of alienation or goals of the absolute abolition of property — as well as liberal fictions of discourse and civility — fail to meet the moment. Robinson's well-intentioned bankers aren't about to dissolve all property; perhaps more crucially, Goldhaber's Dwayne essentially wants land reform, to protect his smallholding from enclosure. In both, we see the languages and logics of exhaustion emerging. The Children of Kali and the unnamed cell exhibit not only Benjamin's "hatred and spirit of self-sacrifice" but also "the confidence, courage, humor, cunning and fortitude"[206] engendered in struggle.

Miéville ends his essay with a rumination on Alexander Cockburn's famous question, "how pure is your hate?" He connects this with תַּכְלִית שִׂנְאָה, or *taklit sinah*, "the utmost or perfect hatred" of Psalm 139. The Talmudic commentary in Tractate Semachot is perhaps even more intense: "eat and drink and rejoice because an enemy of the All-present perished."[207] This goes farther than the famous Quranic injunctions, "fighting is enjoined upon you, though you hate it," or "oppression is worse than killing,"[208] in which physical antagonism is underscored as necessary but is something that people must be convinced of, though they "hate" it. The specter of Carl Schmitt and his political theology hovers here. But Schmitt does not extol hate; rather, he argues hate diminishes "purer" decisionist enmity. Nor does shared hatred or even antagonism necessarily create homogenous identity, as Ngai points out. Miéville's turn to a different way of secularizing theological concepts is instead a specific reaction to *this* moment: "We should hate this hateful and hating and hatemongering system of cruelty, that exhausts and withers and kills us, that stunts our care, makes it so embattled and constrained and local in its scale and effects, where we have the capacity to be greater." This is an intervention in a moment where hate "cannot be the only or main drive to renewal" nor something to celebrate. But it is also an intervention against the suppression of antagonism, the injunction to universality, the uneasy din of a thousand commercials blaring that we are in this together while confirming that we are not, and the fact that ideology as unspoken or unconscious is rending. Disgust, anger, rage, hate — these are instructive in the face of today's catastrophe — not the fear which petrifies or the anxiety which seeks the flight to safety.

In Spinoza's logic of self-preservation, we *will* hate, we *will* destroy; but "joy arising from the fact, that anything we hate is destroyed, or suffers other injury, is never unaccompanied by a certain pain in us." This is the risky gambit that somehow, even if destruction is necessary, the pleasure in the ever-proliferating possibilities of friends is greater than the damaging pain-diminished joy of defeated enemies. Adorno and Horkheimer famously argued that the "dialectic of Enlightenment" is Spinoza's self-preservation taken to its self-defeating and annihilating conclusion within the conditions of capitalism: the domination of nature, of other people, and eventually even the dissolution of the self. In other words: The Rex Position. Right-wing climate realism. But the dialectic of Enlightenment in the Anthropocene cuts both ways. Self-preservation for the vast majority necessitates reconciliation with difference, social and ecological. It requires turning messages in bottles into Molotov cocktails. Self-preservation is just the beginning. Much more is at stake; much more is possible. An "epoch of hate" can still become an "epoch of rest."

Now isn't it time to tell your story of exhaustion? Isn't it time to pick a side? Aren't you sick and tired of being sick and tired? Isn't it time to fight fire with fire?

# Chapter 5

# The Long Now

*It's after the end of the world, don't you know that yet?*
— Sun Ra, 1970

On April 30, 2018, the temperature reached 122.4 °F (50.2 °C) in Nawabshah, Pakistan, a city of 1.1 million people a mere 127 miles from Karachi, the capital city with approximately 15 million residents.[1] Although Pakistan is a large and varied country geographically speaking, it is the fifth most populous country in the world, with just under 230 million residents. This was not only the hottest April day ever in Nawabshah; it was the hottest April day ever in recorded human history.

Many climate scientists who work on "dangerous heat" — in some ways the most straightforward social impact of global warming — talk about "wet-bulb" temperatures, a combination of heat and humidity measures.[2] And for good reason. Wet-bulb thresholds are much lower (around 32–35 °C) and are already being crossed all over the world.[3] But even a "dry-bulb" threshold (around 35 °C) without the compounding issues is "dangerous." For a few days, maybe even a few hours, 50.2 °C is deadly.[4]

There is so much to think through in an example like this, from rising temperatures and their systemic causes to less obvious factors: even with mitigation, these are *millions* of people who can and will move. But what I wish to focus on for the moment are the *temporal* aspects of such blunt empirical realities. Horizons seem to fade, a fog descends, the past is churning geologically in the present, while if there is a "future," it is already here. So many warnings about climate change — even when consciously people know this not to be the case — make appeals as if they are about a future about to arrive "before it's too late" to "leave a better future for our children or grandchildren," and so on. But the Anthropocene, definitionally, has already existed for some time, and the challenges it poses are already here.[5] A whole way of thinking about time changes.

One of the greatest confusions around climate change is temporal. This is a question, in our case, of political-time, though. And that political-time is "out of joint" when one understands the political divide of climate politics, their non-universal nature.[6]

As we've already seen, not everyone will die in a likely scenario — only a rather large number of people. It is far more likely that most people will *live*, just even more miserably so. Across the many sites of the extractive circuit, more and more is extracted, exploited, expropriated, and exhausted for less and less, in the basest material terms, from majorities across the world. Far from tales of market efficiencies, we find a system efficient only at the maintenance of profit, creating institutions and technologies specifically designed to chew through ecological, social, and individual life in pursuit of profitability and the maintenance of existing power structures.

As you can probably already hear, there are different ways of conceiving of time embedded in these different ways of understanding the politics of climate change. Different presents, different futures. Much "green" discourse talks about the relationship of present and future in which the current generation or current interests are robbing "our children" or "our grandchildren." From this point of view, the fundamental issue is some broad register of the "desires" of this moment outweighing the needs of the future. This too is in error. The challenge of climate adaptation and mitigation is about the here and now. It is not a question of doing something today for hopeful results tomorrow; rather, it is a question of the direct intervention and fundamental transformation of the systems of today as the precondition for even thinking about tomorrow. This is not only in terms of the overt characteristics often associated with climate change — atmospheric carbon, soil depletion, ocean acidification — but with the attendant economic, social, and political crises that are internal to our global human ecological niche.[7] Economic, social, and political systems are utterly embedded within this ecological niche. Our "current global socioeconomic system" — in that careful phrase climate science literature uses to describe capitalism[8] — is one of the primary means of extracting from social systems and the biosphere alike. It is not simply that climate impacts exacerbate existing inequalities and inequities. Rather, these inequalities, social and natural, are *drivers* of climate change — both in contributing to overall systemic "overheating" and in the ways in which profits and wealth concentration are principal forms of *political* power. Paradoxically, today's needs and today's desires — at least those of the vast majority of people on Earth — are precisely what we need to fulfill to meet the fundamental *technical* needs of a sustainable global human ecological niche. This is true not only in the populations of the Global South but even within relatively wealthy and well-off Global North states. It is not about what "we" can achieve tomorrow; it is a political struggle *in this moment.*

I call this moment "The Long Now." The ecological challenges so many different ways of understanding time. But I am not interested in moving past these understandings to talk about some "true nature" of time. To pursue more metaphysical or existential questions — to search for a "fundamental ontology" of time — would be to miss the everyday affective structure of time. It is precisely there — in this moment as a "structure of feeling" — that the temporal and political implications of the ecological are most vital.

The Anthropocene is a geological epoch perfectly

understandable within the standard linear frameworks of what Walter Benjamin called "empty, homogenous time."[9] It is *visible* in the geologic record. Such a standard temporal framework is vital for understanding climate change. Conditions like those in Nawabshah are only knowable through these kinds of temporal measurements. But when we think about time in more social terms, we reach a limitation. Take another example from climate science literature. Assuming all carbon emissions magically just stopped, right now, as you're reading, global warming would continue.[10] "Committed warming" is an active cumulative effect of the history of carbon-fueled economic life. The rate and pace of that continuation would make avoiding some of the catastrophic outcomes outlined in the $\Delta1.5$ °C IPCC special report quite probable. But the warming itself would continue. "Committed warming" is our history literally haunting us. The ghost of dead labor living on not only in commodities, but in "externalities." The past is here, whether we're looking stratigraphically at layers of plastics, signs of increased carbon emissions, or even radiological changes. Similarly, Nawabshah is likely unlivable. That future is also already here.

The Long Now attempts to capture both the strangeness and the radical potential of this particular moment. The Anthropocene, as an epoch, is here to stay. If the most optimistic outcomes come to fruition, we will still be living in a world in which human activity is fundamentally organizing the ecological niche.[11] I am far from the first to note that this time *feels* different. As Berlant has argued, before it is an object for contemplation, "the present" is *felt* and that, currently, we exist within an "elongated *durée* of the present moment."[12] Fisher, riffing on Franco "Bifo" Berardi, wrote of the "long slow cancellation of the future" that accompanied the triumphalist spirit of the era of "capitalist realism" and what Francis Fukuyama famously identified as Hegel's "End of History." A time in which the hegemonic commonsense is that "not only is capitalism the only viable political and economic system, but also that it is now impossible even to *imagine* a coherent alternative to it."[13]

Speaking of political instability and a larger sense of a cultural masculinist, chauvinist backlash, Dayna Tortorici writes of "The Long 2016" that seems without end.[14] In both social and ecological terms, Rob Nixon argues, "ours is an age of onrushing turbo-capitalism […] with rare exceptions, in the domain of slow violence, 'yes, but not now, not yet' becomes the modus operandi."[15] At the turn of the millennium, Stuart Hall recapitulated the title of his now classic analysis of the dawn of neoliberal hegemony, "The Great Moving Right Show," as the "The Great Moving Nowhere Show" to characterize its Blair-era solidification.[16] Although there are significant differences in all of these authors and positions, there is a sense in all of them of a future forestalled, stolen, waiting to be reached, rekindled, or reinvented. Time has run out. Not only in these texts but quite visibly in social, economic, and political crises the world over, we can observe the exhaustion of our moment. That *this* time feels like it will never end is often presented as a barrier to some unknown "future." The heavy blanket of capitalist realism — growing more and more threadbare by the day — and the weight of particulate matter — growing more and more palpable by the day — are changing the stakes of the game.

The vertigo we feel as we reconstitute our vision of time as truly geological is not something we should push past, but rather something we should *embrace*. It is inseparable from the stark presentation of empirical reality — like the heat record in Pakistan — and its effect on us. Somewhere between unbridled optimism ("the arc of history is long, and it bends towards justice") and absolute fatalism (the die is already cast, there is nothing you can do) lies The Long Now.[17] "This civilization is over. And everyone knows it," writes McKenzie Wark.[18] The question is, what now? A whole different way of looking not at the future but at the *present* opens up.[19] A flourishing already latent, possible, potential, here — not on some distant temporal horizon.

That proverbial vision on the horizon leads to a series of

analytic and political responses constructed accordingly. A certain morality is preached, and we dutifully place ourselves on Hegel's slaughter-bench to sacrifice ourselves for the promise of a future perfect, of a heavenly end. Theological ideas — in particular a Christian ideal of providential history — suffuse dominant modes of thinking about time.

There's a game I play with students when I'm teaching philosophy of history, and Benjamin's view in particular. During the game, we imagine a coordinate plane, where the y-axis tracks some quantum of "progress" and the x-axis some quantum of time (see Figure 1).

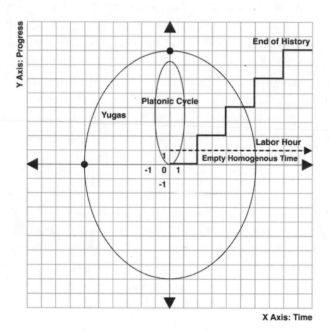

*Figure 1*

With this graph, it's possible to draw a series of figures that roughly conform to famous arguments in the philosophy of history and even broader human systems for understanding time. It's not only Hegel and his dialectical stepladder (or curlicues) proceeding ever upwards to some definite point, the End of History. If we utilize negative coordinates, we can draw Platonic cycles of the instantiation, flourishing, decay, destruction of any material object; we can create great oblong figures from the Brahmanical account of cosmic *Yugas* spanning eons, but, like the Platonic story, fundamentally returning us in a perpetual cycle. I can show with a simple line the very idea of temporal forward motion through ideas from the Hebrew Bible; definite irreversible pinpoints of creation, covenant, flood, revelation at Sinai, and so on. We can paint Augustine's decoupling of the "truth" of Christian theology from the political "success" of the Roman Empire, creating an early prototype of "empty, homogenous time" in which the "City of God" sojourns among a separate "City of Man." In this way, politics has its own distinct virtues but, fundamentally, *ideally* serving the true needs of the "City of God" far beyond our coordinate plane. As with Platonic Forms, this heavenly end is truly not of *this world*. We can talk about how this forward-moving but empty time gets refashioned — secularized and universalized — into capitalism's "labor hour." And we can return to the figure I started with, Hegel's progress of history, in which quite literally the Christian *ideal* of Heaven, static, unchanging, "perfect," ended, over, is brought down to being something that will be historically actualized on Earth.

I can draw other critical figures as well: Reactionary lines sloping downwards (see Figure 2). Or, slightly altering the y-axis, Marx's substantive, and formal transformation of Hegel (see Figure 3).[20] A long series of lines representing classes, ever simplifying towards the classless society, not "heaven on Earth" or any End of History, but rather the end of prehistory.

Marx's transformation is incomplete on its own terms. The story in the *Manifesto* is not quite as neat as it first appears and is not obviously complementary with the history of capital's development, even in Marx's hands.[21] By the time of the collapse of the Second International, "orthodox" Marxism — exemplified by political actors and thinkers like Eduard Bernstein, Karl Kautsky, and the German Social Democratic Party (SPD) in general — had reduced historical materialism to a mechanistic, deterministic "catechism."[22]

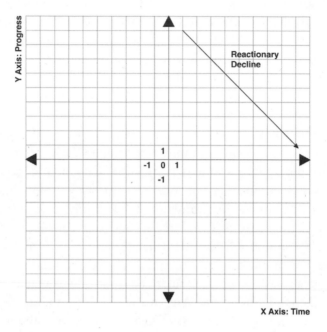

*Figure 2*

While fascism was on the rise, economic conditions cataclysmic, and socialist power riven, Kautsky — with almost unfathomable detachment from the actual conditions surrounding him — posited a faith in technological, social progress, Benjamin was already underlining how the war had demonstrated that there is no such thing. Capitalism and imperialism shaped technology towards destruction and the "domination of nature." In the pursuit of profit "the bridal bed" of the *promise* of technology was turned into a "bloodbath."[23] Communist political struggle was about arresting "an almost calculable moment in economic and technical development (a moment signaled by inflation and poison-gas warfare)."[24] "Before the spark reaches the dynamite," Benjamin wrote, "the lighted fuse must be cut."[25] Given how focused Benjamin is on

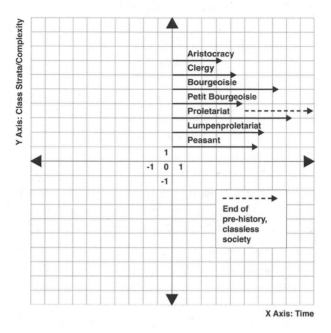

*Figure 3*

capital, technology, nature, and catastrophe, it is easy to see why he is so frequently a theoretical reference point for ecological thought. However, he was hardly the only Marxist who regarded such naïve, deterministic, and "progressive" teleology as absurd. Before and after the interwar period, anticolonial Marxists across the world[26] like Aimé Césaire, M.N. Roy, W.E.B. Du Bois, Amilcar Cabral, Kwame Nkrumah, Frantz Fanon, Claudia Jones, C.L.R James, Mao Zedong, to name just a few, advanced varied critiques and breaks with "orthodox" models in the face of the realities of what capitalist progress was in actuality, and what it necessitated. Closer to home for Benjamin, Luxemburg (whose periodical *Die Internationale* was likely Benjamin's first introduction to Marxism) and Lenin (who Benjamin cites as a political model)[27] grounded their political theories in the clear understanding that capitalist progress had become catastrophic.

For many such thinkers, and Benjamin in particular, Marx's

substantive revision of Hegel's concept of history is insufficiently critical, insufficiently *materialist*. Even if Marx does thoroughly transform Hegel and breaks with stagism in his late thinking, his work still contains traces of the bourgeois concept of progress.[28] These traces are contorted into socialist fantasies, amplified by hegemonic liberal idealism, or rejected in undialectical conservative projections of pure regress. True to the language of catechisms and confessions, such views find us lost again on our coordinate planes, following the idealistic lines of Hegel's stepladder or reactionary cascades from some projected Golden Age. Stripped of their progressive (or reactionary) valence, at best they merely return us to "homogeneous, empty time." This is not the way a good historical materialist views time and matter. Instead, Benjamin offers the view of the "Angel of History." "His eyes are wide, his mouth is open […] his face turned toward the past. Where a chain of events appears before *us, he* sees one single catastrophe, which keeps piling wreckage upon wreckage and hurls it at his feet."[29] The question I ask students as I wind down this exercise is: in the graphs we've been making, in the simple lines and geometric figures we've drawn on our coordinate planes "where a chain of events appears before us," where does this "Angel" go — and what does that do to our understanding of progress and time?

It helps to look at the original Klee painting (see Figure 4) while doing this, since unlike the easy-to-read chart — left-to-right, precise, elegant — this "Angel" looks *out* of the page. Benjamin tells us the "Angel" is facing the past.

*Figure 4*

Just as with drawing out the simplified figures of philosophies of history and theories of time, eventually, after several tries, someone inevitably offers up a solution to the question. If we are placing the "Angel" on our graph, we would have to draw it in *profile*, hovering somewhere just off or all the way to the right of the plane, eyes open, mouth agape, looking backward on all of our various lines (see Figure 5).

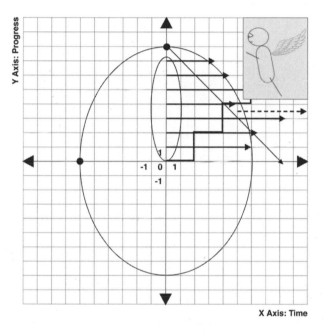

*Figure 5*

Then the question: what does the "Angel" see? The "Angel" sees a gigantic pile of detritus (Figure 6) — "wreckage upon wreckage" — not these linear *theodicies* telling us a story of the Fall or of future redemption, but layer upon layer of the failure to bring about the Messianic condition that, for Benjamin, is complete and utter emancipation and redemption. The Angel sees the present as the accumulation of *ruins* — "hell is not what awaits us but this life, here and now"; "that it goes on like this *is* the catastrophe."[30]

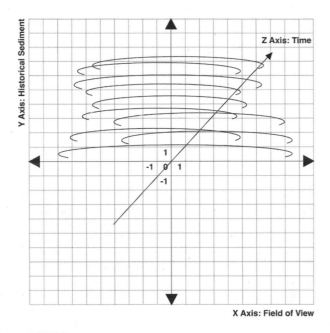

*Figure 6*

Progress as critiqued relies on "empty, homogenous time." This three-dimensional panorama demonstrates the allegory; Christian providence is not replaced with Jewish messianism. Rather, the latter helps historical materialism understand that the possibility for revolution is in the present and its constitutive, accreting past, and paves the way for a *more materialist* view. It "coincides exactly" with the *biological* view that "in relation to the history of all organic life on earth, the paltry fifty-millennia history of *homo sapiens* equates to something like two seconds at the close of a 24-hour day."[31] (Compare now with Figure 7.) Where in our graphical simplifications, we see a fairytale that *justifies* today and whose payoff is tomorrow, the "Angel" sees *stuff*, stratigraphically. The "Angel" sees, well, *a* Long Now. (If not *The* Long Now.)

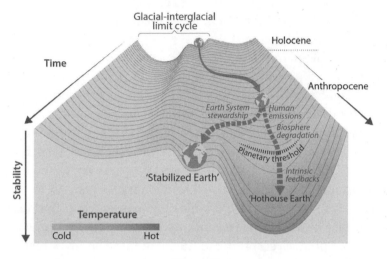

*Figure 7*

The IPCC indicates that to implement any plausible climate adaptation and mitigation scenario "all [mitigation] pathways [must] begin now and involve rapid and unprecedented societal transformation."[32] At this juncture, the horizon is the last place we need to look. Nearly every political-ideological construction around time — our temporal common sense — is pointed, sometimes almost comically, away from the present. There are always brighter tomorrows, tomorrow; never brighter or different or *new* todays, today. Tomorrow is somehow always "today" again in political time. Yet simultaneously, every economic incentive is pointed at maximizing the profit of this very moment. The mainstream macroeconomic approach to climate change — the so-called "discount rate" — perfectly encapsulates *both* these ideas. In trying to "price in" the "externalities" potential future costs are "discounted" against present economic "need." Economic benefit (i.e., profit) is to be maximized, even while a better future is promised. William Nordhaus and the interwar SPD agree: "the fabulous expansion of the capitalist mode of production" is worth it. In language that mirrors common "green" discourse, we'll sacrifice a little today — in the form of a largely meaningless added cost, a very modest carbon tax for example — for an eventual payoff for "our children."

Lee Edelman's radical queer critique of futurity — encapsulated perfectly in his slogan of "No Future" — is in this sense right: the invocation of the symbolic Child is omnipresent; the Child of tomorrow demands sacrifices today. For Edelman, the radical possibility of queer politics is precisely in embracing the "pure" *jouissance* of non-productive sexuality as set against that of "reproductive futurism."[33] The best answer to the deep inscription of providential or reactionary history is to take seriously the seemingly hyperbolic right-wing critique that queer life might be a challenge to nothing less than civilization itself. All other politics, according to Edelman, have an investment in "the Child" and in a vision of the future for that child. Edelman's argument is elegant and he is undeniably correct in diagnosing the overall commitment of existing politics to a *rhetoric* of "reproductive futurism," as well as in his excoriation of the "cult of the Child," which is used both as a normative, reactionary prop and as a hammer against the politics of the present.

But left-wing climate realism *is* a politics of this moment. Whose benefit and whose cost? Whose today and whose tomorrow? One might even ask: whose child? Thinking in terms of the Anthropocene in particular, not all reproductions are equal, and not all politics as actually practiced are particularly concerned with *social* reproduction.[34] It is difficult to square Edelman's reproductive portrait with the actual conditions of most parents in the twenty-first century, even in the overdeveloped North — what Helen Hester describes as "exhaustion, impoverishment, and exploitation."[35] Racialization often marks a site of social abandonment, and certainly no prevailing investment in reproduction. Ruth Wilson Gilmore, for example, paints a bleak portrait, in her thoroughgoing examination of the nexus of the California penal system and political economy, of the "utter abandonment by capital" of racialized "surplus populations."[36] These are not *universal* concerns or conditions in The Long Now. In contrast, as I've previously discussed, there is a reconfiguration and intensification of political conflict — in a word, *repoliticization*. In this aspect, Edelman is wrong. To invest in a specifically *socially* "reproductive futurism" in The Long Now is to abjure the cult of the Child in favor of a radical politics of today. Understanding The Long Now helps break the grip of providential history, of Hegel's liberal belief in the orderly progress of reason.

It is tempting to put a time stamp on The Long Now: it's the next ten years or even less. But The Long Now is *not* synonymous with rough timelines for implementing sustainable pathways. A paper in *Nature Communication* points out that a world where warming is only an additional 1.5 °C is still possible if all "carbon-intensive infrastructure is phased out at the end of its design lifetime from the end of *2018*."[37] Such phasing out starting in 2030, even at a rapidly accelerated rate, does not guarantee such an outcome. The difficulty is, of course, "socioeconomic constraints." Since such a phaseout has not happened, more recent studies, like those synthesized in the most recent IPCC AR6 reports, call for even more dramatic drawdowns to hold to a $\Delta 1.5$ °C world.[38] "Any further delay in concerted anticipatory global action on adaptation and mitigation will miss a brief and rapidly closing window of opportunity to secure a livable and sustainable future for all."[39] Part of the urgency and intensity of this moment is born out of the growing realization of facts likes these. But The Long Now is not only a concept for a different way to understand time; it also describes a widespread *feeling* about time. Social and ecological exhaustion is palpable today. Climate mitigation and adaptation is not about the mythical future or Child. Climate change is *driven* by already existing exhaustions. The Long Now is as much an affective definition as our actual, immediate political-time.

José Esteban Muñoz, in a different imagining of queer futurity that moves away from Edelman's stark oppositions between the reproductive, queerness, and social reproduction, talks of "another world." Following Ernst Bloch, Muñoz wants to linger with and project "concrete utopias" — abjuring the "abstract utopias" of classical utopianism in favor of the "actualized or potential" utopian elements buried in history or coming to be.[40] But what if that other world is not another time? One of the underappreciated aspects of Benjamin's concept of history is that it smuggles in the Judaic concept that "the world to come" [HaOlam HaBa] is "this world" [HaOlam HaZeh] in the Messianic age.[41] It is not bringing heaven to Earth, but allowing what is already here to come about "at any moment." Except there's an anti-utopian rub: only the Messiah redeems, you can't make it happen. "No cookshops for the future." Benjamin's political theology is unsurprisingly well-suited to Marxism. You are not creating the entire world from scratch, but building and transforming from what developed before.[42] The Long Now may not promise the Messianic rupture, but it does still work from the pieces of what is here — all those potentials and possibilities, and that life among the ruins — not to dwell in ruins but to create "another world" right here in this one.

There is a precedent for this talk of potentiality and creating among the ruins of a world that is passing away. That precedent is modernism. Owen Hatherley notes that "modernism, in many (if not all) of its manifestations had no interest in the continuity of our civilization and the uninterrupted parade of progress."[43] To speak of modernism is to conjure images of industrialism at its worst, or perhaps the European avant-garde at its best. Closer to today, perhaps the "supermodernism"[44] of architects like Zaha Hadid or Norman Foster, throwaway Ikea furniture, or the often racially charged and historically muddled visions of public housing blocs, whether in Chicago, Moscow, or London. Modernism in the pages of so many late-twentieth-century texts (for example, the influential political science of James C. Scott[45]) is shorthand for a deeply oppressive, flattening, standardizing, routinizing, and difference-obliterating planning, policy, aesthetic ideology and practice.

As we've seen, so-called "ecomodernism" is even worse.[46]

As it is currently used, it is a worst-of-both-worlds doubling-down on the shortcoming of some of those modernist projects. Coming in left and right varieties, this Promethean techno-mysticism imagines the extremes of the developed bourgeois world — the consumption patterns of the American top 4 or 5% or so — expanded to all human beings everywhere through technological innovation and a true, final domination of the Earth. The left variety is the Climate Lysenkoism we've already explored. This rests on unfounded technological assumptions. Against vast evidence to the contrary and serious time constraints, literally magic technology is thought to be able to overcome what the climate science literature notes as already exceeded "planetary boundaries" and a deep constriction on the most classic question of political philosophy: what constitutes "the good life"? Ironically, so-called "ecomodernists" on the first count are not particularly ecological, and on the second, not particularly modernist. For all their claims to a rigorous reintroduction of science and engineering into the discussion of climate, their belief in (aptly described) technological "miracles" is as dubious as the technocracy they often deride.[47] For all their claims to hew to a radical reintroduction of "the future," they seem to be unable to imagine any life, any culture, beyond a particular vision of an idealized, wealthy American or perhaps European lifestyle, just writ large and universalized. This is precisely that vision of the good life that constitutes "cruel optimism": a holistic vision shot through with emotional resonance, from family to economy to society to state.[48] What so-called "ecomodernists" seem to be missing is that the very people demanding something — anything — different do not seem particularly interested in this "dreamworld" any longer. They are — in a word — *exhausted* by this life.

While some, in understandable contrast, invoke a kind of eco-romanticism, I wish to propose something quite different: a *real* ecomodernism in both the political and aesthetic implications of the term.[49] That is to say, "the minor paradise of the sustainable niche." We do, indeed, need a "concrete utopia," although, depending on local conditions, that "concrete" is possibly "rammed earth" or "compressed stabilized earth blocks" or bamboo, and that "utopianism" is, paradoxically, *anti*-utopian in the best sense of the word.[50] Getting pulled into "the pragmatics of the present" can be a kind of radical realism, particularly at this moment in the Anthropocene. To be truly utopian "concedes [...] everything realistic to the enemy."[51] I have in mind rather all those *other* modernisms; some repressed, some never achieved, and some never dreamt. Think of the Black modernism described by Fred Moten as "one of space-time separated coincidence and migrant imagination, channels of natal prematurity as well as black rebirth, modernism as intranational as well as international relocation."[52] What is more modern and more of a tradition than jazz, one of the objects in question in Moten's formula? Or the oft-misunderstood, when addressed at all, Islamic modernism of Jalal Al-e Ahmad, who argued against the *parochial* nature of European modernism while still holding out that new forms — of literature but also of life, politics, economy, and society — could be fashioned in a rejection of the antinomy of an essentialized traditionalism and the erasing, dominating capitalist modernism that conjured it.[53]

One can find similar modernisms across the Global South (and, in those increasingly large pockets of Global-South-in-Global-North). Not simply in cultural practice or in the built form. The immediate ruins of *our* past also include the dreams of what was once the non-aligned movement for alternative modernisms, from Lagos to Tehran and to "the visions of a domination-free international order that anticolonial worldmakers pursued," in Adom Getachew's phrasing.[54] From the Bandung Conference to the New International Economic Order (NIEO), non-aligned states proposed rather different alternatives to the encapsulation and enclosure of the global neoliberal world we live in today.[55] As Theodor Adorno once said, "in the determinate negation of that which merely is," what we discover is that "it could have been otherwise."[56] This is not only the truth of even the most "sublimated art," it is also a way of telling a history, one less of missed opportunities than of prevented possibilities. We are in many ways living through a specific set of political solutions to a simultaneous economic, social, and political crisis that began in the 1970s (or even since World War I). Whether truly global, as with the NIEO, or parochial, as in the long-forgotten American liberal alternative of a "permanent incomes policy,"[57] such possibilities might be, even if in some sense truly radical alternatives to the present day, shockingly prosaic. It is simultaneously, again with Fisher, "unimaginably stranger than anything Marxism-Leninism had projected."[58] We must understand that the global institutional, national political, economic, cultural, and total socioecological path taken was structurally necessary *for* capital, even as we grasp the potential of the "otherwise" that lies all around us.

This potentiality is perhaps best imagined, experienced, or *felt* in the built form. Upon seeing "Split 3," the last of the great Yugoslavian public housing blocks, in 1981, of all people, Jane Jacobs — the famous American critic of modern urban planning and development — wrote, "Split 3 makes me feel so optimistic, thank you!"[59] Split 3 is less than a hundred miles from the Šerefudin White Mosque, but both are traces of a very different ideal of modernism. Split 3 stands as a brutalist refutation to both the conservative *and* liberal critiques of "public housing," "an exceptional achievement: the transformation of modernist mass housing into a planned environment nevertheless marked by the diversity and spontaneity typical of vital urban neighborhoods."[60] It incorporated space for human solitude and human society. In many ways, Split 3 is the idealization of the design philosophy and process developed in Yugoslavia called the "social standard," where neither market principle (nor party necessity) guided the development but rather a commitment that all people have access — without monetary or physical restriction — to decommodified housing, education, arts, and cultural goods. And in the Yugoslav case in particular, the diversity of the possible expressions of those goods was part of the mode of their expression. The Šerefudin White Mosque —

built as part of an overall dense town center also including a public library and other community spaces — integrates an expansive sense of the "social standard." The spiritual needs of the largely Muslim community are integrated in a design that is utterly modern and yet consonant with theological concepts from a Sunni Islamic tradition. In sharp distinction with, say, the postmodern excesses of contemporary Mecca, the White Mosque utilizes its modernist aesthetic to achieve an enhanced effect of communal spiritual seclusion. "Only after a visitor makes the slow descent from street level to the subterranean level of the Mosque — a descent which removes her not only from the chatter of the surrounding market but also from her everyday preoccupations — can she fully appreciate the dignified austerity of the light-bathed interior."[61] This space elevates (metaphorically) an Islamic spiritual ideal of contemplative and devotional seclusion through an utterly modernist aesthetic and ideal.

My point with these examples — raising the specter of modernism — is not, as some have argued, to underline how a new urbanism will be the end-all, be-all of a "good Anthropocene."[62] It is rather to highlight just how much and in how many different ways a *real* ecomodernism can salvage and develop from the world all around us. Not terribly far from these two sites one can find examples of traditional North African agriculture which are without question part of the portrait of a sustainable global human ecological niche.[63] Thinking on those basic realities of "dangerous heat," the impoverished worldview of the techno-mystic imagines American-style air-conditioning on a *maybe* global scale, powered by endlessly and catastrophically increasing energy consumption, resource use, and sometimes miraculously "cleaned" fossil fuels or rings of nuclear power plants (that each take ten years to build with all the enormous economic, extractive, and emission costs upfront). These are HiFi delusions on Hegel's stairway to heaven that simply *buy time* for capitalism-as-we-know-it. But cooling systems have long been built into architectural design before there ever were such things as wall units or central air. From the Red Fort in India to the Alhambra in Spain, one can see examples — decorative fountains, water channels — of deceptively complex passive, evaporative air-cooling systems that were (and sometimes still are) prevalent in geographies experiencing high temperatures (both dry and humid) as well as water scarcity. Sometimes coupled with wind towers, small water bowls, courtyard design, plant life, specific building materials, *mashrabiya* (ornate carved windows), such technologies developed over considerable time are in many estimates more efficient and more effective than more seemingly "high-tech" alternatives; moreover, these technologies detach cooling from energy needs.[64] As scores of Asian and African engineers and architects have noted, such technologies and designs are frequently more comfortable than contemporary "global" equivalents. They are also considerably more beautiful.

Some of these technologies are quite common in peasant housing. Others — say, the *mashrabiya* — are historically associated with aristocratic privilege, though forms have been adapted already in much broader contexts. "Rationing" within planetary boundaries can be luxurious. In the face of this generalized abundance realized through ecological limits and "the mastery of our mastery over nature," faux-populist defenses of "cheap plastic crap" evaporate and, alongside them, the "green consumption" of "elevated" crap looks like the marketing strategy it always was.[65] There is nothing magical about technology simply because of its age. Contemporary technologies and techniques are as much a part of a sustainable ecological niche for mass flourishing. Some current ideas — like vertical green covering, in which urban buildings are insulated with living plant materials — are complementary with these kinds of so-called "vernacular" ideas, as are commonly understood practices in energy efficiency and retrofits.[66] Nor is there any magic in the "vernacular" itself: Ugandan bricks may make concrete look comparatively sustainable since they have 5.7 times the embodied energy, while mud-brick cities in Yemen look like urban modernism *avant la lettre* but at a fraction of the ecological or economic costs.[67] In Niger, some examples of compressed earth building are relatively recent postcolonial constructions, already an alternative modernism. Or one could look to Cuba's "special period" in which more and better food was produced, with increased soil quality and *less* labor inputs, through cooperative development of agroecological farming.[68] In both classical and modern forms, there is sedimented knowledge born out of generations of doing quite a lot with extraordinarily little under adverse circumstances. Engineers and researchers of all kinds continue to develop these applied sciences, only these are not the sciences that "business-as-usual" or techno-mystics are interested in. A real ecomodernism, "the minor paradise," is both a political horizon and a temporal intervention in the structure of feeling that characterizes The Long Now.

A desire for the future is often the lament of living postmodernism, capitalist realism, or just the accelerating daily exhaustion of the vast majority of people on this planet. The endless recycling of the themes and styles of what have defined capitalist modernity under the hegemonic certainty that there is no alternative; the endless personal and structural "adjustment" to the conditions of one's own exhaustion. The recession of temporal horizons in The Long Now is an opening up of political ones, not least the path to a new political imagination that the cruelest optimism tells us we must wait for. Even some who are trying to grasp the possibilities of this remain trapped with not only that ever-present Hegelian "step-ladder," dancing up that coordinate plane to heaven on Earth, but also mistake the aspiration to modernism with the material fantasies of some "future" as imagined by capitalist modernity. Capitalism-as-we-know-it, as we experience and feel it, is *unrealistic*. Its fantasy life, far from overabundant, decadent, or free, is stulted, pathetic, and *limited*. The "concrete utopia" for The Long Now presents not one single answer but an enticing panoply of real possibilities. It's not air-conditioners vs. asceticism. It's Green

*Proletkult.* Fantastic visions with Mughal cooling systems. Local knowledges linked and spread. It is the desire for the future realized in creating a truly different world built from the materials and dreams, ecological and cultural, of the world we actually live in.

Even on the left, green discourse is often trapped within a question of the limit or limitlessness of desire. Do "we" need some kind of green asceticism or "fully automated luxury communism"? Should "we" want more or less? One can roughly describe the necessary parameters of a sustainable world (and it should always be mentioned that these in no way comport in natural or social scientific terms to what mainstream liberal policy consensus argues in terms of climate action), but who would possibly want to live in such a world? These are the questions that characterize many recent debates concerning climate policy in this moment, like that between the economists Branko Milanović and Kate Raworth. Raworth presents one plausible picture of what a sustainable world might require. Milanović translates that into starker economic terms (extraordinarily high taxes, an extremely limited work week, an effective end to much air travel, dramatic transformation in food systems, etc.). The question that is prompted in debates like these is *not* about technological or economic feasibility but about political desirability. Speaking about the UK specifically and wealthy Global North states in general, he asks why would people *want* such a life? Who would politically support it?

These are fascinating question on two levels. First, on the bare numbers, what Milanović describes is already an improvement for significant numbers in geographies like the UK or the US. Smaller but still relatively ample incomes with guaranteed social goods and services is almost certainly already better for a simple majority in the UK and just under an equivalent number of people in the US. For the 50% of Americans earning $31,000 or less, it's clearly a better deal without even taking into account qualitative questions. The second level, though, is that of desire and ideology. Only 16% of Americans identify "becoming wealthy" as essential to "the American Dream." Americans now value social goods, or what the authors call "amenities" — like parks, libraries, food sources, and schools — far more than they value individual wealth or even homeownership. This survey, from the American Enterprise Institute of all places, reflects pre-existing trends that have been even more sharpened by the pandemic.[69] According to the more recent 2020 General Social Survey, Americans report an all-time low level of happiness and optimism. A scant 14% of Americans describe themselves as very happy, and over 60% are pessimistic about their own future and, where applicable, their children's future, a stunning turnaround from mid-century or even mere decades ago. "A majority felt anxious, depressed, or irritable in the past week" in 2020, increasing over already growing numbers from recent years.[70] The political promise of climate politics, the political possibility of The Long Now is a differently socially and aesthetically rich life of greater economic and social security, less work, greater material and what I like to call "temporal freedom," a slower, less exhausting pace of life, let alone the relief from ever intensifying socioecological pressures.

Milanović, and many of his ecological critics, seem to share a bedrock conviction that it is simply "more" or "less" of *this life* — of "wealthy" capitalist modernity — that defines the boundaries of the politically possible. Capitalist realism and cruel optimism reinforcing each other in what some social psychologists who study exhaustion would call a "loss spiral"; an ever more hopeless expending of resources on an impossible resolution. On this level, Milanović and Raworth (both of whose core work I deeply respect) are actually in much agreement, Milanović coming down on the side of "more" of this life and Raworth on "tolerably less."

But whether one looks to measures of social and psychological stress, ecological economic calculation, social and political instability, or the direct experience of climate catastrophe, the question would not seem to be about the proper distribution of "this life," but rather whether anyone desires it at all. When I speak of the exhaustion of our global human ecological niche, I have this in mind as well. Time has run out on this life. Deleuze and Guattari once argued that capitalism, knowing no external limit, would constantly recast "immanent limits on an ever widening and more comprehensive scale."[71] There is a descriptive power in their understanding of capitalism, but they make an idealist's error in ascribing capitalism a kind of transcendental constant on the world itself. There are external limits on capitalism — its political life-extension program that we call "neoliberalism" stumbles on today, extracting ever greater ecological and social inputs for ever-diminishing returns, its legitimacy in shambles.

Deleuze and Guattari provocatively suggest, of course, that perhaps the answer lies not in opposition to capital but in allowing its internal desire-producing functions to "accelerate" and exceed capitalism itself.[72] But desires for what? What kind of excess? Again, whose desires? As designed and fulfilled by what? In almost every direction we turn in green discourse we find these questions unanswered and rarely addressed. Here again, a different turn to modernism — and its attendant aesthetic, political, and temporal implications — haunts us with experiments and ideas, with lives and desires unfulfilled. The proliferation and freedom of queer desire, sexualities, and spaces discussed by Muñoz are *excessive* in a way that was predicated on the *slowdown* of the 1970s in a place like New York City. Already, we have discussed the political organizational and policy implications of what an example like the Kerala model provides for the possibilities of social need addressed at dramatically lower and different organizations of social wealth.[73] Communist-led participatory democracy not only achieved the living standards of a Portugal on the budget of a Sudan, but it also necessarily fostered one of the world's greatest (in both scale and scope) modernist literary cultures — both popular and avant-garde — among the mere 30 million readers and writers of Malayalam. A far *richer* aesthetic life is possible within the enabling constraints of an emancipatory ecological niche. Here, again, we might think with Moten (as he says with and against

261

Adorno) not of an accelerated modernity-of-the-new-but-same but of syncopated, punctuated, ever-proliferating modernisms, of acoustic possibilities that are seemingly inexhaustible. Not of the desires that must be withheld to make "the future" possible. But of the desires that are cast out to keep the present the same.

Charles Mingus' resistance to electrified jazz was supposedly in part a testament to all the *new* possibilities that had hardly been and never have been exhausted through acoustic music. This should hardly be held up as a rejection of the technological or the digital. Even in new medias, the most compelling creations are *stifled* in obeisance to official progress, in which human desires are ever rechanneled into logics of profitability and reproduction. Through low-power circulation, refurbishment and repair, in technology and tool libraries, opportunities proliferate to further develop nascent, radical, and popular digital cultures from games to media. There are real challenges to how certain contemporary media technologies — not to mention the renewable energy systems they could run on — must be designed quite differently than they are now through cheap, carbon-intensive production, with rare-earth metal batteries, built for planned obsolescence. But how many people *really* want a new iPhone? Or a phone at all. Designing within limits provokes creativity and excitement among artists, designers, engineers, and audiences alike. The possibilities for radical digital cultures — potentially reproducible practically *ad infinitum* at relatively low ecological costs — are also part of a sustainable, flourishing human ecological niche. As are simple individual indulgences, solitary pleasures, or collective, effervescent ones. A real ecomodernism weaves a vivid socioecological tapestry from threads strewn about and cut short in the progress of the extractive circuit, found in the impasse of The Long Now.

Just as the austere buildings that dot the landscape of what was once the built world of failed, forgotten, or suppressed modernisms were often festooned with bright colors, individual qualities, and vital daily life, we *can* imagine a real ecomodernism, "the sustainable niche," characterized by different vectors of desire and excess. One of the fascinating qualities of Split 3 is that it was not built to dominate the natural landscape but to fit snugly within it. The human world realized in its "natural form" as a mere appendage of nature. As Višnja Kukoč observes, Split 3's design is particularly well-adapted for a low-carbon, high-density urban plan.[74] Planetary boundaries are real, but in many ways scarcity is still an ideological fiction.

These are not the markers of some future to come; they're the possibilities already here, all around us. It is *not* that we must wait for some distant time of technological, economic, or cultural maturity to achieve them. Our collective crisis, our general exhaustion, are *caused* by the failure to actualize these potentials. Our collective crisis, our general exhaustion, in turn *causes,* is fuel for the fire which consumes our ecological niche. In the context of the contradictions of capitalism, theorists like Adorno and Marcuse, Fanon and Césaire would return to the point again and again that the material potential for human flourishing had long since been achieved within global capitalist development. The Long Now forces us to confront this not as a question of aesthetic longing and autonomy (Adorno) or as philosophically existential orientation (Marcuse), but as political limit case. That the "dialectic of Enlightenment" might lead to global catastrophe was always on the table. That human life might go on with the material *potential* for mass flourishing *obliterated* was not considered. This is the "immediacy" (Fanon) of socioecological political time.

The ecological pulls the rug out from under so much of the philosophy of history. It's not that the world is ending. It isn't. But rather that the end of the world is so *perfect* for so many of these theories; the apocalypse so sublime. What our thinking about time wasn't prepared for was for the very constitution of that world to change. For it to not become destroyed for everyone (or forever and ever, Amen) but increasingly — and permanently — incommodious for ever more people. Whether time was cyclical or linear, progressive or regressive, it floated comfortably on a world where only, at best, the human bits changed. On that initial diagram it is only Marx, really, who ever thought seriously that conditions in the natural world — the messy, sensuous, material one we actually inhabit — might change. And even then, only in bits and pieces. These ways of thinking are so deeply engrained culturally — and are so frequently repackaged and so hegemonic in their congruity to existing power — that they worm their way back into our thinking time and again. Not only in our thinking about "the world," or rather our ecological niche within it, but also in other concepts that have become suspiciously Platonic over time — "the economy," "the state," capitalism itself. None of these concepts exist outside of time; they have all changed, some so much as to hardly be recognizable. They don't have Forms or Heavenly Bodies. We don't have nineteenth-century states or nineteenth-century markets; why should we have the nineteenth century's past, present, and future?

Cutting the fuse, pulling the break — today, these Benjaminian metaphors mean stopping the "extractive circuit" that lies as the parasitic center of our global human ecological niche. It is fundamentally and radically transforming that system. It is not inventing the future so much as it is building the present. It is a fight for *time*: a temporal freedom, a temporal luxury that stalks about this rather different ecomodernism. Whether and to what degree it's "managed capitalism" or "market socialism" becomes a scholastic question. These coordinates do not turn on a metaphysics of alienation, nor on a redemptive justice. This "minor paradise" is not a way station on the road to "The Revolution," nor a step beyond it. It is a "lateral" project all its own. It may very well be that something like the absolute overcoming of "alienation" — in the strong sense in which many propose it — could actually stand in the way of this revolution.[75] It turns out the "vector of happiness" — Benjamin's own graphic metaphor for the specific politics of the here-and-now — has its own revolutionary needs. Like the Messianic, it is a political rupture, but, to be simple, it is not the Messiah. This rupture is closer to Fanon's material decolonization which "cannot be accomplished by the wave of a magic wand, a natural cataclysm, or a gentleman's agreement. Decolonization, we know, is a historical process [...] captured in a virtually grandiose fashion by the spotlight of History."[76] The constellation one might see from the point of view of the Angel of History is not the constellation of *this* rupture. The spotlight of history shines on unexpected "wreckage upon wreckage" with a more immediate task at hand. For the time being, perhaps the better political theology is that of Imam Mahdi, the twelfth Imam in the Shii tradition. This Messiah is not "gone," merely in occultation. The Messianic hope not overturned, abandoned, or overcome; simply in temporary abeyance, always there to remind us of yet another *imaginable* standard against which we can hold, critique, and transform the present.

But the principles expressed in all those concepts — "concrete utopia," "no future," "materialist history," "Messianic time" — can still serve us in The Long Now. And this "revolution," I would argue, for all its urgency and for all its departure from the Messianic, is no less radiant. In the classroom exercise, the Angel of History sees what we cannot. We can only catch a glimpse of the constellations the Angel must see so clearly. The graphic picture of The Long Now shares in this. As with the Angel, we see the great "pile of debris" that contains history that — like atmospheric carbon — never really goes away. If we try, we can see the potent present possibilities, hidden or submerged but held back, rechanneled, unfulfilled. But not only can we not see the whole picture, the Messianic — and the Angel with it, even as allegory — are in abeyance. We grasp about as if in a maze. We are in the rubble, breathing the past, living the enemy's future already. But we can build from that rubble. We can steal pragmatism and realism from the reigning powers that are anything but pragmatic or realistic. All around us we see a world shot through with the paradigm of Nawabshah. A world where, unexpectedly, the literal grounds of history have shifted. But the longer we look — if we know where to look, in the distance between unrealistic realism and anti-utopian reality — the more we begin to see a different constellation, one that corresponds to a radical portrait of the possibilities latent in the present, waiting to be loosed from the fetters that hold them back *today*. This is the geometry of The Long Now which, we see, is not splayed out on an imaginary coordinate plane but rather in biomes, in built worlds, in systems, in our niche.

And that dream — a world freed from exhaustion and arrayed, organized for human flourishing — is no cheat for revolutions or messiahs. Better — at least as far as I am able to see — than the End of History or its beginning, it is otherworldly in the only sense that can be true. While "we all await the Imam, each in our own way,"[77] we may discover that we never really understood what the Angel was looking at in the first place, what the Messiah is, or what it might bring. The dreamworld — held back not by technical limitation but by political enervation — born from our sensuous ecological life and dreamed for the ruin, for the world, for the possibility of The Long Now, may prove, in all its earthly splendor, more attractive than heaven on Earth could ever be.

# Endnotes

## Chapter 1: We're Not in This Together

1   Matt Simon, "Take a Good Look, America. This is What the Reckoning Looks Like," *Wired*, Nov 13, 2018; John T. Abatzoglou and A. Park Williams, "Impact of anthropogenic climate change on wildfire across western US forests," *PNAS* 113, no. 42 (2016); John T. Abatzoglou, A. Park Williams, and Renaud Barbero, "Global Emergence of Anthropogenic Climate Change in Fire Weather Indices, *AGU* 46, no. 1 (2019), 326–336.
2   Katharine Ricke et al., "Country-level social cost of carbon," *Nature Climate Change 8* (2018), 895–900; David Wallace-Wells, *The Uninhabitable Earth: Life After Warming* (New York: Tim Duggan Books, 2019), 194.
3   McKenzie Wark, *Molecular Red: Theory for the Anthropocene* (New York: Verso, 2016), 149.
4   Oliver Milman, "California fires: what is happening and is climate change to blame?" *The Guardian*, Nov 12, 2018. TMZ, "Kim Kardashian and Kanye West Hire Private Firefighters, Save Neighbors from Wildfires," Nov 12, 2018; West and Kardashian, unlike most recipients of such services, *did* contract directly with "Wildfire Defense Services" a leading private emergency service provider. Salvador Hernandez, "California's Rich Are Protecting Their Homes With Private Firefighters. Officials Say It's Only Making Their Lives Harder," *Buzzfeed News*, Nov 21, 2018. On the commodification of risk, see Christopher Wright and Daniel Nyberg, *Climate Change, Capitalism, and Corporations* (Cambridge, UK: Cambridge University Press, 2015), 62; in Chapters 3 and 4 we will explore different risk paradigms needed for real climate realism.

5    An almost perfect example of Ruth Wilson Gilmore's "workfare–warfare" state. See Gilmore, *Golden Gulag: Prisons, Surplus, Crisis, and Opposition in Globalizing California* (Berkeley: University of California Press, 2007).

6    Leslie Scism, "Wildfire Risk in California Drives Insurers to Pull Policies for Pricey Homes," Jan 19, 2022. AIG did not — as many inferred — exit the high-risk, high-end market; it largely exited the "admitted" or formal, regulated market. First, it trimmed and migrated its "Private Client Group" away from the "admitted" market to largely unregulated "excess and surplus" subsidiaries. While regulatory bodies approved rate increases as high as 17.5%, the new arrangements allow AIG and similar firms to increase rates by 40–60%; see Jason Woleben, "AIG to exit California homeowners insurance market at January-end," *S&P Global Market Intelligence*, Jan 25, 2022. Even relatively well-off client policies have been sold to middle-tier third-party firms with considerably weaker coverage and no luxury services (see Ryan Smith, "AIG to transition upper-middle-market clients to other insurers," *Insurance Business*, Jun 30, 2020) while new surplus lines for the ultra-high-end are sold with phenomenal growth: surplus lines in California increased an astonishing 83% from 2018 to 2021; AIG's surplus lines alone increased 61% (see Tom Jacobs and Hassan Javed, "Use of surplus lines for homeowners coverage surging in California," *S&P Global Market Intelligence*, Apr 4, 2022). To qualify for AIG's exclusive policies, one must bundle a number of expensive assets: homes over $1 million, art collections, yachts, airplanes, even horses. In return, not only do clients receive services like the "Wildfire Protection Unit," they receive bespoke "crisis management" teams in case of kidnapping or extortion, for example, personal liability expansions covering personal injury, slander, and libel against others, as well as discounts for and vetting of private security services (see Policy Genius, "Best high-value home insurance companies of 2023," Feb 6, 2023). As of writing, AIG is restructuring its Private Client Group into a special management general agency called "Private Client Select Insurance Services"

in partnership with Stone Capital, which will formalize the full shift as well as bring additional capital to back the new packages; see "AIG Finalizes Agreement With Stone Point Capital to Form Private Client Select Insurance Services, an Independent MGA," Apr 26, 2023.

7   Dipesh Chakrabarty, "The Climate of History: Four Theses," *Critical Inquiry* 35, no. 2 (2009), 197–222, here 220–222. Discussions about climate change as "apocalyptic" or human extinction are addressed further in this and the next chapters.

8   For data and analysis on climate denial, please see section "Really Unreal?" below.

9   In his short article, "Survival of the Richest," Douglas Rushkoff describes his experiences at a remote, exclusive mini-retreat for the ultrawealthy. During this retreat, at one point he is taken to an even smaller gathering in which issues such as how to pay private security in an economy in which resources have more value than money, or how to effectively prepare for mass rioting, migration, etc. resulting from climate and related crises, came up directly. I discuss both direct and indirect forms of right-wing climate realism in this chapter. Aspects of this are also visible in the details of the AIG insurance arrangements discussed in Endnote 6.

10  Noah Gallagher Shannon, "Climate Chaos is Coming — and the Pinkertons are Ready," *New York Times Magazine*, Apr 10, 2019.

11  The team led by the late atmospheric chemist Will Steffen on the scientific basis for understanding the Anthropocene in general and the accelerating pace of climate change in particular, observe a direct correlation between FDI (among many other indicators) and anthropogenic climate change which establishes that "it is now impossible to view one [the socioeconomic] as separate from the other [anthropogenic climate change]. The Great Acceleration trends provide a dynamic view of the emergent, planetary-scale coupling, via globalisation, between the socio-economic system and the biophysical Earth System." See Steffen et al., "The trajectory of the Anthropocene: The Great Acceleration," *The Anthropocene Review* 2, no. 1 (2015), 1–18; we

will revisit this article later in this and the following chapter. FDI was in single digits in the immediate post-WWII period, and then begins a geometric increase starting roughly in the mid-1980s.

12  On "missing profits," see Thomas Tørsløv, Gabriel Zucman, and Ludvig Wier, "The Missing Profits of Nations," NBER Working Paper no. 24701 (Jun 2018). Their lower bound is 36% and 43% their upper, although, as we'll explore in the next chapter, the very nature of globalized "value chains" make such figures likely *under*estimated. Furthermore, as they record, about 50% of such profits end up (which they underscore does not mean gets accounted for officially) in the US, 25% in the EU, with the remaining 25% split between the remaining OECD states and the rest of the developing world.

13  On luxury survival architecture, see, for example, Proven Partners, "Resilient Design — The Newest Luxury in Real Estate?"; *Icon Architecture*, "Tempo: Cambridge, MA," accessed 7/13/2023; Matt Shaw, "This Luxury Tower Has Everything: Pools. A Juice Bar. And Flood Resilience," *New York Times*, Apr 29, 2020; and, for more extreme examples: Evan Osnos, "Doomsday Prep for the Super-Rich," *The New Yorker*, Jan 22, 2017, and Bradley Garrett, "Doomsday preppers and the architecture of dread," *Geoforum* 127 (2021), 401–411. In a recent study, the Egyptian architects Bassent Adly and Tamir el-Khoury explore exclusive adaptation within "gated communities and enclosed neighbourhoods, gated enclaves, and private cities." Palatial villas are retrofitted for energy efficiency — from insulation and orientation to private photovoltaic power supply (and private security).

14  "Guard Labor: An Essay in Honor of Pranab Bardhan," *Umass Amhurst Economics Department Working Paper Series* (2004).

15  On tax havens see Zucman, Gabriel, et al., *The Hidden Wealth of Nations: The Scourge of Tax Havens*. The University of Chicago Press (2015), 4–24. A team led by the Argentinian ecologist Sandra Díaz finds, among a general account directly linking "[t]he increase in the global production of consumer goods and the

decline in almost all other contributions" to human ecological well-being, that "funds channeled through tax havens support most illegal, unreported, and unregulated" activity in areas like fishing, for example (Díaz et al., "Pervasive human-driven decline of life on Earth points to the need for transformative change," *Science* 366, no. 6471 (2019)). As we'll see in the next chapter, such activity is actually a widespread and *cited* feature of contemporary global capitalism.

16  C.f. Ajay Singh Chaudhary, "Sustaining What? Capitalism, Socialism, and Climate Change," in *Philosophy and Politics — Critical Explorations* (Cham: Springer International Publishing, 2022); Chaudhary, "Emancipation, Domination, and Critical Theory in the Anthropocene," in *Domination and Emancipation: Remaking Critique*, ed. Daniel Benson (Lanham, MD: Rowman Littlefield, 2021); Ajay Singh Chaudhary, "Toward a Critical 'State Theory' for the Twenty-First Century," in *The Future of the State: Philosophy and Politics*, ed. Artemy Magun (Lanham, MD: Rowman Littlefield, 2020); Chaudhary, Ajay Singh. 2018. "It's Already Here." *n+1*. October 10. https:// nplusonemag. com/online-only/online-only/its-already-here/; Chaudhary, Ajay Singh. 2019. "The Climate of Socialism." *Socialist Forum*. Winter. https://socialistforum.dsausa.org/issues/winter-2019/ the-climate-of-socialism/; Chaudhary, Ajay Singh. 2019. "Subjectivity, Affect, and Exhaustion: The Political Theology of the Anthropocene." *Political Theology*. February 25. https:// politicaltheology.com/subjectivity-affect-and-exhaustion/ The Extractive Circuit is the subject of Chapter 2. Throughout this book, I am using an expansive understanding of "extraction" that I think best characterizes social relations in contemporary capitalism, particularly in socioecological terms.

17  See Wright and Nyberg for a wonderful account of how "business-as-usual" discourse proceeds internally within firms. There we also see, alongside knowledgeable action, the choice *not* to know; see also Supran, G., S. Rahmstorf, and N. Oreskes. "Assessing ExxonMobil's Global Warming Projections." *Science* 379, no. 6628 (January 13, 2023) for a full

account of, even with available records, just how clear climate outcomes were within firms like Exxon as far back as 1977. On the lack of urgency to shift to renewables, see Jesse Baron, "How Big Business Is Hedging Against the Apocalypse," *New York Times Magazine*, Apr 11, 2019; as we'll see in the next chapter, Exxon was actually deeply invested in renewables for some time. The problem they discovered is not the unviability of renewables — that they're somehow technically not up to the task — but their likely *unprofitability* vs. traditional extraction. Today, ExxonMobil is investing in "renewables" like CCS (the technical deficiency of which is discussed in Chapter 3). Tillerson quotes are taken from Andreas Malm, *Fossil Capital* (New York: Verso, 2016) and Baron's *New York Times Magazine* reporting.

18   The 2019 Intergovernmental Science-Policy Platform on Biodiversity and Ecosystem Services Report is the most frequently cited for estimates like these; it should be noted this estimate is likely highly conservative. A comprehensive review of existing research by Forest Isbell et al. finds that anthropogenic impact between 1500–1900 has already driven nearly 30% of existing animal and plant life to extinction; continued business-as-usual says an *additional* 37% by 2100. Isbell et al. note that the highest level of consensus is that the ultimate driver of this phenomenon is economic "production and consumption" which, as we'll return to, is cautious climate science language for capitalism. See Isbell et al., "Expert perspectives on global biodiversity loss and its drivers and impacts on people," *Frontiers in Ecology and the Environment* 21, no. 2 (2023), 94–103. There is "overwhelming consensus that global biodiversity loss will likely decrease ecosystem functioning and nature's contributions to people." Díaz et al. (see Endnote 15 of this chapter) — focusing more closely on social and ecological interconnection — find this acceleration in biodiversity loss, among other human impacts, "to have increased sharply since the 1970s. The world is increasingly managed to maximize the flow of material contributions from nature to keep up with

rising demands for food, energy, timber, and more, with global trade increasing the geographic separation between supply and demand. This unparalleled appropriation of nature is causing the fabric of life on which humanity depends to fray and unravel." This unraveling is profoundly uneven, as the authors note: "For example, the European Union, the United States, and Japan together accounted for ~64% of the world imports of fish products in value, whereas developing countries accounted for 59% of the total volume of traded fish."

19 Raymond Geuss, *Philosophy and Real Politics* (Princeton: Princeton University Press, 2008), 8–9. Katrina Forrester makes the case that a kind of synthesis is possible between situated historical political understanding and Rawlsian "ideal theory," but in both cases the limits of liberal institutional and consensus theories would be essentially cast aside; see Forrester, "Liberalism and Social Theory after John Rawls," *Analyse & Kritik* 41, no. 1 (2022), 1–22.

20 The original term was coined by Ludwig von Rochau in his 1853 book *Grundsätze der Realpolitik, angewendet auf die staatlichen Zustände Deutschlands* (*Practical Politics: An Application of its Principles to the Situation of the German States*).

21 Gramsci, Antonio. *Selections from the Prison Notebooks*. Edited and translated by Quintin Hoare and Geoffrey Nowell Smith. (New York: International Publishers, 1971), 171.

22 See, on Machiavelli's unapologetics, Nadia Urbinati across multiple works; Gramsci, *Prison Notebooks*, 134–135.

23 Raymond Geuss, *Philosophy and Real Politics*, 2008. Scholars like Lorna Finlayson and Enzo Rossi, among others in a revival of sorts within academic political science, have explored radical realism in slightly different if largely complimentary ways, particularly in the mode of ideology critique of normative realism. See Finlayson, "With radicals like these, who needs conservatives? Doom, gloom, and realism in political theory," *European Journal of Political Theory* 16, no. 3 (2015); or Rossi, "Being realistic and demanding the impossible," *Constellations* 26, no. 4 (2019); Audre Lorde, "The Master's Tools Will Never

Dismantle the Master's House" in Audre Lorde, *Sister Outsider: Essays and Speeches (Crossing Press Feminist Series)* (Crossing Press, 2007), 110-114. Lorde's work – prose and poetry – drips with an intense political realism, today largely whitewashed and sanitized. On Lincoln, see Carl Schmitt, *Dictatorship* (Polity, 2013); and on the friend–enemy distinction, Carl Schmitt, *The Concept of the Political* (Chicago: University of Chicago Press, 1996), 26. It must be noted that in Schmitt's thinking the friend/enemy distinction and the actuality or possibility of subsequent violent conflict is elevated to an existential ideal whose potential historical passing implies a diminishment of human life. This is of course untrue of even the most radical left political actors and thinkers he appropriates who view such conflict, even if necessary, as damaging and hopefully passing into what, for example, Fanon calls "a new history," or Marx, the actual *beginning* of history. We will return to many of these questions in chapters 3 and 4.

24  Marx, "The Eighteenth Brumaire of Louis Napoleon," in Karl Marx, Friedrich Engels, and Tucker Robert C. (eds.), *The Marx-Engels Reader*, 2nd ed. (New York: Norton, 1978), 595.

25  Frantz Fanon, *The Wretched of the Earth* (New York: Grove Press, 2004), 2.

26  I emphasize structure and material conditions — including and even especially in my discussions of climate, politics, and affect — because many conventional political "realists" instead view political realism as grounded in some transhistorical human nature. Left-populist projects like "post-Marxism" and "radical democracy" often sever a thinker like Gramsci from his Marxist roots and cast "the political" too as a transhistorical condition in which the material world is irrelevant to a purely discursive political "agonism." In contrast, as will become even starker in the next chapter, a left-wing climate realism is structured — if not *determined* — in every dimension by literal, physical, as well as dynamic social structures. (Thinkers like Chantal Mouffe are particularly exemplary of the aforementioned "left-populism." I return to these issues in later

chapters.) See also critical assessment in Kai Bosworth, *Pipeline Populism* (Minneapolis: University of Minnesota Press, 2022).

27 Alyson Krueger, "Climate Change Insurance: Buy Land Somewhere Else," *New York Times,* Nov 30, 2018; Atossa Araxia Abrahamian, "Billionaires like Peter Thiel get citizenship abroad so they can run from the problems they create," *Quartz,* Jan 25, 2017; Julie Satow, "The Generator Is the Machine of the Moment," *New York Times,* Jan 11, 2013.

28 Sarah Miller, "Heaven or High Water," *Popula,* Apr 2, 2019.

29 Shannon Hall, "Exxon Knew about Climate Change almost 40 years ago," *Scientific American,* Oct 26, 2015; Wright and Nyberg, 75.

30 While a review of arguments in the philosophy of science is not particularly relevant here, I will note that this argument plays out roughly the same off competing models, Popperian, Kuhnian, etc. (I have addressed to a minor degree methodological questions in the sciences in "Emancipation, Domination, and Critical Theory in the Anthropocene," *Domination and Emancipation: Remaking Critique,* ed. Daniel Benson (Lanham, MD: Rowan and Littlefield, 2021)). On rushing to consensus, it is also worth noting that an increasing number of climate scientists are no longer accepting the pitfall of this discursive maneuver. The prominent physicist and climate modeler Kate Marvel, for example, responds to questions of climate denial by arguing that "I don't think we have a responsibility to personally convert or change every single person's mind. You're just going to drive yourself crazy if you focus on that one person you just can't get to." See Bridget O'Brian in conversation with Marvel, "Scientist Kate Marvel Provides Some Answers on Climate Change and Sustainability," *Columbia News,* Nov 13, 2017.

31 On denialism, see Kari Marie Norgaard, *Living in Denial: Climate Change, Emotions, and Everyday Life* (Cambridge, MA: MIT Press, 2011); Norgaard is adapting earlier categorizations from Cohen and others to climate denial. On measures of "belief," see Matthew Motta et al., "An experimental examination of measurement disparities in public climate change beliefs,"

*Climatic Change* 154 (2019), 37–47; *ScienceDaily*, "Do most Americans believe in human-caused climate change?" May 9, 2019. Here, I am bracketing the much more radical critiques found in Critical Theory. See Chaudhary, "Emancipation, Domination, and Critical Theory" for further discussion and sources on such methodological questions. On the *palpable* experience of climate change, the Yale survey, for example, finds that 65% of Americans think that climate change already seriously or moderately impacts themselves, their families, or Americans in general. The Pew survey finds approximately 70% of Americans supporting renewable energy (wind and solar) over expansion of fossil fuels (oil, coal, and natural gas), including 44% of Republicans and — speaking to palpable sense or feeling, the "structures of feeling" we will explore in Chapter 3 — 67% of Republicans under the age of 30 (see Alec Tyson, Carey Funk, and Brian Kennedy, "What the data says about Americans' views of climate change," Pew Research Center, Apr 18, 2023). As a parallel Pew survey put it, "the study reveals a growing sense of personal threat from climate change among many of the publics polled" (see James Bell et al., "In Response to Climate Change, Citizens in Advanced Economies Are Willing to Alter How They Live and Work," Pew Research Center, Sep 14, 2021).

32  See Kate Aronoff, "Call the Fossil Fuel Industry's Net-Zero Bluff," *The New Republic*, Mar 5, 2021. On "magic": one of the reasons I prefer "techno-mysticism" to "techno-optimism" is precisely because these claims are "magic" or, in the Marxist sense, "mystified" at best. In Chapter 3 we will examine claims around CCS and other negative emissions technologies. On "carbon offsets," see, for example, Thales A.P. West et al., "Overstated carbon emissions reductions from voluntary REDD+ projects in the Brazilian Amazon," *PNAS* 117, no. 39 (2020): "the weight of the evidence suggests that these projects caused less reduction in deforestation than claimed and that few projects actually achieved emission reductions. Suspicion about the environmental integrity of carbon offsets is not restricted

to REDD+ or voluntary interventions. A series of reports on other market-based initiatives for climate change mitigation, i.e., the Joint Implementation (JI) and the Clean Development Mechanism (CDM) of the Kyoto Protocol, also raised concerns about the true climatic contributions from certified carbon offsets."

33 Mark Fisher, *Capitalist Realism* (Winchester, UK: Zer0 Books, 2009), 2.

34 Max Horkheimer, *Critical Theory: Selected Essays* (New York: Continuum Pub. Corp., 1982), 207. See also "Notes on Science and the Crisis" in that collection; and Max Horkheimer, Theodor W. Adorno, *Dialectic of Enlightenment*, ed. Gunzelin Schmid Noeri (Stanford University Press, 2002).

35 "The truth that [the culture industry is] nothing but business is used as an ideology to legitimize the crash they intentionally produce. They call themselves industries, and the published figures for their directors' incomes quell any doubts about the social necessity of their finished products." (Theodor Adorno and Max Horkheimer, *Dialectic of Enlightenment* 95). The "more or less phenomenal box-office success" of a given film serves as "evidence" that production merely follows consumption. Commercial success proves the efficacy of the production by the very criteria of commercial success — sales and salary. Such logic is comically tautological but, as we will explore further in Chapter 3, is surprisingly found in some parts of the "left".

36 Gramsci, *Prison Notebooks*, 276.

37 Thomas Stocker, "The Closing Door of Climate Targets," *Science* 339, no. 6117 (November 29, 2012).

38 "Greenhouse Gases Continued to Increase Rapidly in 2022," *National Oceanic and Atmospheric Administration*, accessed September 18, 2023, https://www.noaa.gov/news-release/greenhouse-gases-continued-to-increase-rapidly-in-2022: "Levels of carbon dioxide ($CO_2$), methane and nitrous oxide, the three greenhouse gases emitted by human activity that are the most significant contributors to climate change, continued their historically high rates of growth in the atmosphere during

2022." The NOAA scientists also observe "2022 was the 11th consecutive year $CO_2$ increased by more than 2 ppm, the highest sustained rate of $CO_2$ increases in the 65 years since monitoring began. Prior to 2013, three consecutive years of $CO_2$ growth of 2 ppm or more had never been recorded." As of 2023 the rate was still *higher*; see Rasmussen, Carl Edward. "Atmospheric Carbon Dioxide Growth Rate," *University of Cambridge, Engineering Department, Machine Learning Group*, accessed September 18, 2023, https://mlg.eng.cam.ac.uk/carl/words/carbon.html. Rasmussen points out common statistical errors and distortions, which either mistake or utilize extremely short-term cyclical fluctuations to argue for slowing rates or even plateaus. This is further discussed in Chapter 2.

39 Yangyang Xu and Veerabhadran Ramanathan, "Well below 2 °C: Mitigation strategies for avoiding dangerous to catastrophic climate changes," *PNAS* 114, no. 39 (2017), 10,315–10,323. See also Timothy M. Lenton et al., "Quantifying the human cost of global warming," *Nature Sustainability* (2023) for some of the most recent estimations; in pathways based not on worst case scenarios but "existing policies." 1.5 °C is crossed around 2030. A recent *optimistic* analysis (Dirk-Han van de Ven et al., "A multimodel analysis of post-Glasgow climate targets and feasibility challenges," *Nature Climate Change* 13 (2023), 570–578) finds that initial Paris commitments fall far short of either goal, with existing revised "ambitions" indicated by existing policy and implementation resulting in Δ2.1–2.4 °C, only modestly improved by further revised NDC commitments, to 2.0–2.2 °C. By bracketing technological questions whose significance the study confirms, the authors argue, "if countries also comply with their stated LTTs [Long Term Targets] after meeting their current NDC pledges in 2030, temperature increase will be further limited and stabilize around 1.7–1.8 °C, which is arguably in line with a future 'well below 2 °C.'" Even with all the stipulations, the closest they can come to *with* high tech ambitions, and without simply ignoring physics, is redefining "well below" as a rounding error.

40   IPCC, 2018: Summary for Policymakers, in *Global Warming of
     1.5 °C. An IPCC Special Report*, Masson-Delmotte et al. (eds).

41   For more on carbon capture and storage (CCS), please see
     Chapter 3.

42   On 2100 goals, see Katsumasa Tanaka and Brian C. O'Neill,
     "The Paris Agreement zero-emissions goal is not always
     consistent with the 1.5 °C and 2 °C temperature targets," *Nature
     Climate Change* 8 (2018), 319–324; Joeri Rogelj et al., "A new
     scenario logic for the Paris Agreement long-term temperature
     goal," *Nature* 573 (2019), 357–363, here 357; Carl-Friedrich
     Schleussner et al., "Science and policy characteristics of the
     Paris Agreement temperature goal," *Nature Climate Change* 6,
     no. 9 (July 25, 2016): "flexibility disappears when aiming
     for more stringent goals such as a hold below 2 °C goal with
     likely or higher probabilities, a 1.5 °C limit, or when taking
     a more precautionary approach assuming limited negative
     emissions potential."; and IPCC AR6 WGIII (2022), 385. (IPCC
     reports are vital but particularly cautious and conservative
     due to their subordination to political pressures, especially
     from wealthy states, as well as issues like modeling and other
     frameworks which we will return to several times.) On Article
     6, see UNFCCC Paris Agreement; the establishment of the
     "internationally traded mitigation outcomes" is simply the
     establishment of market-based carbon tradition as the principle
     mechanism, although markets are not explicitly mentioned.
     There is no controversy here. If the discussion of the use
     of "proceeds" and "expenses" in the Article were not direct
     enough, the 2022 Glasgow Climate Pact explicitly identifies the
     mechanisms of Article 6 with "market-based mechanisms." Kate
     Aronoff records Shell executives boasting the ease with which
     they had basically written Article 6 whole cloth; see Aronoff,
     "Shell Oil Executive Boasts that His Company Influence the
     Paris Agreement," *The Intercept*, Dec 8, 2018. See also Felipe
     Duarte Santos et al., "The Climate Change Challenge: A Review
     of Barriers and Solutions to Deliver a Paris Solution," *Climate* 10,
     no. 5 (2022), 75. On the Glasgow Climate Pact, see Report of the

Conference of the Parties serving as the meeting of the Parties to the Paris Agreement on its third session, held in Glasgow from 31 October to 13 November 2021, "Outcomes of the UN Climate Change Conference in Glasgow," *Center for Climate Change and Energy Solutions*, 7: "aviation, shipping, steel, and trucking, which collectively make up at least 20% of global emissions," were left out of consideration and taken up in a separate ad hoc market-based group announced by John Kerry. Some of these issues like sectoral difficulties and scope 3 emissions are further addressed in Chapters 2 and 3. The UNFCCC language on open international trade has not been challenged through any of the revisions. This language is also remarkably underappreciated — discussed principally in legal briefs, reviews, and individual firm reports. The major exception is its discussion in Naomi Klein's *This Changes Everything* (although I must admit, I encountered it first in the actual document archive and could not believe it was so blatant). See Isak Stoddard et al., "Three Decades of Climate Mitigation: Why Haven't We Bent the Global Emissions Curve?" *Annual Review of Environment and Resources* 46 (2021), 653–689; it is remarkable that one of the authors of the *optimistic* account mentioned in endnote 39 — Glen Peters — is also an author here.

43  On climate science producing its own critique of capitalism, see Chaudhary, "Emancipation…," where I call this "intuitive critical theory." Will Steffen et al., "Trajectories of the Earth System in the Anthropocene," *PNAS* 115, no. 33 (2018).

44  A.T.C. Jérôme Dangerman and Hans Joachim Schellnhuber, "Energy Systems Transformation," *PNAS*, 110, no. 7 (2013); Timon McPhearson et al., "Radical changes are needed for transformations to a good Anthropocene," *npj Urban Sustainability* 1 (2021); Tiffany H. Morrison et al., "Radical interventions for climate-impacted systems," *Nature Climate Change* 12 (2022); Rogelj et al., 2019; Jean-Louis Martin, Virginie Maris, and Daniel S. Simberloff, "The need to respect nature and its limits challenges society and conservation science," *PNAS* 113, no. 22 (2016); and Diaz et al., 2019. There

are, of course, many climate scientists, like Michael Mann, who endorse market mechanisms to deliver the unprecedented change, misunderstanding socioeconomic realities, political economy and power, and how both structure the metabolic relation with a sustainable ecological niche. This is precisely the place for critical intervention.

45  On climate financialization, see Aaron Clark and Taiga Uranaka, "Here Are Five Ways Finance Is Trying to De-Risk Heat Waves," *Bloomberg*, May 2, 2023; on carbon capture credits, see Camilla Hodgson, "Banks and oil groups place bets on carbon capture schemes," *Financial Times*, April 30, 2023. As we'll see in Chapter 3, pseudo-left claims that these "miracle" technologies wither on the vine due to lack of direct or indirect investment and development are absolutely unfounded.

46  Wright and Nyberg, *Climate Change, Capitalism, and Corporations*, xi.

47  Scare quotes because, since no longer *technically* capital, no longer particularly connected to a dynamic process of investment and accumulation, one might better imagine this as simple wealth.

48  Philip Alston, "Climate Change and Poverty," UNHCR A/HRC/41/39 (2019); the discrepancies between arguments from different UN bodies is exceptional.

49  As of 2014, a quarter of the American workforce was employed in one form or another of what Samuel Bowles and Arjun Jayadev describe as "guard labor" (Bowles and Jayadev, "Guard Labor: An Essay in Honor of Pranab Bardhan," *Umass Amhurst Economics Department Working Paer Series* (2004)). See also Ajay Singh Chaudhary and Raphaële Chappe, "The Supermanagerial Reich," *Los Angeles Review of Books*, Nov 17, 2016, on the already existing role of "supermanagers" as key privatized governance agents and complementary arguments in Brooke Harrington, *Capital Without Borders* (Cambridge, MA: Harvard University Press, 2016) about the specific role of "wealth managers" as an example of such agents.

50  See Sam Moore and Alex Roberts, *The Rise of EcoFascism*

(Cambridge, UK: Polity Press, 2022); Mukul Sharma *Green and Saffron* (Ranikhet: Permanent Black, 2011); Anwesh Dutta and Kenneth Bo Nielson "Autocratic environmental governance in India," in *Routledge Handbook of Autocratization in South Asia* (Abingdon/New York: Routledge, 2021); almost all studies — academic and journalistic — treat such politics as deception or hypocrisy, which is fundamentally an error.

51  On Austria, see Benjamin Opratko, "Austria's Green Party will pay a high price for its dangerous alliance with the right," *The Guardian*, Jan 9, 2020, and Oliver Noyen, "Secret deals with conservatives expose Austrian Greens," *Euraktiv*, Feb 1, 2022. On France, see Aude Mazoue, "Le Pen's National Rally goes green in bid for European election votes," *France24*, Apr 20, 2019. On Switzerland, see Joe Turner and Dan Bailey, "'Ecobordering': casting immigration control as environmental protection," *Environmental Politics* 31, no. 1 (2022), 110–131.

52  https://heartiste.org/2018/08/05/full-text-of-alleged-manifesto-of-el-paso-shooter/; Thomas L. Friedman, "Trump Is Wasting Our Immigration Crisis," *New York Times*, Apr 23, 2019.

53  Bowles and Jayadev, "Guard Labor."

54  Wright and Nyberg, *Climate Change, Capitalism, and Corporations*, 63.

55  Luke Kemp et al., "Climate Endgame: Exploring catastrophic climate change scenarios," *PNAS* 199, no. 34 (2022); Steffen et al., 2018.

56  On "existential risk," see Kemp et al., "Climate Endgame"; for Kate Marvel's remarks, see John Schwartz, "Will We Survive Climate Change?" *New York Times*, Nov 19, 2018.

57  Christian Parenti, *Tropic of Chaos: Climate Change and the New Geography of Violence* (New York: Nation Books, 2011), 11.

58  See DARA, and Wearebold.es, *Climate Vulnerability Monitor 2nd Edition. A Guide to the Cold Calculus of a Hot Planet* (DARA and the Climate Vulnerable Forum, 2012); see also Lenton et al., "Quantifying the human cost of global warming."

59  See William Dalrymple, "The East India Company: The Original Corporate Raiders," *The Guardian*, Mar 4, 2015; Angus

Maddison, *The World Economy: A Millennial Perspective* (OECD Development Centre, 2001).

60 Mike Davis, *Late Victorian Holocausts* (New York: Verso, 2000) generally should be emphasized here. Famines of such extreme nature were unheard of before, although of course shortages — and measures to anticipate and ameliorate them — were in place long before.

61 See Dalrymple, "The East India Company," as well as Davis, *Late Victorian Holocausts*, 42, 44, and 35.

62 See Vimal Mishra et al., "Drought and Famine in India, 1870–2016," *Geophysical Research Letters* 46, no. 4 (February 15, 2019); "Churchill's real Darkest Hour: new evidence confirms British leader's role in murdering 3 million Bengalis," *South China Morning Post*, April 12, 2019; Amartya Sen, "Imperial Illusions," *The New Republic*, Dec 31, 2007.

63 See Cedric Robinson, *Black Marxism* (Chapel Hill: University of North Carolina Press, 1983), 144; C.L.R. James, *The Black Jacobins* (New York: Vintage Book, 1989), 374.

64 I address debates on naming the "Anthropocene" in the chapter 5 endnotes.

65 On climate mortality, see DARA 2012; on climate-related economic and social deprivation, see Ricke et al., "Country-level social cost of carbon"; on Australian fires, see Sharon Zhang, "Australia's Fires Show How Wealth Inequality Compounds Climate Disasters," *Truthout*, Jan 18, 2020.

66 Walter Benjamin, *Selected Writings, 4: 1938–1940* (Cambridge, MA: Belknap Press, 2006), 396.

67 Walter Benjamin, *The Arcades Project* (Cambridge, MA: Belknap Press, 2002), 458.

68 William Nordhaus. *A Question of Balance: Weighing the Options on Global Warming Policies.* (New Haven: Yale University Press, 2008). Originally Nordhaus endorsed Δ3.5 as optimal. He has more recently "revised" his position to Δ3.

69 On the discount rate model and Stern Report, see William Nordhaus, "Estimates of the Social Cost of Carbon: Concepts and Results from the DICE-2013R Model and Alternative

Approaches," *Journal of the Association of Environmental and Resource Economists* 1, no. 1/2 (March 2014), 284; see also IPCC (2018) Full Report, 152; and Elizabeth A. Stanton, "Negishi welfare weights in integrated assessment models: the mathematics of global inequality," *Climatic Change* 107, no. 3–4 (December 16, 2010). As of IPCC AR6, Negishi weights are still included in most models, despite increasing evidence that reducing inequality — and redistributing wealth — maximizes not simply welfare considerations but ecological ones as well (something we will return to in Chapters 3 and 4). As Lenzi et al. note, "integrated models currently include Negishi weights, a technical means to ensure that models do not produce any 'additional' global redistribution of wealth. While not explicitly justified, this choice appears again to rely upon unstated beliefs about political feasibility. Yet such a choice ignores much research arguing for decreasing inequality on moral, political, economic, and environmental grounds." Lenzi, D., & Kowarsch, M. (2021). "Integrating justice in climate policy assessments: Towards a deliberative transformation of feasibility," in Kenehan, S. & Katz, C. (eds.), *Principles of Justice and Real-World Climate Politics* (pp. 15–33). Rowman & Littlefield. This should not be construed as a dismissal of models, but rather a point for interpretive intervention (one suggested by AR6's own inclusion of "alternative sustainability models" such as degrowth, post-growth, and Kate Raworth's "donut economics"). However, core assumptions like Negishi weighting and similar functions also help explain discrepancies between ideal models and more granular findings, even within IPCC reports.

70 Even though Nordhaus is technically a "liberal" (in the American sense) few of the "liberals" within the spectrum of "right-wing climate realism" base their arguments on his pure economic gains models (at least directly). Beyond the stray comment about learning to live with Δ3 at meetings and conferences, Δ3 is normalized in mainstream circles in a few other ways. First, by technological investments (akin

to those we'll examine in Chapter 3) that allow for overshoot past $\Delta 2$. Second, by policy incrementalism. Perhaps the most interesting example of this is the case of the Obama-era fuel efficiency standards. There was decent skepticism even early among climate analysts but attention to them reached a peak when the Trump administration moved to roll them back. Although Trump is a climate skeptic in the classic sense, his administration produced one of the most interesting documents in the history of American environmental policy. The 500-page "Draft Environmental Impact Statement for the Safer Affordable Fuel-Efficient (SAFE) Vehicles Rule for Model Year 2021–2026 Passenger Cars and Light Trucks" (2018) justified rolling back the previous standards on the grounds that they were trivial in the face of actual climate conditions which would lead to a $\Delta 3$–$\Delta 4$ world (and whose impacts the report thoroughly explored). While media reports covered this as tragic, contradictory, or hypocritical (see Juliet Eilperin, Brady Dennis, and Chris Mooney, "Trump administration sees a 7°[F] rise in global temperatures by 2100," *Washington Post*, September 28, 2018,), upon closer examination the document was largely correct, an early example of a form of right-wing climate realism (see Ajay Singh Chaudhary, "It's Already Here," *n+1*, August 10, 2018). The difference between the two policies was in fact trivial. Although some reports claimed that the change would increase US GHG emissions by billions of metric tons, the 2018 impact statement and a post-hoc analysis by Obama policy advisors largely agreed that the outcomes are negligible by both ecological and economic standards, despite the Trump policies employing a lower discount rate. (Bordoff, Jason E., Joshua Linn and Akos Losz. "Making Sense of the Trump Administration's Fuel Economy Standard Rollback." (2018)). The main difference between the two is that the "liberal" document said relatively nothing about climate change at all, while the *right*-wing document was, despite some errors, both detailed and unvarnished in presenting climate change realities, not only based on the latest IPCC findings (against the wishes of the Cato Institute) but even sharpening them.

In the end, the policy differences had no actual effect, and neither has Biden's partial reversion thus far. While both technological and policy incrementalist normalization still occur — yesterday's "clean coal" and fracked gas "bridge fuel" are today's CCS and geoengineering — the third mode today, ironically, is eco-"realism" (here of the ideological variety), which cautions that Δ2 is long gone but that's ok, if not great (c.f. David Wallace-Wells, "Here's Some Good News on Climate Change: Worst-Case Scenario Looks Unrealistic," New York Magazine, December 20, 2019) to dictated optimism as with *Our World in Data*'s Hannah Ritchie propounding the absolutely baseless claim that Δ1.5 and Δ2 are arbitrary with no basis in science (Hannah Ritchie, "Stop Telling Kids They'll Die From Climate Change," *Wired*, November 1, 2021). *Our World in Data* is a fascinating case of manipulating data to fit exactly any narrative desired. As Ritchie herself explains, her principal inspiration for her optimism prescription is from Peter Thiel (Hannah Ritchie, "We need the right kind of climate optimism," *Vox*, March 13, 2023), and *Our World in Data* reflects his inspiration. Throughout writing this book, I've tried to reckon with competing claims about empirical data through a critical and synthetic lens, but *Our World in Data* has always proved itself to be a particularly egregious outlier. For example, they have made much publicity about how deaths from natural disasters are declining steadily even as some events (excess heat, extreme weather) multiply. If you use their interactive data tool, you find, for example, a total number of deaths from all "natural" disasters between 1998–2017 to be roughly 60,000. And yet the ever-cautious IPCC cites 526,000 deaths from *just* extreme weather in AR6 WGII. OWID claims all American excess temperature deaths between 1980–1998 as 1281. But the EPA records for the same period 11,000 (and notes a likely significant undercount). One of the best regarded studies on heat deaths (Y. Guo, A. Gasparrini, S. Li, F. Sera, A.M. Vicedo-Cabrera, M. de Sousa Zanotti Stagliorio Coelho, et al. (2018) "Quantifying excess deaths related to heatwaves under climate

change scenarios: A multicountry time series modelling study".
*PLoS Med* 15(7): e1002629) finds a staggering 79,287,540 deaths
from heatwaves between roughly 1971–2020. And increasing.
OWID draws most of their mortality statistics from EM-DAT,
who's methodology is prone to undercounting. And yet, EM-
DAT's own reporting counts 1.19 million deaths from natural
disasters between 1980–1999, and an *increased* number of
1.23 between 2000–2019. These numbers are not explained by
over-reporting (one of their and Ritchie's favorite stories) nor
do they remotely match the OWID presentation, even if you
remove some of the more obvious methodological tricks (like
decadal averaging). As the EM-DAT report states, this is clearly
increasing death rates due to climate change. *Our World in Data*
is heavily entwined with the pseudosciences and philosophies
of effective altruism and longtermism which, not incidentally
argue for a very low discount rate. *Our World in Data* — while
not always wrong — often just wraps the story of unidirectional
market-based technological progress in charming if wildly
inaccurate infographics.

71 See Ajay Singh Chaudhary, "The Long Now," *Late Light*.
November 2022. Questions of temporality are addressed through
the coming chapters but are fully explicated in Chapter 5.

## Chapter 2: The Extractive Circuit

1 Trading Economics, "Philippines — Remittance Inflows to
GDP," 2023 Data 2024 Forecast, accessed 8/9/23.

2 On migrant labor statistics in Saudi Arabia, see IMF Staff
Country Reports, Saudi Arabia: 2022 Article IV Consultation-
Press Release; and Staff Report, accessed 8/9/23; this is actually
*down* from recent highs due to COVID-19 and initiatives to
increase women's participation in the workforce. Regarding
"weak states," Saudi Arabia has, for example, tried and failed
to implement extremely low but basic taxation on its citizens
multiple times over the course of the past two decades;
such a measure would be vital to shift from being a deeply

dependent rentier state — although, as I will argue here, what we increasingly contend with in the political landscape in the twenty-first century are *layers* of weak states or *networks* of weak states contained within in a loose global governance regime attuned to the needs of capital accumulation. It should be noted, as Quinn Slobodian does in his excellent economic history *Globalists* (Harvard, 2018), that this international system dates back at least to the years following the close of World War I and always had at its center the problem of how to "contain" decolonizing states. *Some* post-colonial states are particularly weak by design. As *many* post-, de-, and anti-colonial scholars have noted, what gets called the "neoliberal" state-form in the Global North is, in many ways, a classic case of a *technology* produced for control in the colonized world later imported into the "metropole" for controlling workers and other subaltern classes "at home." See also Ajay Singh Chaudhary, "Toward a Critical 'State Theory' for the Twenty-First Century," in *The Future of the State: Philosophy and Politics*, ed. Artemy Magun (Lanham, MD: Rowman Littlefield, 2020). On oil availability and pricing, see Timothy Mitchell, *Carbon Democracy: Political Power in the Age of Oil* (New York: Verso, 2013), 206–207. On intensifying climate change in relation to migration, see Chandra Segaran Thirukanthan et al., "The Evolution of Coral Reef under Changing Climate: A Scientometric Review," *Animals* 13, no. 5 (2023): "The increasing global temperatures are posing a significant threat to coral reefs, leading to widespread coral bleaching and mortality. Moreover, changes in ocean chemistry brought on by an increase in carbon dioxide levels lead to ocean acidification, which can worsen the effects of rising temperatures on corals. In addition, sea level rise and coastal development are transforming the physical structure of coral reef ecosystems, exacerbating the negative effects of the other stressors. These changes are harmful not only to the coral reefs but also to the plethora of species that rely on them for survival and the communities that rely on them for livelihoods and for protecting the coast. Future challenges for developing countries like those within the coral reef triangle initiative (i.e., Indonesia, Malaysia, the Philippines, Papua New Guinea, Timor-Leste, and the Solomon Islands) will center on access to funding for conservation-restoration efforts and continued monitoring studies"; see also Yi Guan et al., "Vulnerability of global coral reef habitat suitability to ocean warming, acidification and eutrophication," *Global Change Biology*, Jul 25, 2020: "We show that 61% of the total number of reefs in the six regions will likely be affected by both local and global threats under changing environmental conditions. These reefs are found in all six major regions we considered. Local management can act to reduce the anthropogenic pressures in these regions. However, although essential, local management programmes may produce little benefits [*sic*] in areas subject to global threats because the impacts of warming and ocean acidification may outweigh the benefits from local management"; and Mikhael Clotilda S. Tañedo et al., "Individual and Interactive Effects of Ocean Warming and Acidification on Adult *Favites colemani*," *Frontiers in Marine Science* 8 (2021): "Coral reef ecosystems are of immense biological, economical, and societal importance. Philippine reefs are among the most productive and diverse in the world. The country's 25,000 km$^2$ of total reef area contribute about US $4.94 billion to *The Nation*'s economy in the form of fisheries, tourism, and coastal protection. A recent nationwide coral reef assessment (2014–2018) estimated that over the past decade, the country lost about a third of its coral reefs." On gendered perceptions of migrant labor, see Melissa Wright, *Disposable Women and Other Myths of Global Capitalism* (New York/London: Routledge, 2007), and John Chalcoft, "Monarchy, Migration and Hegemony in the Arabian Peninsula," Kuwait Programme on Development, Governance and Globalisation in the Gulf States, LSE Global Governance, 2010.

3   As we'll examine further, this can be understood through the mainstream macroeconomic concept of secular stagnation, as discussed by economists like Robert Gordon (demand-side) and Larry Summers (supply-side), or through Marxian accounts of

industrial overcapacity, as most recently explicated in Aaron Benanav's *Automation and the Future of Work* (Verso, 2020); as is a common theme in this chapter — and book — different analytic perspectives alight on remarkably similar accounts. One also finds a constant and desperate focus in the business press on long-term productivity growth decline, which is visible not only in the wealthiest economies like the US, EU, and Japan, but even in former global powerhouses like China (see, for example, Chris Giles in the *Financial Times* from Apr 13, 2023, "The dire outlook for global growth — and for forecasters"). It is important to distinguish *labor* productivity (whose rate has increased) and total or multifactor productivity, which attempts to account for all or at least multiple inputs in the production process. (Although rarely including energy and other physical inputs). Giles, for example, correctly points out that "residual" factors (using Robert Solow's terminology) are economic "pixie dust" (i.e., unexplained) but mistakenly lapses back into labor productivity questions. Almost all such investigations posit some kind of magic solution, from technological innovation to various forms of stimulus and regulation. But the global productivity growth slowdown persists, in fact outpacing already pessimistic predictions. As Benanav puts it in a recent article, "in the world economy today, unlike during the postwar boom, some economies are able to expand rapidly, but only at the expense of others. That is because, as the world economy grows more slowly, any one country can achieve rapid export-led expansion only by taking market share away from others," resulting in "an increasingly zero-sum game" (see Benanav, "A Dissipating Glut?" *New Left Review* 140/141 (2023)). In ecological terms, which we will revisit, as global total or multifactor productivity growth *slowed*, material intensity — the tonnes of material inputs per $1 of production — actually *increased*, particularly in the last 20 years. Not only does this disaggregate from population growth rates, but it also demonstrates a complex relationship that outstrips the economic models and debates about growth (see IPCC AR6 WGI (2022), 1,169, for example).

4    One of the clearest demonstrations of this phenomenon is
     presented, perhaps surprisingly, by the liberal, market-oriented
     German economist and physicist Andreas Siemoneit, who
     points out that consumption of time-saving services and
     products like consumer electronics become compelled aspects
     of overall economic life (consumers must keep up in order
     to simply continue earning income), facilitating flexibility
     and overall economic efficiency. Echoing his study's title,
     he concludes "Efficiency is an offer you can't refuse," and
     that, furthermore, efficiency becomes one of the key growth
     imperatives (Siemoneit, "An offer you can't refuse: Enhancing
     personal productivity through 'efficiency consumption,'"
     *Technology in Society* 59 (2019)). This echoes, on the one
     hand, earlier ideas in Critical Theory about the blending not
     only of individual work and "free time" but of the supposedly
     separate economic and cultural spheres (see, for example, Max
     Horkheimer, "Notes on Science and the Crisis," or Theodor
     Adorno, *Minima Moralia*, "Free Time," among other essays),
     and, on the other, more contemporary related investigations
     like Jonathan Crary's *24/7* (Verso, 2013) or Tithi Bhattacharya's
     work on social reproduction. Siemoneit's demonstration draws
     on Gary Becker's human capital theory, particularly his 1965 "A
     Theory of Time Allocation," while the Critical Theoretical and
     other Marxian accounts draw on and expand Marx's discussions
     of "productive consumption" and "consumptive production"
     across the *Grundrisse* and all three volumes of *Capital* (see
     David Harvey, *The Limits to Capital* (Basil Blackwell, 1982)).
     Remarkably, both accounts challenge the simple division
     between production and consumption in capitalist society with
     an emphasis on acceleration, speed, velocity, etc. as fundamental
     to productivity in capitalism.

5    Statistics here are derived from Lucas Chancel, Thomas Piketty
     et al., *World Inequality Report 2022* and the related World
     Inequality database using numbers at personal median levels
     measured through PPP (purchasing power parity). Income
     decile thresholds at national and global levels can also be found

through the database. Almost unbelievably — and in shocking contrast to political commentary — most Americans fall well below the top *global* income decile, despite the extraordinary wealth of the United States. The US is the most unequal country in the OECD world, not even accounting for the US being uniquely more "precarious" (a term with different meanings across regions; by European standards, most Americans are by definition precarious). GDP is notoriously poor at capturing genuine social development and/or quality-of-life issues, as perhaps most famously explored by the literatures on communist-led Kerala, India, which achieved human development levels on par with many wealthy countries with an incredibly low GDP. It is remarkable, though, that in discussions of climate and economics, massive redistribution from the top down globally is often painted as unpalatable in the North. But even with the extremely *limited* quantitative tools at hand, it seems to escape scrutiny that such redistribution effectively aimed at eliminating the top 10% or even 20% of highest incomes to raise socioecological standards for the bottom 80–90% would clearly benefit *majorities,* even in the US. Putting aside ecological and qualitative questions, which I will return to in later chapters, even comparing quantitatively across countries is difficult. For example, most US incomes reported in rough PPP comparisons with many other OECD incomes show Americans spending approximately 64% of income on "basics," while Europeans spend on average 44% (see James Manyika et al., "The Social Contract in the 21st Century," *McKinsey Global Institute Report,* Feb 5, 2020), and with dramatically different outcomes. US inequality is so profound across dimensions of class and race as well geography that such numbers are compounded for most Americans by a large margin. Thus, US incomes, particularly in lower deciles, are largely unreflective of lived reality. Chancel et al. provide a useful example comparing the bottom 50% of American and Chinese wealth (i.e., upper middle income): even though overall the US is many times wealthier than China, "the bottom 50% of the US population owns less wealth than the Chinese bottom 50%, in purchasing power parity terms" (52). This is consistent with findings that the US, among OECD countries, is at or near the bottom of a host of actual well-being outcomes and measures (see Carlotta Balestra, Donald Hirsch, and Daniel Vaugh-Whitehead, "Living wages in context: A comparative analysis for OECD countries," *OECD Papers on Well-being and Inequalities* 13 (2023)). Less than 60% of Americans earn enough to live at OECD living wage standards; the US has poverty rates (measured in the OECD's balanced cross-country terms) comparable to countries like Chile and Mexico.

There are a host of systems to try to adjust GDP — among them, the UN's human development index (HDI), multidimensional poverty index (MPI), and others. HDI gives heavy weight to national economic indicators, but even as such the US lands with an HDI rank of 21st and falling. MPI, developed by Oxford economists Sabina Alkire and James Foster, attempts to capture "deprivations" otherwise missed even in living wage or HDI calculations. It is far more useful than the World Bank's highly flawed attempts to measure absolute poverty lines. While World Bank poverty measures focus largely on income, MPI looks to standard-of-living, health, and education outcomes. In this light, Pakistan's poverty rate of about 5% by World Bank accounting in 2018 is more realistically measured by real outcomes and conditions closer to 38% by MPI calculation. (There are many other such frameworks but a full review is beyond the scope here.) MPI has largely been used to understand low- and middle-income living standards, although its theoretical foundations made the case for broader application. The US census bureau began issuing a roughly comparable MPI for the United States in 2019. By World Bank measures, there is about 1% poverty in the US. By MDI, the US poverty rate exceeds the official poverty rate, around 13%, placing real US poverty rates roughly between Honduras and Gabon (see Shatakshee Dhongde and Robert Haveman, "Spatial and Temporal Trends in Multidimensional Poverty in the United

States over the Last Decade," *Social Indicators Research* 163 (2022) and MPI Datasets from 2023). Meanwhile, US median per capita wealth is $79,274 in 2021, *far* below the $100,000 that comprises the top wealth decile and even the $82,000 for the top 20%. This is not to disregard how US per capita is rich by global standards, but rather to show in even the most basic terms how much ground there is for converging interests in radical climate politics across seemingly unlikely global positions, while at the same time underlining how income and wealth at US median levels does not translate to standard-of-living or quality-of-life conditions, a general problem with GDP as a concept and measure well reflected in critical development literatures.

6   There are many approaches to "neoliberalism." This area — the internalization of principles of optimal capitalist performance as a kind of post-Fordism, endlessly flexible, endlessly adaptable, and "resilient" — is where theorists focused more on culture are the most adept (see, for example, Lauren Berlant, "Risky Bigness," in *Against Health* (NYU Press, 2010) or Wendy Brown, *Undoing the Demos* (Zone Books, 2015)). We will look more closely at this kind of embodied intervention in Chapter 4. For details on technologies for labor intensification, see discussions and notes in related sections of Chapter 3, and for pharmacology in particular, those in Chapter 4.

7   My treatment of desire and self-preservation here obviously follows Baruch Spinoza's *Ethics* and in particular the concept of the *conatus*, at its most basic the power or desire of any relatively discrete thing to preserve itself. I return to questions of affect and desire in Chapter 3. In a radically different framework, the American economist Robert Allen — much in line with the kinds of arguments we saw in Endnote 5 — views poverty levels as the threshold between pure "necessity" and "desire." While I disagree with pure GDP metrics (like the World Bank's essentially meaningless absolute measures), I also find, despite agreeing in large part with many of their components and aims, many common counter-metrics to be theoretically and sometimes empirically lacking (particularly in relation to global

political economy and other areas of social complexity). Further discussion of these questions can be found in Chapters 3 and 4. Allen's Basic Needs Poverty Level model has a theoretical elegance expressed in his title "Absolute Poverty: When Necessity Displaces Desire." Implicitly, Allen alights on the quite radical notion that desire is largely trapped or negated if all or most of life is dictated by the quantitative compulsion of mere survival imposed by the market. If and when the relationship is overcome or reversed though, desire is unbound and moves from the limitations of quantitative calculation into an open field of *qualitive* freedom. (An implication remarkably akin to Marx's incomplete arguments about the realms of freedom and necessity which will be addressed again in Chapters 3 and 4.) While it is more parsimonious than even many highly restrictive ecological models, and not as thorough as those discussed in Endnote 5 (and in later chapters), and misses many factors including the technological consumption *for* production discussed in these sections, its focus on *outcomes* in basic goods — food, clothing, housing — has an extraordinary power to correct geographic and *historical* errors. For the former, it shows how a single poverty line even today does not match reality. Historically, though, it is powerful in refuting economic claims from many schools which, using poverty measures based on units of income, have little meaning in pre-capitalist, pre-industrial societies and imply that, for example, poverty was universal before the nineteenth century globally. This is hard to square with historical and geographical and other accounts of most of the subsequently colonized world. It is clear from many accounts (including economic histories that use other approaches and/or draw on non-European sources) that societies like pre-colonial Mughal India or Qing China, to use two well-known examples, were prosperous in terms of their era but also in comparison to their subsequent colonized periods. It is clearly obvious that such economic claims cannot apply (but it still argued) in places where colonization wiped out entire civilizations (e.g., in the Americas). Such questions (on

which Allen is hardly alone or even predominant but is useful
as described here for a starting point) are addressed across
by so many different scholars from different disciplines that
unfortunately further discussion is beyond the scope here.

8   Both emissions and material accounting are confounded by the
reorganization of GVCs in ways that escape standard production
*and* consumption measures. In business accounting, these are
usually described as "scope 1" (direct emissions production
by sector), "scope 2" (indirect emissions in final consumption
or use), and "scope 3" (indirect emissions from intermediary
services and process). As of the most recent AR6 WGIII report,
most "scope 3" emissions are not accounted due to this difficulty
(see IPCC AR6 WGIII (2022), 236–244; further elaborated in
the supplemental material). Existing scope 3 and value chain
estimates are largely produced through firms self-reporting for
ESG investment purposes (and almost certainly undercounted;
see, for example, Maida Hadziosmanovic, Kian Rahimi,
and Pankaj Bhatia, "Trends Show Companies are Ready for
Scope 3 Reporting with US Climate Disclosure Rule," *World
Resources Institute*, Jun 24, 2022). Scope 3 GVC emissions
account for vast and increasing amounts of carbon emissions
over time: "Between 1995 and 2015, global scopes 1, 2, and 3
emissions grew by 47%, 78%, and 84%, to 32, 10, and 45 Pg
$CO_2$, respectively" (Edgar Hertwich and Richard Wood, "The
growing importance of scope 3 greenhouse gas emissions from
industry," *Environmental Research Letters* 13, no. 10 (2018));
these too are underestimates, as they only account for some
sectors. WRI finds a massive range, from 99.8% in financial
sectors and 99% in capital goods (an extraordinary amount) to,
say, 44% in apparel. Aurel Stenzel and Israel Waichman discuss
the ramifications for undercounting (for example through
shifting standard accounts into "scope 3" through digitization
and unaccounted "carbon leakage"; see Stenzel and Waichman,
"Supply-chain data sharing for scope 3 emissions," *npl Climate
Action* 2 (2023)). The EU's own reporting indicates it "is the
world's largest importer of virtual $CO_2$-emissions: its net imports

of goods and services contain more than 700 million tons of $CO_2$ emitted outside of the EU's territory. This is more than 20% of the EU's own territorial $CO_2$ emissions" (Think Tank European Parliament, "Economic assessment of Carbon Leakage and Carbon Border Adjustment," Apr 14, 2020). This too is likely an undercount, as Hertwich and Wood note that downstream scope 3 (indirect emissions) in the EU standard are optional. As we will examine again, here and in Chapter 4, this has profound impacts for the now common claim that "carbon leakage" is no longer an issue, particularly in relation to upper-middle income countries. Also, like the discussion of material intensity above, it helps account for accurate total measures of atmospheric carbon and warming but discrepancies between measured territorial accounts. GVCs are *explicitly* designed to escape such knowledge in parallel to the "missing profits" discussed in Chapter 1.

9   See longer discussion of the Green Revolution in Chapter 3, in particular the long scientific documentation on the failures of the Green Revolution. See also pointed critiques from a range of Marxian perspectives in Jason W. Moore, *Capitalism in the Web of Life* (Verso, 2015); Raj Patel and Jason W. Moore, *A History of the World in Seven Cheap Things* (University of California, 2018); and Max Ajl, *A People's Green New Deal* (Pluto Press, 2021).

10  Statistics in this section are from Deloitte, *Technology, Media, and Telecommunications Predictions 2022*; as noted in earlier, there is limited public (scientific or governmental) information available about these aspects of production, so it is largely in trade publications and voluntary firm disclosures that one finds data. The total count — as already discussed — is likely *deflated*, but as these are averaged from several producers, they would have little interest in skewing the percentages against their favor (i.e., in claiming such high environmental impacts in the value chains they "coordinate" and not in, say, domestic energy, which they do not).

11  Richard Baldwin uses the construction of a cellphone as the quintessential example of a GVC and how they function economically. If we think of every process separated along

such a value chain — raw materials, components of degrees of complexity, middleware, design, marketing, retailing, etc. — the profit is concentrated at the governing firm level. But even if the best paying labor in the process, like the design being discussed here, occurs in "headquarters states," the nominal home of the TNC, the vast majority of the actual wealth is transnationally mobile. Even relatively complex parts of the process — say, the manufacturing of a high-resolution camera lens — do indeed "pay out" to some degree at the far-flung value chain sites, but at an extraordinarily lower level than they previously did within the "national" system. This is in some ways obvious. It otherwise would hardly be worth the trouble. What is truly fascinating is that even the nominally well-remunerated position within the "headquarters economy" is fundamentally detached from the overall profitability of the firm (Richard Baldwin, "Global supply chains: Why they emerged, why they matter, and where they are going," *CEPR Discussion Papers* 9103, 2012). The example of Apple is illustrative here: extractive sites across the world, component construction of chips (for example) in Taiwan, final manufacturing largely in China, coding spread out over South Asia and the US, design largely concentrated in California — but profits are recorded, tax-free, in Ireland.

12  The concept of "planetary boundaries" was proposed by the environmental scientist Johan Rockström et al. in 2009 as a way to measure "a safe operating space for humanity with respect to the functioning of the Earth System." They would answer the question prompted by the Anthropocene: "What are the non-negotiable planetary preconditions that humanity needs to respect in order to avoid the risk of deleterious or even catastrophic environmental change at continental to global scales?" As later clarified and expanded by chemist Will Steffen et al. in 2015, the concept of planetary boundaries does not draw lines at thresholds or tipping points where "a biophysical threshold" is "likely to exist" but "rather upstream of it — i.e., well before reaching the threshold." Planetary boundaries are not the only model of such a space, but any real sustainable

model requires this fundamental shift in conceptualizing and dealing with socioecological risk in ways entirely antithetical to the financialized, just-in-time, high-speed world of twenty-first-century capitalism. See Rockström et al., "A safe operating space for humanity," *Nature* 461 (2009), 472–475; and Steffen et al., "The trajectory of the Anthropocene."

13 Cobalt is not technically a "rare earth metal," even though it is rarer than many rare earth metals. It is often dubbed a critical rare metal (as I've done here) or a critical resource for its economic importance. Cobalt is, though, highly *concentrated*. The 2023 US Geological Survey estimates 8.3 million tons in total reserves, almost 50% of which are in the DRC. This is fairly unique (it is even more concentrated than lithium, which is also more abundant). The rarity of "rare earth metals" can be overstated and often involves geopolitical maneuvering as much as actual geology. Furthermore, "even under the most technologically optimistic scenario," as a recent study put it, cobalt will still remain indispensable up to the point of potentially exceeding reserves (see Anqi Zeng et al., "Battery technology and recycling alone will not save the electric mobility transition from future cobalt shortages," *Nature Communications* 13 (2022)). For more on "rare earth metals" and their geopolitical dimensions, see Julie M. Klinger, *Rare Earth Frontiers: From Terrestrial Subsoils to Lunar Landscapes* (Ithaca: Cornell University Press, 2017).

14 On planned or forced obsolescence, see, for example, Lieselot Bisschop, Yogi Hendlin, and Jelle Jaspers, "Designed to break: planned obsolescence as corporate environmental crime," *Crime, Law and Social Change* 78 (2022), 271–293. The factuality of the process — obvious to users — has largely been established in court cases and, as such, the most synthetic analyses are found in legal reviews and in journalism. Bisschop et al. not only review case history from electronics to fashion to transport, they connect the increasing prevalence of "planned obsolescence as a core business strategy" to "bolster private profit at the expense of consumer interests and environmental sustainability" in the

context of markets characterized by "zero-sum competition and innovation, as well as the systemic drivers of the environmental and waste issues surrounding this topic" by drawing on business, economic, engineering, climate, eco-Marxist, and historical literatures.

15 On relative surplus population (RSP), see David Neilson and Thomas Stubbs, "Relative surplus population and uneven development in the neoliberal era: Theory and empirical application," *Capital & Class* 35, no. 3 (2011). Neilson and Stubbs build on Marx's concept of "relative surplus population," but adjusted for the unanticipated empirical conditions of contemporary capitalism. For their calculations, the RSP are roughly those — *beyond* simple unemployment — who need formal exploitation to live but are permanently consigned to un- or under-employment or informal labor and are neither part of an active agrarian peasantry nor part of "classical" unwaged social reproduction, as with the following example of unwaged domestic work. They cite Mike Davis's accounts of the billions of slum dwellers around the world as a non-exhaustive jumping off point. One could also look to Ruth Wilson Gilmore's work on "organized abandonment." I return to these questions in Chapters 3 and 4. See also Chong Soo Pyun, "The Monetary Value of a Housewife: An Economic Analysis for Use in Litigation," *The American Journal of Economics and Sociology* 28, no. 3 (1969), 271–284; it is also an extremely (I assume unintentionally again) *comedic* article. Pyun begins his investigation by noting that the "monetary value of a housewife" is "imponderable." And then proceeds to ponder it in rather precise terms. (Note: both inflation adjustments in this section are my own done using US Bureau of Labor Statistics methodology.) And Alyssa Battistoni, "Bringing in the Work of Nature: From Natural Capital to Hybrid Labor," *Political Theory*, 45, no. 1 (2016). For an analysis that demonstrates refugee "capitalization" through data extraction and circulation, see Martina Tazzioli, "Extract, Datafy and Disrupt: Refugees' Subjectivities between Data Abundance and Data Disregard,"

*Geopolitics* 27, no. 1 (2022); alternatively, for an analysis that follows a Marxian labor theory of value for refugees and migrants, see Prem Kumar Rajaram, "Refugees as Surplus Population: Race, Migration and Capitalist Value Regimes," *New Political Economy* 23, no. 5 (2018), 627–639. Davis's work itself highlights a number of additional sources of profitability in the global surplus population, from simple "slumlordism" to complex real estate valuations and fees for toilets, in addition to modes of informal labor and even "enforced entrepreneurialism" (Davis, *Planet of Slums* (New York: Verso, 2006)).

16 Malm, *Fossil Capital*.

17 These emissions statistics may be an underestimate given current rate increase and modest rounding of the 1980 level. 1980 is not the beginning of the neoliberal period as already discussed. Rather, it is roughly when one can observe a number of key inflections in economic indicators of the effects of neoliberalism: a rapid rise in corporate profits (see Economic Research, FRED Economic Data, Corporate Profits After Tax), in stock valuations, the continuous decline in labor share (Thomas Piketty, *Capital in the Twenty-First Century*, 2014), etc. As I discuss further, we can also see these same inflections in ecological measures. As regards "corporate social responsibility" (CSR), both proponents and critics find, with only slight differences, CSR is essentially ineffective or "greenwashing" (see Gereffi and Lee, 2016, and Bischopp et al., 2022). ESG metrics are, simply put, updated branding of greenwashing.

18 See Steffen et al., "The trajectory of the Anthropocene"; see also Thomas Wiedmann et al., "Scientists' warning on affluence," *Nature Communications* 11 (2020).

19 Else Pirgmaier and Julia K. Steinberger, "Roots, Riots, and Radical Change — A Road Less Travelled for Ecological Economics," *Sustainability* 11, no. 7 (2019).

20 Robert J. Gordon, *The Rise and Fall of American Growth* (Princeton University Press, 2017), 547.

21 See, for example, the United Nations, Department of Economic and Social Affairs, Statistics Division's Sustainable Development

Goals, section 12 on Responsible Consumption and Production from 2019: "the global material footprint is increasing at a faster rate than both population and economic output. In other words, at the global level, there has been no decoupling of material footprint growth from either population growth or GDP growth." Subsequent reports no longer provide these key comparisons. However, 2023's report does provide the total global material footprint number for 2019 as 95.9 billion metric tons. 2019's analysis is based on a total global material footprint of 92 billion metric tons for 2017. This corresponds with the trajectory of the earlier report (and similarly the emissions and atmospheric carbon concentration data discussed in Chapter 1). Economic growth rates declined globally between 2017 and 2019, confirming the 2019 report's more thorough and accurate presentation.

22  For carbon growth rates, see Carl Edward Rasmussen, "Atmospheric Carbon Dioxide Growth Rate," https://mlg.eng. cam.ac.uk/carl/words/carbon.html (accessed 8/10/23); the long-term decline in economic growth rates at the aggregate global level is well documented.

23  Benanav, *Automation and the Future of Work*.

24  Piketty, *Capital in the Twenty-First Century*, 95.

25  Jürgen Habermas, *Legitimation Crisis* (Beacon Press, 1975), 42 and 46. Remarkably, Habermas notes the empirical and theoretical shakiness of the famous *Limits to Growth* report from the Club of Rome in much the way subsequent skeptics and proponents would challenge and revise its claims. At the same time, he underscores that the global warming dimension is "even on optimistic assumptions" unavoidable, drawing on German physicist and philosopher of science Klaus M. Meyer-Abich (not to be confused with the elder German biologist Adolf Meyer-Abich). Unfortunately, much of Meyer-Abich's work, including this paper (even in German), is currently inaccessible (at least to me). However, Meyer-Abich is widely recognized as an important contributor to both early climate modeling as well as the politics of climate change. Already in the 1970s,

he argued that the most reasonable course of action (what we would today call mitigation) would be blocked by political, not technical or scientific limitations. There would be "winners and losers" relationally and even in toto, namely shaped around questions of profits and costs, geography and wealth. And he did not foresee "winners" voluntarily relinquishing wealth or power. As he rephrased it in 1993: "Traditional political rationality has become unreasonable at the present stage of global interrelations" (see Wolfgang Sachs (ed), *Global Ecology: A New Arena of Political Conflict* (Zed Books, 1993)). There were in fact already a host of analyses around climate change by the early 1970s; the US and USSR signed an extremely loose (but surprisingly fruitful) climate agreement in 1972 based on already existing analyses and public concerns with environmental issues. Although the history of climate science has increasingly narrowed, it was difficult to write out Soviet contributions since many modern frameworks (not least the IPCC) were built out of US-USSR cooperation. At the same time, other global climate science was largely incorporated into increasingly American and European research, sometimes simply a matter of credit and others cutting out vital observations. So while names like Verdanskii and Budyov are — at least in passing — still discussed, their third world counterparts are not. For example, the work of meteorologist Duzheng Ye is hardly discussed but is highly influential on a host of well-known climate science figures like "C.-G. Rossby, J.G. Charney, B. Bolin, E.N. Lorenz, N.A. Phillips" (see Jianhua Lu, "From General Circulation to Global Change: The Evolution, Achievements, and Influences of Duzheng Ye's Scientific Research," *Atmosphere* 14, no. 8 (2023)). Figures like Ye, Tu Changwang, or Chen-chao Koo (among others) played both acknowledged and unacknowledged roles in pivotal journals like *Tellus* and in climate science more broadly, including, probably most famously, both learning and contributing to the rise of science oriented towards socialist ends in, for example, Britain with J.D. Bernal and Joseph Needham.

26  See Odette K. Lawler et al., "The Covid-19 pandemic is

intricately linked to biodiversity loss and ecosystem health,"
*The Lancet Planetary Health* 5, no. 11 (2021); Ashley Smith,
"Competing with Nature: Covid-19 as a Capitalist Virus,"
*Spectre*, Oct 16, 2020; Andreas Malm, *Corona, Climate, Chronic
Emergency: War Communism in the Twenty-First Century* (New
York: Verso, 2020); and Rob Wallace, *Dead Epidemiologists: On
the Origins of COVID-19* (Monthly Review Press, 2020).

27  Laleh Kalili, "A World Built on Sand and Oil," *Lapham's
Quarterly*, accessed 8/10/23; see also Walter Leal Filho et
al., "The Unsustainable Use of Sand: Reporting on a Global
Problem," *Sustainability* 13, no. 6 (2021).

28  See Klinger, *Rare Earth Frontiers*; I have put her principle
geopolitical focus into complementary political economic and
ecological terms here. The political implications of such divides
are addressed in the coming chapters.

29  See Thea Riofrancos, *Resource Radicals* (Duke University Press,
2020) for an excellent analysis of not only the environmental
questions at hand in, for example, resource extraction in South
America, but the political limitations and binds it imposes on
local governments, regardless of intent. In a different register,
see Eliza Griswold's longform narrative non-fiction *Amity and
Prosperity* (Farrar, Straus and Giroux, 2018) for a thorough
exploration at the personal and community level of decisions
and effects of fracking in rural Pennsylvania in the United States.
I return to these questions in Chapter 4.

30  See Klinger, 14; Anna Tsing, *The Mushroom at the End of the
World* (Princeton University Press, 2021 Gary Gereffi, *Global
Value Chains and Development: Redefining the Contours of 21$^{st}$
Century Capitalism* (Cambridge University Press, 2018); Richard
Baldwin, "Global Supply Chains: Why They Emerged, Why They
Matter, and Where They Are Going" in *Global Value Chains
in a Changing World* (WTO, 2013) — many of whom see the
Japanese *keiretsu* as the origin of the modern GVC (beyond the
earliest cases in fossil fuel extraction). Tsing focuses in particular
on the first transnational form of such production in the
manufacture of furniture, which was created precisely to profit

off of illegal logging, local social conflict, and precarious labor in Indonesia.

31  Speed is a *constant* emphasis in both mainstream and critical literatures on GVCs. In many natural and non-critical social scientific literature, even when GVCs are addressed, they are still separated into Fordist sectors (energy, industry, transport, etc.). Peter Dicken demonstrates how flawed this is in his relatively straightforward presentation of how production processes, once discrete and easily identified, are fundamentally all imbricated with globalized production (almost all sectors involved at sometimes *thousands* of locations). See Dicken, *Global Shift: Mapping the Changing Contours of the World Economy* (Sage, 2014), Chapter 8. The closest the IPCC came to examining this dynamic explicitly was in the AR4 report, WGIII, Chapter 5: "Industrialization and globalization have also stimulated freight transport, which now consumes 35% of all transport energy [...] Freight transport is considerably more conscious of energy efficiency considerations than passenger travel because of pressure on shippers to cut costs, however this can be offset by pressure to increase speeds and reliability and provide smaller 'just-in-time' shipments. The result has been that, although the energy-efficiency of specific modes has been increasing, there has been an ongoing movement to the faster and more energy-intensive modes. Consequently, rail and domestic waterways' shares of total freight movement have been declining, while highway's share has been increasing and air freight, though it remains a small share, has been growing rapidly." Even this limited presentation has been *weakened* practically to the point of non-existence in later reports. At the same time, it is only in relatively scant climate science literature that mechanisms to take not only GVC (or GPN, etc.) into account but specifically the questions of speed have been explored. For example, Ugarte et al. show how just-in-time speeds "significantly increase GHG emissions" in production-related transport, while Muñoz-Villamizar et al. demonstrate how just-in-time speeds in final delivery dramatically increase emissions (see Gustavo Ugarte,

Jay Golden, and Kevin Dooley, "Lean versus green: The impact of lean logistics on greenhouse gas emissions in consumer goods supply chains," *Journal of Purchasing and Supply Management* 22, no. 2 (2016); Andrés Muñoz-Villamizar et al., "The environmental impact of fast shipping ecommerce in inbound logistics operations: A case study in Mexico," *Journal of Cleaner Production* 283 (2021)). Still, these are quite isolated (due in no small part to necessarily cautious methodology) focusing only on transport. While GVCs are not discussed in Moore and Patel, I find several of their concepts highly adaptable to it (as I've done here). Despite their differences, Malm also discussed this in *Fossil Capital* by reference to the general theoretical ideas in Crary's *24/7*. What is interesting though is that especially within industry and pro-industry bodies like the IEA or UN, reviews on maritime transport — although again within a narrow framework — slowing is noted as the most obvious way of drastically and immediately lowering emissions, which will also decrease profitability by raising transport and logistics costs (see, for example, United Nations Conference on Trade and Development, "Review of Maritime Shipping," 2021, 88). The IEA — not known for being tech averse — even mentions that wind-powered shipping, *literally modern sail boats*, is the only 100% "clean" solution despite, again, slower speed (see IEA Energy System, Transport report: https://www.iea.org/reports/transport, accessed 2023, or 2001 IEA report on "just-in-time" speed dilemmas). Carl Folke et al also discuss related questions on globalization, but much more generally (see Folke et al, "Our future in the Anthropocene biosphere," *Ambio* 50 (2021)). As Intan Suwandi has observed, earlier theorization of GVCs was based in the general frameworks of world-systems and Marxian analysis. A good example in my own research is Erica Schoenberg's contribution "Competition, Time, and Space in Industrial Change" in: Gereffi et al (eds.), *Commodity Chains and Global Capitalism* (Praeger, 1994).

32    While some hallucinate a "good" Amazon or "socialist" Walmart as the backbone of a "Good Anthropocene," a 2019 public letter

signed by nearly nine thousand Amazon employees detail the
ways in which the company is in no way currently compatible
with even the simplest necessities of climate mitigation and
adaptation (see Amazon Employees for Climate Justice, "Open
letter to Jeff Bezos and Amazon Board of Directors," *Medium*
post, Apr 10, 2019). For all its stark and brave honesty, the
letter actually underplays Amazon's ecological unsustainability
in terms of climate and society. It also does not proceed —
although most analyses rarely do — into thinking what is
*facilitated* by the speed of Amazon. Amazon is paradigmatic
of the extractive circuit; not only in terms of the structure of
the firm, but also in the ways it has shaped many intrinsically
socially and ecologically catastrophic technologies.

33  See Baldwin, "Global Supply Chains," 32; as most authors note,
global supply chains, in a trivial sense, have existed for centuries
or longer. As Gary Gereffi puts it: "The value chain describes
the full range of activities that firms and workers perform to
bring a product from its conception to end use and beyond. This
includes activities such as research and development (R&D),
design, production, marketing, distribution, and support to
the final consumer" (see Gereffi, *Global Value Chains and
Development: Redefining the Contours of 21ˢᵗ Century Capitalism*
(Cambridge University Press, 2018)). Furthermore, the idea of
"unbundling" such processes it itself hardly novel. However,
as we will examine more closely in Chapter 3, the speed,
distance, and degree of "unbundling" has profound, qualitative
implications for questions of economy, ecology, and governance.

34  C.f. Gereffi. *Global Value Chains and Development*; Richard
Baldwin, 2012. "Global Supply Chains: Why They Emerged,
Why They Matter, and Where They Are Going." *CEPR
Discussion Papers* 9103.

35  See Glencore and G4S information from "Violations and the
Security Nexus," *Rights and Accountability in Development*, Mar
15, 2019, and the companies' own websites.

36  Patricia Callahan, "Amazon Pushes Fast Shipping but Avoids
Responsibility for the Human Cost," *New York Times*, Sep 5,
2019.

37 Gereffi et al, *Commodity Chains.*

38 Respectively: Baldwin, "Global Supply Chains"; Herman Mark Schwartz, "The Dollar and Empire," *Phenomenal World*, Jul 16, 2020; and Ajay Singh Chaudhary, "The Amazon Drama," *The Baffler*, Mar 15, 2019.

39 It is also largely mythical. Reshoring, "just-in-case," and regionalization largely mean simply creating alternative far-flung, high-speed networks. See, for example, a theoretical (and notably incomplete) re-mapping of a more ethical, "modular" smartphone-based GVC for German smartphone coordination (see Anna Schomburg et al., "Environmental footprints show the savings potential of high reparability through modular smartphone design," *Research Square* preprint, Apr 5, 2023). Repairability is a real issue, but modularity is largely code for yet another process black box (see Miriam Posner, "See No Evil," *Logic(s)* 4, Apr 1, 2018).

40 See Julie Ray, "Americans' Stress, Worry And Anger Intensified in 2018," *Gallup*, Apr 25, 2019.

41 Fanon, *The Wretched of the Earth.*

42 Fisher, *Capitalist Realism*, 19; emphasis mine.

43 On the group of affects collected under "exhaustion", see COVID-19 Mental Disorders Collaborators, "Global prevalence and burden of depressive and anxiety disorders in 204 countries and territories in 2020 due to the COVID-19 pandemic," *The Lancet* 398, no. 10312; I address this in greater depth — as well as the critical limitations of such studies — in Chapter 4. On global suicide rates, see World Health Organization, "Suicide: facts and figures globally," Sep 6, 2022; suicide was the second leading cause in 2019 but has been crowded out by several other distinctly socioecological phenomena: infectious disease, war and conflict, and increasing maternal mortality; youth is defined here as ages 15–29. On the psychological impacts of climate change, see Paolo Cianconi, Sophia Betrò, and Luigi Janiri, "The Impact of Climate Change on Mental Health: A Systematic Descriptive Review," *Frontiers in Psychiatry* 11 (2020); and Miriam V. Thoma, Nicolas Rohleder, and Shawna L. Rohner,

"Clinical Ecopsychology: The Mental Health Impacts and
Underlying Pathways of the Climate and Environmental Crisis,"
*Frontiers in Psychiatry* 12 (2021).

44  See, for example, Anson Rabinbach, *The Human Motor*
(Berkeley: University of California Press, 1992) and Teresa
Brennan *The Transmission of Affect* (Ithaca: Cornell University
Press, 2004); all these materials are addressed further in Chapter
4.

45  On "imperialism rent," see Samir Amin, *The Law of Worldwide
Value* (New York: Monthly Review Press, 2010); and Amin, "The
New Imperialist Structure," *Monthly Review*, Jul 1, 2019. On G7
share of world GDP, see Baldwin, "Global Supply Chains," 54.
On the "citizenship premium," see Branko Milanović, *Global
Inequality* (Harvard University Press, 2018), 131. On phenomena
of inequality beyond income levels, see, among many, Yufeng
Yan et al., "Carbon endowment and trade-embodied carbon
emissions in global value chains," *Applied Energy* 277 (2020);
Lixia Guo and Xiaoming Ma "Research on global carbon
emission flow and unequal environmental exchanges among
regions," *IOP Conference Series: Earth and Environmental
Science* 781, no. 2 (2021); Xinsheng Zhou et al., "Trade and
Embodied $CO_2$ Emissions: Analysis from a Global Input–
Output Perspective," *International Journal of Environmental
Research and Public Health* 19, no. 21 (2022); Jason Hickel et al.,
"Imperialist appropriation in the world economy: Drain from
the global South through unequal exchange, 1990–2015," *Global
Environmental Change* 73 (2022).

46  Rosa Luxemburg, "The Accumulation of Capital: A Contribution
to the Economic Theory of Imperialism," in: *The Complete Works
of Rosa Luxemburg Volume II, Economic Writings 2* (New York:
Verso, 2016), chapter 27.

47  See Chaudhary, "Toward a Critical 'State Theory'..."

48  Du Bois, W. E. B., *Souls of Black Folk* (Dover, 1994), 9.

49  Walter Benjamin, *The Arcades Project* (Cambridge, MA: Belknap
Press, 2002), Convolute N.

50  On the Cuban example, see Peter Rosset and Miguel Altieri,

*Agroecology: Science and Politics* (Bourton-on-Dunsmore: Practical Action Publishing Ltd, 2017), 79–81. On "temporal luxury," see Ajay Singh Chaudhary, "It's Already Here: Left-wing climate realism and the Trump climate change memo," *n+1*, 2018, and Chaudhary, "The Long Now," *Late Light*, Nov 2022. On "less work," see Sarah Jaffe, "Automatic for the People: Are Robots Coming to Take Our Jobs?" *Bookforum* Feb/Mar 2020.

51  Oliver Natchwey, *Germany's Hidden Crisis* (New York: Verso, 2018), 56. On energy intensity and "hyperbolic" growth, Vaclav Smil, *Growth: From Microorganisms to Megacities* (MIT Press, 2019), 507. On the Keeling Curve, see Rasmussen, "Atmospheric Carbon Dioxide Growth Rate." On profit as driver, see Jennifer Hinton, "Limits to Profit? A conceptual framework for understanding profit and sustainability," *Postgrowth Economics Network Working Paper*, 2022. Hinton covers both approaches and additionally addresses the limitations of pure growth theories in separate articles. Her analysis is particularly salient as she demonstrates both profit as the fundamental driver but also potential flexibility within a zero-sum, profitless, or indeed, receding managed capitalism. However, Hinton is too rosy on non-profit models as opposed to the highly managed scenarios I will address in Chapters 3 and 4. On pricing in the "externalities," see Jesse Ausubel, "Economics in the air: an introduction to issues of the atmosphere and climate," *IIASA Working Paper*, 1980. Ausubel's is a good early example from a mainstream perspective. For a non-exhaustive range of contemporary heterodox arguments, see Elke Pirgmaier, *Value, Capital, and Nature*, diss. University of Leeds, 2018; Moore, *Capitalism in the Web of Life*; and Malm, *Fossil Capital*.

52  Ajay Singh Chaudhary, "It's Already Here: Left-wing climate realism and the Trump climate change memo," *n+1*, 2018.

53  Adam Tooze paraphrases Shin in *Crashed: How a Decade of Financial Crises Changed the World* (New York: Viking/Random House, 2018). See also Hyun Song Shin, "Globalisation: real and financial," presentation for the 87th General Meeting of the Bank of International Settlements (2017); see also Chaudhary,

"Toward a Critical State Theory…" for a full discussion of the state theory I am using throughout this book.

54  Frantz Fanon, *A Dying Colonialism* (Grove Press, 1965), 128.

55  Frantz Fanon, *Black Skin, White Masks* (New York: Grove Press, 2008).

56  See Sen and Dreze, *India: Economic Development and Social Opportunity* (Clarendon Press, 1999).

57  Patrick Heller, *The Labor of Development: Workers and the Transformation of Capitalism in Kerala, India* (Ithaca: Cornell University Press, 2000), 8. It's worth noting the GDP here, not only for comparative analysis but also because atmospheric carbon stood at approximately 350 ppm in 1995.

58  Heller, *The Labor of Development*, 8.

59  "A Review of Public Enterprises in Kerala 2004–2005," Bureau of Public Enterprises, Government of Kerala.

60  See Kali Akuno and Ajamu Nangwaya (eds.), *Jackson Rising: The Struggle for Economic Democracy and Black Self-Determination in Jackson, Mississippi* (Daraja Press, 2017).

61  Sonia Faleiro, "What the world can learn from Kerala about how to fight covid-19," *MIT Technology Review*, Apr 13, 2020.

62  Richard Sandbrook, Marc Edelman, Patrick Heller et al., *Social Democracy in the Global Periphery* (Cambridge University Press, 2007).

## Chapter 3: Climate Lysenkoism; Or, How I Learned to Stop Worrying and Rescue Class Analysis

1  See Nikolai Krementsov, *Stalinist Science* (Princeton University Press, 1996); the actual Michurin's ideas had nothing in common with Lysenko's.

2  Ibid., 171. Targeted opponents included Ivan Shmal'gauzen, Petr Zhukovskii, Boris Zavadovsky, Mikhail Zavadovsky, Aleksandr Paramonov, Efim Lukin, Iurii Polianskii, Ilya Polyakov, Anton Zhebrak, Sos Alikhanian, and Nikolai Dubinin, among others across the biological sciences.

3  For a more thorough and nuanced account of the historical

Lysenko and Lysenkoist movement, please see Richard Levins and Richard Lewontin, *The Dialectical Biologist* (Harvard University Press, 1985).

4   A further discussion of European Social Democracy and specifically Karl Kautsky and the limitations of his thought can be found in my concluding chapter.

5   Not to be confused with Prof. Matthew Huber, the atmospheric chemist.

6   Karl Marx, *Capital Volume I* (Penguin Classics, 1990), 165.

7   See Christopher Lewis and Adaner Usmani, "The Injustice of Under-Policing in America," *American Journal of Law and Equality* 2 (2022), 85–106.

8   On anti-police activism, see Phillips, *Austerity Ecology & the Collapse-Porn Addicts* (Zer0 Books, 2015). On anti-Islamic sentiment, see Phillips, "Lost in Translation: Charlie Hebdo, free speech and the unilingual left," *Ricochet*, Jan 13, 2015. On traditionalist natalism, see Phillips, "Hurrah for 8 Billion Humans," *Compact*, Dec 2, 2022. See also Phillips, "Let's Talk about the COVID-19 Lab Leak Theory," *Jacobin*, Mar 19, 2023. Phillips's opening broadside about the Enlightenment, "science," and progress is so comically overstated that even Karl Popper must be flipping in his grave.

9   Kari Marie Norgaard, "Implicatory Denial: The Sociology of Climate Inaction," *Sydney Environment Institute*, Nov 15, 2017.

10  Andreas Malm, "The Future Is the Termination Shock: On the Antinomies and Psychopathologies of Geoengineering," Parts 1 and 2, *Historical Materialism*, Dec 28, 2022.

11  McKenzie Wark, *Molecular Red*. For those interested in the fine nuances of such relationships, please see my comprehensive overviews in "Sustaining What? Capitalism, Socialism, and Climate Change" in *Capitalism, Democracy, Socialism: Critical Debates* (Springer, 2022), and "Emancipation, Domination, and Critical Theory in the Anthropocene," in *Domination and Emancipation: Remaking Critique* (Rowan and Littlefield, 2021). In these articles I provide several examples of the ways in which climate scientists are adopting Marxian and other critiques

driven by the *data* and its clear irreconcilability with given conditions.

12 Caroline Haskins, "Jeff Bezos Is a Post-Earth Capitalist," *Vice*, Oct 5, 2019; these paragraphs adapted from Chaudhary, "Sustaining What?"

13 See Daniel W. O'Neill et al, "A good life for all within planetary boundaries" (*Nature Sustainability* 1 (2018)) and its adaptation by Smil in *Growth* (MIT Press, 2019).

14 *Ecomodernist Manifesto*, Apr 2015, ecomodernism.org, accessed 8/22/23.

15 Armin Grunwald, "Diverging pathways to overcoming the environmental crisis: A critique of eco-modernism from a technology assessment perspective," *Journal of Cleaner Production* 197, no. 1 (2018).

16 On US support for wealth redistribution, see Frank Newport, "Average American Remains OK With Higher Taxes on Rich," *Gallup*, Aug 12, 2022; on international consensus, see Efraín García-Sánchez et al., "The Two Faces of Support for Redistribution in Colombia: Taxing the Wealthy or Assisting People in Need," *Frontiers in Sociology* 7 (2022).

17 See Thomas Piketty, "Redistributing wealth to save the planet," *ZNet Articles*, Nov 10, 2022; I return to this argument in Chapters 4 and 5.

18 See, among others, Narasimha D. Rao and Jihoon Min, "Less global inequality can improve climate outcomes," *WIREs Climate Change* 9, no. 2 (2018); Fergus Green and Noel Healy, "How inequality fuels climate change," *One Earth* 5, no. 6 (2022); Shuai Zhang et al., "Urbanization, Human Inequality, and Material Consumption," *International Journal of Environmental Research and Public Health* 20, no. 5 (2023).

19 See Arpit H. Bhatt, "Evaluation of performance variables to accelerate the deployment of sustainable aviation fuels at a regional scale," *Energy Conversion and Management* 275 (2023); Alexander Barke et al., "Are Sustainable Aviation Fuels a Viable Option for Decarbonizing Air Transport in Europe? An Environmental and Economic Sustainability Assessment,"

*Applied Sciences* 12, no. 2 (2022); and Derek R. Vardon et al., "Realizing 'net-zero-carbon' sustainable aviation fuel," *Future Energy* 6, no. 1 (2022).

20  Frederico Afonso et al., "Strategies towards a more sustainable aviation: A systematic review," *Progress in Aerospace Science* 137, no. 5 (2023).

21  See again Vardon et al., "Realizing 'net-zero-carbon'....".

22  See, respectively, Rosa Cuéllar-Franca et al., "Utilising carbon dioxide for transport fuels: The economic and environmental sustainability of different Fischer-Tropsch process designs," *Applied Energy* 253 (2019); Ikenna J. Okeke et al., "Life cycle assessment of renewable diesel production via anaerobic digestion and Fischer-Tropsch synthesis from miscanthus grown in strip-mined soils," *Journal of Cleaner Production* 249 (2020); Khaoula Ben Hnich et al., "Life cycle sustainability assessment of synthetic fuels from date palm waste," *Science of the Total Environment*, Nov 20, 2021; and Candelario Bergero et al., "Pathways to net-zero emissions from aviation," *Nature Sustainability* 6 (2023).

23  Phillips, *Austerity Ecology*.

24  Phillips, "The degrowth delusion," *openDemocracy*, Aug 30, 2019; and Phillips, "Ban Private Jets? Or Give Private Jets to All?" *The Breakthrough Institute*, Jul 23, 2023.

25  Phillips, *Austerity Ecology*.

26  Phillips, "The degrowth delusion."

27  Jeffrey M. Masters, "The Skeptics vs. the Ozone Hole," *Weather Underground* (2004).

28  D.J. Dudek et al., "Cutting the Cost of Environmental Policy: Lessons from Business Response to CFC Regulation," *Ambio* 19, no. 19 (1990).

29  Phillips, *Austerity Ecology*.

30  Public statement by the United Nations Special Rapporteur on the right to food, Michael Fakhri, 2022.

31  On hidden hunger, see Kadambot Siddique et al., "Rediscovering Asia's forgotten crops to fight chronic and hidden hunger," *Nature Plants* 7 (2021); on globesity, see, for example, UN World

Food Programme, "WFP Nutrition: Saving Lives, Changing Lives" (2023) or UN Food and Agriculture Organization reports; see also IPCC, "Special Report on Climate Change and Land," Chapter 5: Food Security (2019).

32  See Joan Martinez-Alier, "The EROI of agriculture and its use by the Vía Campesina," *Journal of Peasant Studies* 38 (2011); see also Smil, *Growth*, 389: from the early nineteenth century to 2017 "anthropogenic subsidies" (i.e., fossil-fuel inputs, synthesized fertilizers) "increased nearly 130-fold, from just 0.1 to almost 13 EJ" [Exajoule per year].

33  See IPCC 2019; also M. Crippa et al., "Food systems are responsible for a third of global anthropogenic GHG emissions," *Nature Food* 2 (2021).

34  See Farooq Shah and Wei Wu, "Soil and Crop Management Strategies to Ensure Higher Crop Productivity within Sustainable Environments," *Sustainability* 11, no. 5 (2019); Tandzi Ngoune Liliane and Mutengwa Shelton Charles, "Factors Affecting Yield of Crops," in *Agronomy: Climate Change and Food Security* (Intechopen, 2020); Ann R.L.E. Nelson et al., "The impact of the Green Revolution on indigenous crops of India," *Journal of Ethnic Foods* 6 (2019); Tim Benton and Rob Bailey, "The paradox of productivity: agricultural productivity promotes food system inefficiency," *Global Sustainability* 2 (2019); https://www.nature.com/articles/s41598-022-12686-4S.K. Kakraliya et al., "Energy and economic efficiency of climate-smart agriculture practices in a rice–wheat cropping system of India," *Scientific Reports* 12 (2022): "This energy-intensive system has started suffering from other production fatigue owing to over mining of nutrients, declining factor productivity, increasing production cost, reducing farm profitability, deteriorating soil health and labour shortage causing concern about its sustainability"; Deepak K. Ray et al., "Recent patterns of crop yield growth and stagnation," *Nature Communications* 3 (2012); Daisy John and Giridhara Babu, "Lessons From the Aftermath of Green Revolution on Food System and Health," *Frontiers in Sustainable Food Systems* 5

(2021): "To meet the needs of new kinds of seeds, farmers used increasing fertilizers as and when the soil quality deteriorated [...] Although for around 30 years there was an increase in the production of crops, the rice yield became stagnant and further dropped to 1.13% in the period from 1995 to 1996. Similarly with wheat, production declined from the 1950s due to the decrease in its genetic potential and monoculture cropping pattern. The productivity of potato, cotton, and sugarcane also became stagnant. Globally, agriculture is on an unsustainable track and has a high ecological footprint now"; V. Chhabra, "Studies On Use Of Biofertilizers in Agricultural Production," *European Journal of Molecular Clinical Medicine* 7 (2020): "in spite of outsized use of fertilizers, the trend of yield went down due to less soil fertility and reducing soil flora and fauna"; see also: J.G. Conijn et al., "Can our global food system meet food demand within planetary boundaries?" *Agriculture, Ecosystems & Environment* 251 (2018). In the current 6th IPCC report, none of the potential sustainable agriculture paths recommend continuing Green Revolution practices (IPCC 2022 WGIII).

35 Levins and Lewontin, *Dialectical Biologist*, 235–236.

36 In keeping with Phillips' long tradition of carrying water for business in the name of progress, he cites a report from "the International Food Policy Research Institute, an independent agricultural research institute that has been sharply critical of multinationals" for this bizarre claim. The IFPRI is actually a multinational-funded think tank for the promotion of agricultural technologies. Equally bizarre is that the author he cites for his second claim has been adamant to this day about the numbers and rates being undercounted, if anything: see P. Sainath, "Farmers' suicide rates soar above the rest," *Agrarian Crisis*, https://psainath.org/farmers-suicide-rates-soar-above-the-rest/ (accessed 8/23/23).

37 On actual suicide rates, see K. Nagaraj, "Farmers' Suicides in India: Magnitudes, Trends, and Spatial Patterns," *Madras Institute of Development Studies* (2008). For Phillips's misconstrual of Nagaraj's study, see "Frankenpolitics: The Left

defence of GMOs," Jun 1, 2014, on Phillips' personal website; see also *Austerity Ecology*.

38  Perhaps the simplest illustration of non-neutral technology is the long history of technological developments which use a light-skinned model as their baseline assumption, resulting in technologies which fail to work or work more poorly for dark-skinned users: Vanessa Volpe et al., "Anti-Black Structural Racism Goes Online: A Conceptual Model for Racial Health Disparities Research," *Ethnicity & Disease* 31 (2021). Probably the most famous cases of intrinsic racist design in modern technology are in imaging. Many film stocks, designed by and for white constituencies, simply could not properly reproduce black faces and features. The physical materials chosen and processes used were matched to one population; the technology required serious revision to achieve even a modicum of equality. When Microsoft released its Kinect motion capture cameras for its Xbox gaming consoles in 2010, the same error was notoriously repeated: the Kinect could not "recognize dark-skinned users" (Meredith Broussard, *Artificial Unintelligence: How Computers Misunderstand the World* (MIT Press, 2019), 157). Attempts to "debunk" that the technology was racist required highly unique and unusual lighting situations to achieve "universal" results. I want to thank my colleague Danya Glabau for introducing me to this kind of technological racism.

39  Bryan Newman, "A Bitter Harvest: Farmer suicide and the unforeseen social, environmental and economic impacts of the Green Revolution in Punjab, India," *FOODFIRST Institute for Food and Development Policy*, Development Report no. 15 (2007).

40  In "Food sovereignty and farmer suicides: bridging political ecologies of health and education" (*Journal of Peasant Studies* 49 (2022)), David Meek and Ashlesha Khadse observe that this process was clear as day to farmers across a series of interviews.

41  See Ann R.L.E. Nelson et al., "The impact of the Green Revolution…" and Siddique et al., "Rediscovering Asia's forgotten crops…".

42  And not without good reason. China's early and post-Great
Leap Forward record on food was astonishing. American fears
that India and other Asian countries would pursue Chinese
development began as early as the 1950s when Chinese land
reform and collectivization were yielding agricultural advances
which already impressed countries like India (see John H.
Perkins, *Geopolitics and the Green Revolution* (Oxford University
Press, 2004)). While most are only familiar with the disastrous
famine during the GLP period, China, largely cut-off from the
United States and wary of emulating the Soviet Union, was
proving quite capable of massively increasing its food supply
through its own red-green revolution. The international "Green
Revolution" had little to do with China's version, although
the two are often added together. The Chinese started using
hybridized rice early in the 1960s, and the Chinese red-green
revolution was as much based on the synthesis of land reform,
unique ownership structures, and the adaptation of tradition
multicropping and biological pest control, as high-yield varieties.
To this day, Chinese agriculture remains largely dominated by
smallholders, and the Chinese red-green revolution emphasized
actual production-for-use (namely the expansion of the chief
staple crop, rice, as opposed to expanding cash-crops for
export). Narratives have been adjusted both in the West and
China to emphasize post-Deng reforms but, just as with Chinese
economic reform, this red-green revolution flourished on the
foundations of earlier agricultural developments, including
the post-GLF commune period. The political and economic
structure of China was already separating from the Soviets and,
by the early 1960s, it had become the mix of local autonomy
and national party authority that broadly still exists. Chinese
rice production remained higher than all other examples but,
just as with its Western counterparts, it required prodigious
chemical and mechanical inputs — even if, especially pre-
1980s, these were sometimes traditional fertilizers like manure
(see Benedict Stavis, *Making Green Revolution: The Politics of
Agricultural Development in China* (Cornell University Press,

1974); Stavis, "How China Is Solving Its Food Problem," *Bulletin of Concerned Asian Scholars* 7, no. 3 (1975); Joshua Eisenman "Building China's 1970s Green Revolution: Responding to Population Growth, Decreasing Arable Land, and Capital Depreciation," in *China, Hong Kong, and the Long 1970s: Global Perspectives* (Palgrave Macmillan, 2017); Sigrid Schmalzer *Red Revolution, Green Revolution* (University of Chicago Press, 2015)). As of 2023, further development of Yang Longping's groundbreaking high-yield rice varietal (produced earlier, independently, and with dramatically different techniques than the breed popularized in the Philippines through American and Swiss anti-Communist aid efforts) had resulted in "perennial rice" — a non-GMO hybrid of existing Chinese varietals and a Nigerian perennial. Perennial rice is already proving successful across southern China as well as several other countries like Uganda and the Ivory Coast. "After more than 20 years of effort, the cultivation of PR has become a reality. This represents a cropping system that simultaneously achieves grain production, labour reduction and ecological security, especially for terraced and fragile farmland. PR has demonstrated good yield potential, and agronomic traits for four years and eight cropping seasons from a single planting can enhance soil fertility and reduce requirements for inputs through ecological intensification" (Shilai Zhang et al., "Sustained productivity and agronomic potential of perennial rice," *Nature Sustainability* 6 (2023); and, on connections to earlier Chinese agronomy, see Jian-Guo Gao and Xin-Guang Zhu, "The legacies of the 'Father of Hybrid Rice' and the seven representative achievements of Chinese rice research: A pioneering perspective towards sustainable development," *Frontiers in Plant Science* 14 (2023)). These developments — combining "high" and "low" technologies — fit within the sustainability and agroecological goals of the 14th Five Year Plan, although, as an incredibly recent development, they go beyond the scope of this analysis. However, to return to the calamity of the Great Leap Forward, as Amartya Sen and Jean Drèze observed in 1989, India has excess deaths from ordinary "deprivation" on the scale of the GLF every eight years or so (Sen and Drèze, *Hunger and Public Action* (Oxford University Press, 1991)).

43 See Benjamin K. Sovacool et al., "Differences in carbon emissions reduction between countries pursuing renewable electricity versus nuclear power," *Nature Energy* 5 (2020), and Sovacool et al., "Reply to: Nuclear power and renewable energy are both associated with national decarbonization" *Nature Energy* 7 (2022).

44 Renewables included, as Thea Riofrancos has argued vigorously (see Alyssa Battistoni, "The Lithium Problem: an Interview with Thea Riofrancos," *Dissent*, Spring 2023).

45 Jessica Wang, Ellie Zhu, and Taylor Umlauf, "How China Built Two Coronavirus Hospitals in Just Over a Week," *Wall Street Journal*, Feb 6, 2020.

46 On the 13th Five Year Plan, see Arendse Huld, "The Status of China's Energy Transition and Decarbonization Commitments," *China Briefing*, Apr 22, 2022. On Chinese wind and solar, see Energy Foundation China, "Electrification in China's Carbon Neutrality Pathways" (2022). On the 14th Five Year Plan targets, see Ivy Yin and Eric Yep, "China could exceed renewables generation target of 33% by 2025," *S&P Global Commodity Insights*, Sep 23, 2022; and, again, Huld, "The Status of China's Energy Transition…". On the wastefulness, difficulty, and expense of building nuclear, see Xinyang Guo et al., "Grid integration feasibility and investment planning of offshore wind power under carbon-neutral transition in China," *Nature Communications* 14 (2023). And on storage and grid transmission, see again "Electrification in China's carbon neutrality pathways" and Guo et al., "Grid integration…".

47 On the nuclear share of energy production, see, for example, International Energy Agency, "Net Zero by 2050: A Roadmap for the Global Energy Sector" (2021). On nuclear power shrinking, see IPCC Special Report: Global Warming of 1.5 °C, Chapter 2; the wording of the report can be confusing since, even as nuclear maintains or decreases, it's *share* goes up in some

scenarios because (a) fossil fuels are quickly eliminated and (b) total energy production *decreases* or grows as only a tiny (less than 1%) amount at in *all* 1.5 °C scenarios (see table 2.6 for relevant reference). On the impossibility of a both/and approach, see again Sovacool et al., "Differences in carbon emissions reduction…" and ibid., "Reply to…". On a "massive worldwide buildout," listen to Phillips on the Dead Pundits Society podcast, "Science, the Labor Movement, and the Green New Deal w/ Leigh Phillips" on Soundcloud (2019). And on supportive estimates, see Syed Tauseef Hassan et al., "Is nuclear energy a better alternative for mitigating $CO_2$ emissions in BRICS countries? An empirical analysis," *Nuclear Engineering and Technology* 52, no. 12 (2020); and Harrison Fell et al., "Nuclear power and renewable Energy are both associated with national decarbonization," *Naure Energy* 7 (2022).

48  On small modular reactors, see Amory B. Lovins, "US nuclear power: Status, prospects, and climate implications," *The Electricity Journal* 35, no. 4 (2022). For IEA numbers, see "Net Zero by 2050…," 115; as of this writing, two SMRs have finally come on line — one in Russia and one in China. However, there is little evidence for them being any more efficient or having potential for climate mitigation (see Paul Hockenos, "The Big Problem with Small Nuclear Reactors," *Undark*, Jul 20, 2023). On the drawbacks of SMRs, see Lindsay Krall et al., "Nuclear waste from small modular reactors," *PNAS* 119, no. 23 (2022).

49  Nikit Abhyankar et al., "Achieving and 80% carbon-free electricity system in China by 2023," *iScience* 25, no. 10 (2022).

50  Phillips' ignorance truly knows no bounds. For example, he insists in a tweet from August 2022 that the PRC's economic structure under Mao was the same as the Soviet Union under Stalinism: "Stalinism is any polity employing the political economic mode pioneered by Stalin; *not* just the period of the USSR when Stalin was leader. Thus Mao's China was Stalinist, yet there was no person named Stalin. This should be ABCs for anyone commenting on politics or economics" (https://twitter. com/Leigh_Phillips/status/1557953327152762884). This is

absolutely incorrect. The CPC tried Stalinist planning for a few years but found it underperforming and ill-suited to Chinese conditions. Phillips could have just looked it up in the standard undergrad introduction to comparative political economy: *Comparative Economics in a Transforming World Economy* by Marina and Barkley Rosser.

51  See EPA Center for Corporate Climate Leadership, GHG Inventory Development Process & Guidance, Scope 3 Inventory Guidance.

52  For an extended discussion of this, see Chaudhary "Sustaining What?…".

53  What is vital in Marxism is principally methodological, not articles of faith to be incanted transhistorically; I do not find Marxological debates to be of much political use. That said, of the contemporary Marxologists, Kohei Saito's work is the most thorough and compelling in showing Marx's engagement with the natural sciences and ecology in particular. On connections with the Golem, see Gad Yair and Michaela Soyer, "The Ghost is Back, Again: Karl Marx and the Golem Narrative," *Journal of Classical Sociology* 8, no. 3 (2008).

54  Marx, *Capital*, Chapter 15.

55  See Phillips, "Let's Talk about…" and Phillips, "Tough on Anti-Vaxx Nonsense, Tough on the Causes of Anti-Vaxx Nonsense," *Jacobin*, Oct, 16, 2020. It's worth noting that there were several stages of xenophobic and racist explanations for COVID. Original zoonotic spillover was depicted as the result of "barbaric" food consumption and procurement. In a surprise to no one, Phillips does not ever discuss how it is the industrial food system bequeathed by the "Green Revolution," alongside climate change itself, that leads to the proliferation of epidemic disease, particularly diseases of zoonotic origins. For Marx's views on positivism, see his letter to Engels, dated Jul 7, 1866 (https://megadigital.bbaw.de/briefe/detail.xql?id=M0000133#, accessed 8/23/23).

56  I have addressed some questions on science and Marxism in "Emancipation, Domination, and Critical Theory in the Anthropocene".

57  On global farmer suicides, see Debbie Weingarten, "Why are America's farmers killing themselves?" *The Guardian*, Dec 11, 2018. See also Qi Wu, "Economic and Climatic Determinants of Farmer Suicide in the United States," diss. UC Davis, 2022.

58  Lauren Berlant, "Risky Bigness," in *Against Health* (NYU Press, 2010). See also Berlant, *Cruel Optimism* (Duke University Press, 2011).

59  Phillips, *Austerity Ecology*.

60  Vladimir Lenin, "Kommunismus," in *Lenin's Collected Works*, 4th English edition, volume 31 (Moscow: Progress Publishers, 1965).

61  Mike Davis, *Old Gods, New Enigmas* (Verso, 2018); see pages 21–22, xviii, and 22, respectively.

62  Although Fukuyama is fundamentally a Hegelian, he of course does not share Hegel's admittedly bizarre idealization of a perfected constitutional monarchy. Rather he sees liberal capitalist "democracy" now triumphant against any conceivable systemic challenge, and subject only to revanchists' outlashes against it. This last part is often excised by Fukuyama's critics from the left but also overblown by leftist attempts at recuperation. As with Hegel, Fukuyama's *Geist* (Reason, Sprit) has a borrowed Christian providentialism. It is true that Fukuyama does not actually argue that *events* will stop. However, he argues, we have reached a clear end point and the best that any countermovement can achieve is destruction, not competition. This claim is obviously untrue but also (unlike some other examples of right-wing thought) has very little analytic value for Marxists or the left writ large.

63  Though many contemporary Anglophone Marxists distance themselves from postcolonial theory — and subaltern studies in particular — there remains probably no greater critique of this phenomenon than Gayatri Spivak's evisceration of post-Foucauldian French Theory in her famous essay, "Can the Subaltern Speak?" As Spivak points out, long before the fall of the Soviet Union, it was the historical appearance of subjects who challenged existing power relations — understood in Marxist terms — that moved theorists like Deleuze and Guattari to abandon the subject altogether.

64 Benhabib, "Modernity and the Aporias of Critical Theory" *Telos* 49 (1981).

65 Ellen Meiksins Wood, *The Retreat from Class: A New True Socialism* (Verso, 1999), 15; Davis quotes Wood seemingly from memory but inverts her negative critique of post-Marxists who ignore these qualities into positive ones.

66 György Lukács: "the proletariat has the opportunity to turn events in another direction by the conscious exploitation of existing trends. This other direction is the conscious regulation of the productive forces of society. To desire this consciously, is to desire the 'realm of freedom' and to take the first conscious step towards its realisation" (*History and Class Consciousness* (MIT Press, 1972)). Lukács is more reserved than Marx himself on desire, which he describes as materially, historically produced on both biological and social foundations. Strangely, it is rarely observed that in the *Manifesto* itself, Marx describes Communist activity as actively drawing on all the most "advanced" revolutionary desires across the entirety of a broadly and loosely defined working class. One of the clearest — although least explored — threads through the entirety of Marx's work is the political necessity of demystifying grievances and desires, looking at a given time to understand "its own struggle and its own desires" toward a "ruthless criticism of everything existing" (*Letter to Ruge*, 1843). The role of Communists in the *Manifesto* (1848) is to capture, engender, and nurture the most radical "desires" for social transformation; in *Capital Volume 1* (1867) "the desire for social transformation" is inseparable from "class antagonisms" (Marx, *Capital*, 637; although the translation in the standard *Marx-Engels Reader* (ed. Robert Tucker) is more accurate in this case. See Tucker, 416). Marx also explicitly condemns "bourgeois socialists" who "desire the existing state of society minus its revolutionary and disintegrating elements" (*Manifesto* in Tucker, 496).

67 Murat Arsel, "Climate change and class conflict in the Anthropocene: sink or swim together?" *Journal of Peasant Studies* 50, no. 1 (2023), 24.

68  Davis, *Old Gods*, 7.

69  See, for example, not only the points of view mentioned here but varied arguments from Marxists like Daniel Bensaïd and Olivier Schwartz (both cited by Davis as well) that class theory cannot be reduced to simple workplace exploitation (Bensaïd, *Marx for Our Times* (Verso, 2009), 109). This is in Marx himself, again in the *Manifesto*, in particular in its brief comments on actual politics, where he argues that

> Finally, in times when the class struggle nears the decisive hour, the process of dissolution going on within the ruling class, in fact within the whole range of society, assumes such a violent, glaring character, that a small section of the ruling class cuts itself adrift, and joins the revolutionary class, the class that holds the future in its hands. Just as, therefore, at an earlier period, a section of the nobility went over to the bourgeoisie, so now a portion of the bourgeoisie goes over to the proletariat, and in particular, a portion of the bourgeois ideologists, who have raised themselves to the level of comprehending theoretically the historical movement as a whole (cited in Tucker, 481).

70  Adam Przeworski, *Capitalism and Social Democracy* (Cambridge University Press, 1985), 23.

71  The rejection of classic Marxist theories of imperialism and colonialism is another sad dimension of contemporary Anglophone socialist revivalism — in both economic and ecological terms. As the most thorough recent review finds: "economic growth and technological progress in 'core areas' of the world-system occurs at the expense of the peripheries, i.e., growth is fundamentally a matter of appropriation. In fact, modern technological systems may, in part, be driven by differences in how human time and natural space are compensated in different parts of the world. High resource consumption is enabled by globally prolonged supply chains, favoring countries with high-value added processes" (Christian Dorniger et al., "Global patterns of ecologically unequal

exchange: Implications for sustainability in the 21st century,"
*Ecological Economics* 179 (2021)). In yet another sleight of hand,
Huber deflects such considerations, not through any kind of
empirical or historical analysis, but by impugning that such
documents do not care whether these flows are to "capital or
labor." However, it is Huber, following the even more extreme
case of Phillips, who effaces this dimension by not distinguishing
the rather dramatic differences in consumption that reflect
the dramatic economic inequality in a country like the United
States.

72  David Neilson and Thomas Stubbs, "Relative surplus population
and uneven development in the neoliberal era: Theory and
empirical application," *Capital & Class* 35, no. 3 (2011), 435–453;
here 444 and 450, respectively. See Chapter 2 notes for
explication of the concept of "relative surplus population."

73  Ruth Wilson Gilmore, *Golden Gulag: Prisons, Surplus, Crisis, and
Opposition in Globalizing California* (University of California
Press, 2007), 70 and 68, respectively.

74  Davis, *Old Gods*, 8.

75  "The problem with class abstractionism thus lies in its poor
specification of the structural primacy of class, which, rather
than providing theoretical warrant for its political primacy, is
instead a function of a prior assumption of political primacy
[…] 'Class abstractionism' dodges any consideration of how class
formation relates to other forms of collective identification and
action at the political level. The class structure is conceptualized
in the image of a class formation whose form is already assumed,
and the question of structural determination is reduced to a
binary outcome in which all that matters is whether this image is
realized or not." See Michael A. McCarthy and Mathieu Hikaru
Desan, "The Problem of Class Abstractionism," *Sociological
Theory* 41, no. 1 (2023). McCarthy cites Vivek Chibber as a
paradigmatic example.

76  See Huber, *Climate Change as Class War* (Verso, 2022), part II.
In addition to the absolutely bizarre nature of this assertion,
Huber seems unaware that his argument about pathological

projection of social guilt requires some kind of Critical Theoretical or psychoanalytic account.

77  I return to the question of progress in my concluding chapter. For those who want a more Marxological approach, please see Kohei Saito's recent *Marx in the Anthropocene* (Cambridge University Press, 2023). Marxological debates — incredibly prolific in eco-Marxist literature — aren't particularly germane to climate politics. As Saito himself notes:

> It is surely too naïve to believe that the further development of productive forces in Western capitalism could function as an emancipatory driver of history in the face of the global ecological crisis. In fact, the situation today differs decisively compared with that of 1848: capitalism is no longer progressive. It rather destroys the general conditions of production and reproduction and even subjects human and non-human beings to serious existential threat (2).

Even though Saito goes on to demonstrate that both positions can be found in Marx — and Saito's works are probably the most thorough account of Marx's ecological investigations — the above would still be true.

78  Mike Davis, "*Planet of Slums*," *New Left Review* 26 (2004).

79  Huber, *Climate Change*, 38; Huber extrapolates his estimates from Kim Moody's 2017 data, but he conveniently ignores that Moody justifies his numbers on grounds of simple labor necessity for capital (waged and unwaged) *and* by explicitly emphasizing that it is not in fact the "point-of-production" that matters (see Moody, *On New Terrain: How Capital is Reshaping the Battleground of Class War* (Haymarket Books, 2017), and also Moody, "The New Terrain of Class Conflict In The United States," *Catalyst* 1, no. 2 (2017)). Of course, Huber could have used more recent numbers and analysis by Moody, but that would deviate from his a priori theoretical commitments and his a posteriori conservatism.

80  See Huber, *Climate Change*, 35; interestingly, the ILO Social

Dialogue Report (2022, here 120) emphasizes that almost all recent unionization is of precisely the kinds of people Huber wishes to ignore: "waste pickers, translators, journalists, actors, musicians, interpreters and some other professions (such as social care workers in some countries)."

81  A bowdlerization of Hal Draper's class model.

82  Gilmore, *Golden Gulag*, 72.

83  Wood, *Democracy Against Capitalism* (Cambridge University Press, 1995), 95.

84  Huber, "Ecology at the point of production: climate change and class struggle," *Polygraph* 28 (2020).

85  Chibber, "Why We Still Talk About the Working Class," *Jacobin*, (Mar 15, 2017), emphasis added. Chibber explicitly defines workers through the formal wage relation and the labor movement. Almost unbelievably, he continues: "[Workers] are the ones that suffer the most under capitalism, and hence they not only have a *capacity*, but also an *interest* in coming together and struggling towards those ends which we think would generate more just social arrangements," (emphasis in the original). This claim — extended into *The Class Matrix* (Harvard University Press, 2022) — is ridiculous in every dimension.

86  Huber, "Ecology at the point…," 24.

87  Ibid., 1.

88  Luxemburg, *The Letters of Rosa Luxemburg* (Verso, 2013), 115.

89  See, for example, a fairly thorough review of the literature by Dan La Botz, again hardly an *anti-labor* or anti-union thinker: "The Marxist View of the Labor Unions: Complex and Critical," *Journal of Labor and Society* 16, no. 1 (2013).

90  Przeworski, *Capitalism and Social Democracy*, 43.

91  Gabriel Winant, "Strike Wave," *NLR Sidecar*, Nov 25, 2021. See also Winant's discussion here of the "bargaining for the common good" framework in contrast with Huber and in conversation with the further discussion with other texts referenced throughout this section.

92  Huber's pathic projection onto his apparently hopelessly bourgeois colleagues bears more than a whiff of his own "reaction formation," in the language of psychoanalysis.

93 Alex Press, "This Machine Kills Fascists," *Bookforum*, Feb/Mar 2020; for an intra-labor approach, see Moody, "Reversing the 'Model': Thoughts on Jane McAlevey's Plan for Union Power," *Spectre Journal*, Nov 8, 2020. One of the genuinely enjoyable aspects of Huber's book is his seeming enchantment with the concept of "power structure analysis" which he picks up from McAlevey and correctly attributes (through her) to Frances Fox Piven. Unless you are Glenn Beck, though, Piven's theory is *not Marxist*. As the theory has been adapted over the years, it's come closer to the kind of analysis one finds in post-Marxists like Chantal Mouffe. Although Piven's original work retained some elements of structural critique, current employment assumes a roughly discursive, not structural field. Piven's original theorizations are unobjectionable, if limited (she rightly argues quite against Huber's style of analysis as stuck in the nineteenth century, unable to cope with real world changes that were not predicted in "orthodox" class theory). But "power structure analysis" as it stands today is the methodology used by the *very* types of organizations Huber critiques. Piven herself cites the Obama-era ACORN as a key example (see Piven's introduction to *Roots to Power: A Manual for Grassroots Organizing*, 2nd edition, ed. Lee Staples (Praeger 2004)). I have had the opportunity to work with many social and political activist organizations (and labor ones) over the years and was absolutely floored when I first witnessed this kind of exercise, to which I desperately tried to add actual, Marxian analysis in order to demonstrate how simplified "power structure" analyses didn't actually account for much real-world power.

94 Davis, *Old Gods*, 6.

95 See Nachtwey, *Germany's Hidden Crisis* (Verso, 2018).

96 As Davis notes: "Organized labor also opposed introduction of contributions from general income to fund social security (the usual arrangement in other countries); given the regressive character of the social security tax (same rate for all, with an upper limit on taxed income), this ensured that the burden was disproportionately concentrated at the lower income end" (Davis, *Prisoners of the American Dream* (Verso, 1986)).

97  Davis, *Old Gods,* 218.

98  Ibid., 213.

99  Davis, "The Monster Enters," *New Left Review,* Mar/Apr 2020; strangely, Davis both blurbed Huber's book (despite its clear departure from his work) and at the same time endorsed Schmelzer, Vansintjan, and Vetter's *The Future is Degrowth* (Verso, 2022), which has a diametrically opposed view. I can only assume this has little to do with either book's content, but rather with the mundane ecosystem of endorsements in publishing.

100  Huber, *Climate Change,* 199.

101  See Emily Grubert and Frances Sawyer, "US power sector carbon capture and storage under the Inflation Reduction Act could be costly with limited or negative abatement potential," *Environmental Research Infrastructure and Sustainability* 3 (2023); Jasmin Cooper et al., "The life cycle environmental impacts of negative emission technologies in North America," *Sustainable Production and Consumption* 32 (2022). As chemical engineer Cooper et al. note, on a 20-year timeline "all [CCS] scenarios which use grid electricity and/or natural gas have no net carbon removal." Grubert and Sawyer similarly note that CCS implemented immediately only works in scenarios in which the facilities will not have their lifetime extended. And furthermore that *all* CCS power-generation, even assuming best possible variables, fails in comparison to a phaseout of only coal in 2030 or all fossil fuels in 2035.

102  Another study tracking real-world plants, and stipulating "that captured carbon dioxide can be stored indefinitely, an optimistic and unproven assumption," finds that CCS used for hydrogen production increases GHG emissions by as much as 20% (see Robert Howarth and Mark Jacobson, "How green is blue hydrogen?" *Energy Science & Engineering* 9, no. 10 (2021)). In a rebuttal to Howarth, researchers funded by ExxonMobil, BP, and industry lobbying groups cite an IAEA report which actually confirms the initial analysis, an industry-funded study not subject to peer-review, and a *cartoon* from an oil

and gas trade group website. See both Matteo Romano et al.,
"Comment on 'How green is blue hydrogen?'" *Energy Science &
Engineering* 10, no. 7 (2022) and Howarth and Jacobson, "Reply
to comment on..." in the same issue. More recent research only
confirms the original argument: see Matteo Bertagni et al., "Risk
of the hydrogen economy for atmospheric methane," *Nature
Communications* 13 (2022). It is worth noting yet again that
possible use cases are seen for these immature technologies —
but only *after* an energy transition to a genuinely clean,
renewable power system.

103 The IEA states blankly that CCS-produced fuel "does not
necessarily lead to emissions reduction" (see IEA report "$CO_2$
Capture and Utilisation"); or, as June Sekera and Andreas
Lichtenberger put it: "In sum, the ICR effort globally is
miniscule in relation to the scale of the problem. For DAC
to operate at climate-significant scale, the amount of energy
required is massive and vast amounts of land are required. There
are no plans presently for a pathway for addressing resource
needs or for scaling up operations to a scale that would make
any practical difference to the problem of excess atmospheric
$CO_2$. Moreover, most of the literature, and all of the evident
policymaking dialogue, on ICR ignore the biophysical impacts
of operating an ICR process at a climate-significant scale. These
include emissions from material and infrastructure supply and
the biophysical impacts from the $CO_2$ removal process and from
transport, injection, and storage at scale" ("Assessing Carbon
Capture: Public Policy, Science, and Societal Need," *BioPhysical
Economics and Sustainability* 5 (2020)).

104 Arne Kätelhön et al., "Climate change mitigation potential of
carbon capture and utilization in the chemical industry," *PNAS*
116, no. 23 (2019).

105 Chelsea Harvey, "Cement Producers Are Developing a Plan to
Reduce $CO_2$ Emissions," *Scientific American*, Jul 9, 2018; in his
standard bait-and-switch pattern, Huber vaguely mentions "new
material processes" and then proceeds into the full-throated
case for CCS. The materials aren't always "new", and "processes"

is vague, for example, in discussing construction materials that have long histories of use or recent alternative technical development that does not fit the Climate Lysenkoist vision of "technology." Incidentally, Williams, like many CCS scholars, is not against the technology per se but finds its use limited and largely future-relegated (see Sunxiang Sean Zheng et al., "Ca-Based Layered Double Hydroxides for Environmentally Sustainable Carbon Capture," *Joule* (preprint, 2022)).

106 The reference in Fengming Xi et al., "Substantial global carbon uptake by cement carbonation," (*Nature Geoscience* 9 (2016)) to Pete Smith et al.'s "Biophysical and economic limits to negative $CO_2$ emissions" (*Nature Climate Change* 6 (2015)) is about the fundamental *limitations* of CCS and other negative emission technologies (NETs); the Chelsea Harvey article Huber cites also references a 2018 study involving Stephen J. Davis ("Net-zero emissions energy systems,'" *Science* 360, no. 6396 (2018)) which *also* does not state the necessity of CCS for cement production; its arguments *for* CCS are based on the reticence of firms to pursue *non-CCS* options. Davis, now a research consultant in the private sector, is considerably more gung-ho concerning CCS today.

107 See again Pete Smith et al., "Biophysical and economic limits…"; even in some of the most positive assessments of CCS and NETs, these realities are noted: "In comparison with all hydrogen-based CCU technologies considered in this work, e-mobility and heat pumps reduce the climate impact more strongly per kilowatt hour of electricity used. Hence, from a climate perspective, the implementation of these technologies should be prioritized over the hydrogen-based CCU technologies considered in this work until their demand for renewable energy is fully exhausted" (Kätelhön et al., "Climate change mitigation potential…").

108 See, respectively: Ravikumar et al., "Carbon dioxide utilization in concrete curing or mixing might not produce a net climate benefit," *Nature Communications* 12 (2021); Li et al., "Carbon capture in power sector of China towards carbon neutrality and

its comparison to renewable power," *Fundamental Research*, Jul 4, 2022; Ho, "Carbon dioxide removal is an ineffective time machine," *Nature* 616, Apr 6, 2023. On the comparison with renewables, this does not even get into the issues of geography, land-use, storage, and transportation that still other proponents cite as "critical issues."

109 Huber, *Climate Change*, 52.

110 Marianna Cerini, "Oil giant Shell is the leading the way on this $4 trillion climate change solution some say won't work," *Fortune*, Sep 26, 2022.

111 See Makoto Susaki, "How carbon capture is helping clean up the cement industry, *Spectra*, Oct 20, 2022; and Heidelberg Materials publication on "Carbon Capture and Storage (CCS)."

112 See Aaron Eisenberg's review of Huber's *Climate Change as Class War* ("There is No One Trick to Overcoming the Climate Crisis," *Science for the People* 25, no. 2). In 2022, Huber and co-author Fred Stafford (a retired industry consultant, NSF director, and University of Chicago professor and administrator) could barely contain their joy at the initial failures to pass the BPRA. In their article, they chastised activists for being "disruptive," claiming the act was anti-labor and anti-public, calling for SMRs, DACs, and other unspecified clean energy technologies and chiding the bill's proponents for failing to consult "industry." They even added some anti-Chinese xenophobia for good measure: "cheap solar panels from China made by a combination of forced labor and coal." They also noted — and here I agree — that one of the coalition partners was funded by Jeff Bezos's philanthropy outfit (See Huber and Stafford, "The Problem with New York's Public Power Campaign — and How to Fix It," *The Intercept*, Jul 23, 2022). But one of their key citations, presented as if just an ordinary union member, came from the UA's international representation speaking on behalf of the bizarre, national corporatist (in the classic early twentieth-century sense) "Climate Energy Jobs Coalition — NY" (See the coalition's website: https://www.cleanenergyjobsny.com/who-are-we, accessed 8/24/23). The "Coalition" is quite literally unions allied

with industry/management, advocating not just for an "all of
the above" class compromise strategy but for nuclear-powered
cryptocurrency alongside clean "natural gas" and so on. Of
course, they were dismayed when the bill actually passed. And
probably even more so since it showed that what they criticized
as out-of-touch activists from NYC DSA and non-profits
(despite being quite happy with their own billionaire-funded
sponsors) actually turned out to be a massive coalition including
quite a bit of labor support (See Luca Goldmansour, "How New
York's Democratic Socialists Brought Unions Around to Public
Renewables," *The American Prospect*, Jun 19, 2023). For all their
praise of labor, it turns out that most utility workers deeply
dislike and distrust the New York Power Authority (the center
of Huber and Stafford's arguments), which has long since ceased
to be a New Deal style public agency and is actually now run
largely by a private, right-wing banker. In place of the largely
private administration, the BPRA legislates that utility workers
will get seats on the governing board.

113 Du Bois, *Black Reconstruction* (Free Press, 1998), 216 (above)
and 700.

114 Du Bois, *Writings: The Suppression of the African Slave Trade/The
Souls of Black Folk/Dust of Dawn/Essays and Articles* (Library of
America, 1987), 935.

115 Ruth Wilson Gilmore, "Prisons and Class Warfare: an Interview
with Ruth Wilson Gilmore," *Verso Blog*, Aug 2, 2018.

116 Jodi Dean, "Four Theses on the Comrade," *e-flux Journal* 86
(2017).

117 See, respectively, Silpa Satheesh, "Fighting in the Name of
Workers," in: *The Palgrave Handbook of Environmental Labour
Studies* (Palgrave Macmillan, 2021), 210; Vishwas Satgar, "A
Trade Union Approach to Climate Justice," *Global Labour
Journal* 6, no. 3 (2015; B. Özkaynak et al., "Mining conflicts
around the world: Common grounds from an Environmental
Justice perspective," *EJOLT Report* 7 (2012); Michael Hess et
al., "Establishing the Bougainville Mining Workers' Union,
1969–1976," *Journal of Pacific History* 51, no. 1 (2015); Michael

Klare, *Resource Wars* (Henry Holt, 2001); Stefania Barca and Emanuele Leonardi, "Working class ecology and union politics: a conceptual topology," *Globalizations* 15, no. 4 (2018); UE press release, "UE Endorses Green New Deal, Climate Strike," Aug 29, 2019; and UE news report, "Convention Hears Youth Climate Leader, Endorses Green New Deal and Climate Strike," Oct 23, 2019.

118 See Eisenberg, "There is No One Trick...".

119 For a useful case we already began looking at in Chapter 1, see Michelle Williams, *The Roots of Participatory Democracy* (Palgrave Macmillan, 2008). The NFF within the broad Periyar Malineekarana Virudha Samithi (PMVS) organization is interesting on both these accounts. The PMVS not only contains the NFF and other agricultural unions, its members grow out of and are closely affiliated with Kerala's communist movement, the CPI(M), "the KSSP [People's Science Movement in Kerala], and other organizations affiliated to the political left." The CPI(M) itself has a unique combination of a horizontal participatory democratic network and a traditional vertical party structure. The CPI and CPI(M) split as the more sectoral and business union organizations drifted rightward toward the Congress Party in familiar hopes of greater access to better contracts. The more trade-union-centric CPI was essentially absorbed into the neoliberalization of India, while the CPI(M) has managed to maintain communist rule within the broader now Fascist BJP India, but not without difficulty and loss. Most of the CPI(M)'s successes — from women-led village councils, to massive literacy and publication campaigns, to unparalleled human development success, followed this break. The CPI(M), though, is committed broadly to the industrialization of Kerala and is not particularly ecological in orientation. Both organizations discussed here are closely tied to it.

120 Alyssa Battistoni, "Ways of Making a Living: Revaluing the Work of Social and Ecological Reproduction," *Socialist Register* 56 (2020); Murat Arsel, Bengi Akbulut, and Fikret Adaman, "Environmentalism of the malcontent: anatomy of an

anti-coal power plant struggle in Turkey," *The Journal of Peasant Studies* 42, no. 2 (2015).

121 Stefania Barca, "Labour and the ecological crisis: The ecomodernist dilemma in western Marxism(s) (1970s–2000s)," *Geoforum* 98 (2019); in one of the strangest moments in a simply bewildering book, Huber quotes this very section from Barca approvingly, not seeming to understand that she is condemning *precisely* his version of politics for transforming workers from "agents" into "victims."

122 See Jörg Nowak, *Mass Strikes and Social Movements in Brazil and India* (Palgrave Macmillan, 2019); Aris Martinelli "Why workers' determination is not enough: A case study of workers' struggles in Swiss machinery GVCs," *CLGP Working Paper* (2020); Şahan Savaş Karataşlı, "Surplus Populations, Working Class Struggles and Crises of Capitalism: A World-Historical Materialist Reconception," in *Marxism, Social Movements and Collective Action* (Palgrave Macmillan, 2022); Marcel Paret, "The community strike: From precarity to militant organizing," *International Journal of Comparative Sociology* 61, no. 2–3 (2020). Recall that the only modern value chain structure that preceded the Japanese keiretsu was in the fossil and extractive industries.

123 To her credit, McAlevey seems to understand this dimension well. It is interesting to note that perhaps the world's most active *labor movement* (which is not synonymous with a *union* movement) is in the People's Republic of China. If Huber or Chibber took their arguments seriously, they would be principally focused both on these Chinese workers and the Chinese Communist Party, since labor militancy is strong in China and the Party is likely the most popularly responsive on Earth (see Democracy Perception Index 2022 and Edward Cunningham, Tony Saich, and Jessie Turiel, "Understanding CCP Resilience: Surveying Chinese Public Opinion Through Time" (Harvard: Ash Center for Democratic Governance and Innovation, 2020)). Meanwhile, Huber fails to notice how it is teachers, not his prized industrial unions, who seem to be

capable of winning (even on ecological action) concessions through the state, although his own writing records the fact over and over (see Keith Brower Brown, Jeremy Gong, Matt Huber, and Jamie Munro, "A Real Green New Deal Means Class Struggle," *Jacobin*, Mar 21, 2019).

124 See Kai Heron, "The Great Unfettering," *NLR Sidecar*, Sep 7, 2022.

125 See Roger Moody, *The Gulliver File* (Minewatch, 1992); Gretchen Bauer, *Labor and Democracy in Namibia 1971–1996* (Ohio University Press, 1998); Brian Wood (ed.), *Namibia 1884–1984. Readings on Namibia's History and Society* (London: Namibia Support Committee, 1988); Henning Melber (ed.), *Transitions in Namibia: Which Changes for Whom?* (Nordic Africa Institute, 2007). Although certainly in part due to my own limited access, this history seems largely buried — both because of the relative disinterest in cataloguing native Namibian works, particularly from the pre-independence era, but also due to (a) the effective messaging campaign of Rio Tinto; (b) the pervasive anti-communism of even left-leaning commentators of the era who demonize anti-colonial liberation with support from the Soviet Union, Cuba, etc., and thus try desperately to separate the tightly bound labor and liberation movements into either a story of astroturfing or cooptation — it is quite clearly neither from the records despite these attempts; and (c) the desire of contemporary *progressive* scholars to downplay violence in labor or social movements. This kind of respectability politics is at play in Climate Lysenkoism, as I discuss here, but it is equally vital, as I address in the next chapter, for many of their opponents. Between these sources and fragments of newspaper articles, I was able to get the story straight despite incredibly varying accounts of organizational names, dates, tactics, etc.

126 See Hess et al., "Establishing the Bougainville Mining Workers' Union…"; Colin Kahl, "Plight or Plunder? Natural Resources and Civil War," in *Guns and Butter: The Political Economy of International Security* (Lynn Rienner Publishers, 2005).

127 Isabel Ortiz et al., "An Analysis of World Protests 2006–2020," in *World Protest* (Palgrave Macmillan, 2022).

128 Katie Toth, Jane McAlevey, "Jane McAlevey on how faith can build successful social movements," *Broadview*, Mar 1, 2017.

129 Cedric Durand, "Zero-Sum Game," *NLR Sidecar*, Nov 17, 2021.

130 See Intan Suwandi, *Value Chains: The New Economic Imperialism* (Monthly Review Press, 2019) and "Labour-Value Commodity Chains — The Hidden Abode of Global Production," *The Jus Semper Global Alliance* (2020): "Multinational clients gain advantage from management policies and practices conducted by the bosses in the dependent companies, ranging from specific measurements of workers' performance to forms of direct control on the factory floor. All these practices are enabled by the deskilling of work that has transformed workers into 'mere executors' of work and thus made them vulnerable [...] Remember that those who have suffered the most — workers and peasants in the global South, minorities in the global North, working class women everywhere — are going to lead struggles." See also Grace Reinke, *Extraction: Geographic, Theoretical, and Normative Dimensions*, diss. University of Washington (2022); Paul Apostolidis, "Introduction: On the Timeliness of Precarity-Critique Today," *Emancipations* 1, no. 3 (2022); Benjamin McKean, "Supply Chains and Organizing Against Precarity," *Emancipations* 1, no. 3 (2022).

131 Benanav, *Automation and the Future of Work* (Verso, 2020).

132 Huber, *Climate Change*, 199.

133 C.L.R. James, *The Black Jacobins* (Vintage, 1989), 240.

134 Huber, *Climate Change*, 199; although the positive psychology tics are pervasive through the book. The way he divides affective states is *precisely* the division outlined in normative positive psychology.

135 As a recent massive literature review in the *Journal of Positive Psychology* (yes, really) puts it: "The findings suggested that positive psychology (a) lacked proper theorizing and conceptual thinking, (b) was problematic as far as measurement and methodologies were concerned, (c) was seen as a pseudoscience that lacked evidence and had poor replication, (d) lacked novelty and self-isolated itself from mainstream psychology,

(e) was a decontextualized neoliberalist ideology that caused harm, and (f) was a capitalistic venture." The critical literature was drawn only from academic *positive psychologists*. The authors found these criticisms largely correct, if potentially fixable. Again, hilariously, this kind of pseudoscience is what is used to back up the very sustainable lifestyle, well-being, and other individualized paradigms that Huber *rightly* wishes to criticize — and yet replicates. As I discuss in the next chapter, this kind of discourse and related fields have wormed their way into common climate literature, including IPCC reports (see, for example, IPCC Sixth Assessment Report, Chapter 7: "Health, Wellbeing, and the Changing Structure of Communities"). I will address this in the next chapter in terms of resilience and return to it here concerning risk.

136 Adapted from Chaudhary, "Subjectivity, Affect, and Exhaustion: The Political Theology of the Anthropocene," *Political Theology Network*, Feb 25, 2019. Remarkably, Huber's earlier, better research, *was* built around fossil fuels not only a source of energy and engine for profits but as defining an American "structure of feeling."

137 Berlant, *Cruel Optimism*, 64; their italics. I do not here have the space to address the vast debates within affect theory about difference between feelings, emotions, subjects, non-subjects, etc. I follow here Berlant, Sianne Ngai, Sara Ahmed, and others.

138 For Marx, think of, for example, the "*new* forces and *new passions* spring up from the bosom of society, forces and passions which *feel* themselves to be fettered by that society" in his discussions of the moment of the expropriating of the expropriators (*Capital Vol. 1*, 928; emphasis added). See as well examples from Endnote 66, although these are hardly exhaustive.

139 Lenin, *What is to be Done?* Section A, in *Collected Works, Vol. 5* (Foreign Languages Publishing House, 1961). It is remarkable that both Marx (in his glowing appraisals of the Commune) and Lenin, here in what is regarded as his more "orthodox" period, cite, respectively, police abolition as a crowning achievement,

and police violence as an example. Contemporary red-brown formations (like those clustered around *Compact*, *Damage*, and related centers of reactionary theory including but not limited to Climate Lyenkoists) in contrast, with little historical justification (i.e., without demonstrating what has changed that would require modifying such positions), line up with their liberal and right-wing comrades in celebrating law and order. Regarding understanding and feeling as two sides of the same coin, Lenin writes: "The consciousness of the masses of the workers cannot be genuine class consciousness, unless the workers learn to observe from concrete, and above all from topical, political facts and events, every other social class and all the manifestations of the intellectual, ethical and political life of these classes; unless they learn to apply practically the materialist analysis and the materialist estimate of all aspects of the life and activity of all classes, strata and groups of the population [...] When we do that (and we must and can do it), the most backward worker will understand, or will feel, that the students and religious sects, the muzhiks and the authors are being abused and outraged by the very same dark forces that are oppressing and crushing him at every step of his life, and, feeling that, he himself will be filled with an irresistible desire to respond to these things."

140 Haiyan Lee, "Class Feeling," in Christian Sorace et al. (eds.), *Afterlives of Chinese Communism* (Made in China Journal, 2019); Sorace, "The Chinese Communist Party's Nervous System: Affective Governance from Mao to Xi," *The China Quarterly* 248 (2021).

141 Perry Anderson, *Considerations on Western Marxism* (Verso, 1979), 123.

142 Luxemburg, "The Mass Strike," in *The Essential Rosa Luxemburg* (Haymarket Books, 2007).

143 Gilmore, *Abolition Geography* (Verso, 2022).

144 Raymond Williams, *Marxism and Literature* (Oxford University Press, 1978), 128–135.

145 E.P. Thompson, *The Making of the English Working Class* (Vintage, 1966); unlike McCarthy and Desan, I view Thompson's

understanding as compatible with accounts of socioeconomic structure.

146 Bosworth, *Pipeline Populism* (University of Minnesota Press, 2022), 2.

147 Ibid., 38.

148 Ibid., 32.

149 See, for example, Ngai's introduction to *Ugly Feelings* (Harvard University Press, 2005) as well as *Theory of the Gimmick* (Harvard University Press, 2020). Both situate affect within the "totality" — following Adorno — of capitalism — in particular, as "suturing" pure sense perception and linguistic articulation. In *Ugly Feelings,* Ngai notes that Williams is probably the first contemporary thinker to argue the sociality and materiality of feelings (25). At the same time, in a long footnote, she vacillates on whether Williams is addressing affect, feeling, or emotion at all. This unease seems to have dissipated as the argument at the beginning of *Theory of the Gimmick* is even more in line with Williams than that in *Ugly Feelings* (and Williams's concepts are cited positively in Our Aesthetic Categories (Harvard University Press, 2015)).

150 I am using affect, emotion, feeling, and passion largely interchangeably in line with Ngai, Ahmed, and, in reality, Spinoza, with only relative degrees of difference. One of the great shortcomings of social and political theory derived from Deleuze and Guattari is the recasting of emancipation as a kind of antinomian removal of all restriction. It is not considered in these theories that such a removal can, in many historical instances, be the *condition* of oppression, freeing actually existing formal power to flow far more freely than nascent distributed or minor powers. See Brian Massumi's introduction to Deleuze and Guattari's *A Thousand Plateaus* (University of Minnesota Press, 1987).

151 Ngai, *Ugly Feelings*, 354.

152 Thompson, "Romanticism, Utopianism, and Moralism: The Case of William Morris," *New Left Review*, Sep/Oct 1976.

153 Susan Buck-Morss, *Revolution Today* (Haymarket Books, 2019).

154 *Encyclopedia of Ecology*, "ecological niche," see: https://www. sciencedirect.com/topics/earth-and-planetary-sciences/ ecological-niche, accessed 8/24/23.

155 Ibid.

156 In a Spinozist sense (as well as Marx's), it is of course *all* nature.

157 IPCC AR6 WIII Summary for Policymakers, 32; the calculation is based on the 1–4 trillion per annum estimate on the faster timetable, although this is likely an underestimation for several reasons I discuss here.

158 Arthur Rempel and Joyeeta Gupta, "Fossil fuels, stranded assets and COVID-19: Imagining an inclusive & transformative recovery," *World Development* 146 (2021), 2.

159 Joel Millward-Hopkins et al., "Providing decent living with minimum energy: A global scenario," *Global Environmental Change* 65 (2020); Y. Oswald et al., "Global redistribution of income and household energy footprints: a computational thought experiment," *Global Sustainability* 4 (2021).

160 Shuai Zhang and Dajian Zhu, "Incorporating 'relative' ecological impacts into human development evaluation: Planetary Boundaries–adjusted HDI," *Ecological Indicators* 137 (2022) 2022; Zhu, "Green Consumption: Research Based on Material Flow and Consumption Efficiency," *Bulletin of Chinese Academy of Sciences* 32, no. 6 (2017). It should be noted that when Zhu discusses decoupling, he is explicit that this must be a discussion about material throughput use and not GDP per se; thus for him and the broader Chinese "ecological civilization" model, absolute decoupling means a post-growth economy (he cites Raworth as a model) at sufficiency.

161 Shi Yi, "The 14th Five Year Plan sends mixed message about China's near-term climate trajectory," *China Dialogue*, Mar 8, 2021.

162 Yifan Gu et al., "Ecological civilization and government administrative system reform in China," *Resources, Conservation and Recycling* 155 (2020).

163 Although it is widely agreed (see, for example, Johanna Coenen et al., "Environmental Governance of China's Belt and Road

Initiative," *Environmental Policy and Governance* 31, no. 1 (2021)) that this is the goal of Chinese policy (outside American national security influenced reports), empirical measures are still relatively hard to find. One recent investigation by Zhizhong Liu et al. ("Does the Belt and Road Initiative increase green technology spillover of China's OFDI? An empirical analysis based on the DID model," *Frontiers in Environmental Science* 10 (2022)) finds: "The main conclusions are as follows: First, the BRI has significantly increased the green technology spillover of OFDI into countries along the routes. The heterogeneity study shows that this effect is significant in middle- or low-income countries with high institutional quality or poor environmental performance but not obvious in other countries. Second, the BRI promotes green technology spillover through the mechanism of increasing R&D investment, improving the environmental system, and accelerating the flow of production factors." While another finds: "the results of the analysis and especially the econometric analysis show that BRI infrastructure development contribute technology transfer, in particular in terms of increasing total patent grants, which once again confirms the widespread aspect in professional circles that there are interactions between infrastructural development and innovation activity" (Atom S. Margaryan et al., "Belt and Road Initiative as an innovative platform for technology transfer: Opportunities for Armenia," *Ordnungspolitische Portal*, no. 2022–6 (2022)).

164 Marta Baltruszewicz et al., "Social outcomes of energy use in the United Kingdom: Household energy footprints and their links to well-being," *Ecological Economics* 205 (2023).

165 Porte, *Theses on the Philosophy of Paradise* (unpublished manuscript).

166 Sophie Lewis, *Abolish the Family* (Verso, 2022); Silvia Federici, *Wages Against Housework* (Falling Wall Press/Power of Women Collective, 1975).

167 Almut Grüntuch-Ernst (ed.), *Hortitecture: The Power of Architecture and Plants* (JOVIS, 2019), 73–75.

168 See Xiuli Wang et al., "Vertical greenery systems: from plants to trees with self-growing interconnections," *European Journal of Wood and Wood Products* 78 (2020). See also Viet Nam News, "HCM City housing developments see green spaces shrink," Jan 2, 2022, for evidence of how desired green space is in conflict with the market.

169 We'll return to these questions in a broader discussion of modernism in the final chapter.

170 See Beverly Ann Tan et al., "Nature-Based Solutions for *Urban Sustainability*: An Ecosystem Services Assessment of Plans for Singapore's First 'Forest Town,'" *Frontiers in Environmental Science* 9 (2021). See also Teh Shi Ning, "Green Buildings: Why the benefits of retrofitting outweigh the costs," *The Straits Times*, Aug 15, 2021, for an argument for green space in spite of the market.

171 William Menking (ed.), *The Vienna Model 2: Housing for the City of the 21st Century* (JOVIS, 2019), 171.

172 See, for example, Stefano Boeri's Prato Urban Jungle and Aler public housing redevelopment projects (https://www.stefanoboeriarchitetti.net/en/project/prato-urban-jungle/; https://www.stefanoboeriarchitetti.net/en/project/monza-aler/, accessed 8/24/23).

173 IPCC AR6 WG3, Chapter 8: "three overarching mitigation strategies with the largest potential to decrease current, and avoid future, urban emissions: (i) reducing or changing urban energy and material use towards more sustainable production and consumption across all sectors including through spatial planning and infrastructure that supports compact, walkable urban form (Section 8.4.2); (ii) decarbonise through electrification of the urban energy system, and switch to net-zero-emissions resources (i.e., low-carbon infrastructure) (Section 8.4.3); and (iii) enhance carbon sequestration through urban green and blue infrastructure (e.g., green roofs, urban forests and street trees), which can also offer multiple co-benefits like reducing ground temperatures and supporting public health and well-being (Section 8.4.4)." GHG emissions

344

across plans like these can synergistically add up to 70–90% reductions with *existing* technologies. "While resource efficiency measures are estimated to reduce GHG emissions, land use, water consumption, and metal use impacts from a lifecycle assessment perspective by 24–47% over a baseline, combining resource efficiency with strategic densification can increase this range to about 36–54% over the baseline…"; "green and blue infrastructure…urban forests, street trees, green roofs, green walls, blue spaces, greenways, and urban agriculture" provides in particular many benefits as it sequesters and stores carbon, reduces building energy use, reduces municipal water use, facilitates active mobility, reduces heat stress, mitigates flooding, improves health, improves air quality, promotes biodiversity.

174 Lisa-Marie Hemerijckx et al., "Mapping the consumer foodshed of the Kampala city region shows the importance of urban agriculture," *npj Urban Sustainability* 3 (2023).

175 See Ajl, "The Hypertrophic City vs the Planet of Fields," in *Implosions/Explosions: Towards a Study of Planetary Urbanization* (JOVIS, 2014).

176 See Buck-Morss, *Dreamworld and Catastrophe* (MIT Press, 2022).

177 Eve Blau, *The Architecture of Red Vienna, 1919–1934* (MIT Press, 1999).

178 For this section, please see Matteo Kries et al. (eds.), *Balkrishna Doshi: Architecture for the People* (Vitra Design Museum, 2019); Manon Mollard, "Revisit: Aranya low-cost housing, Indore, Balkrishna Doshi," *The Architectural Review*, Aug 14, 2019; and Menking, *The Vienna Model 2*; strangely, many of the architecture catalogue texts simultaneously claim the project as an unmitigated success and a failure within pages.

179 Buck-Morss, *Dreamworld and Catastrophe*, 115.

180 See Kate Wagner, "Coronagrifting: A Design Phenomenon," on her her blog *McMansionHell*: https://mcmansionhell.com/post/618938984050147328/coronagrifting-a-design-phenomenon, accessed 8/24/23.

181 See, among many others, the proceedings of the 33rd National

Convention of Aerospace Engineers and National Conference on "Emerging Technologies in Aerospace Structures, Materials and Propulsion Systems," *Institute of Engineers India*, Pune Local Centre, Pune, India, Nov 16-17, 2019; Julian David Hunt et al., "Using the jet stream for sustainable airship and balloon transportation of cargo and hydrogen," *Energy Conversion and Management: X* 3 (2019); Mark Piesing, "How airships could return to our crowded skies," BBC - Homepage, November 8, 2019; Christoph Pflaum, Tim Riffelmacher, and Agnes Jocher, "Design and route optimisation for an airship with onboard solar energy harvesting," *International Journal of Sustainable Energy* 42, no. 1 (March 20, 2023). The principle former barrier and cost was *safety*. The early test flights of the "Airlander 10," from which the emissions metrics were measured, were put on hold when one flight partially crashed. The new version is considered extraordinarily safe and modeled to be even more efficient. Actually efficient airships – unlike SAFs – are the technology in danger of being "captured" for exclusive high-end use.

182 Expert Review Comments on the IPCC WGIII AR5 First Order Draft — Chapter 8, 73; Robock is best known as a vociferous critic of solar geoengineering, but through that research is also quite familiar with airship technology, which is often proposed for spraying reflective aerosols in the stratosphere in such projects.

183 Marx, *Capital Vol. III* (Penguin, 1993),197; Leslie, "'The murderous, meaningless caprices of fashion': Marx on capital, clothing and fashion," *Culture Matters*, May 4, 2018.

184 Peláez, "Sustainable Fashion in the Soviet Union? A Historical Perspective," *Paradigme Mode*, Mar 28, 2023.

185 Christina Kiaer, cited in blogpost: https://measures319.rssing.com/chan-25935166/latest.php, accessed 8/24/23.

186 Rempel and Gupta, "Fossil fuels, stranded assets…".

187 See Abbas Jong, "World Risk Society and Constructing Cosmopolitan Realities: A Bourdieusian Critique of Risk Society," *Frontiers in Sociology* 7 (2022) for a fairly concise overview.

188 This literature is where the next chapter begins.

189 "In the case of global risks, side effects are so dangerous that everyone has to adopt survival strategies (not in the military sense) — the imperative of survival of humanity. Some social and political theorists are trying to capture the coming future with old concepts. They use old lines of demarcations ('friend and foe') to put boundaries between earnest and worthy on the one side and those disingenuous and dangerous on the other side. Using those distinctions to understand the conflict dynamic of climate politics is actually not helpful. Those who can be labelled 'foes' as, for example, industries, are actually 'both-and'. They are producing side effects and thereby push the metamorphosis they are trying to hold back" (Ulrich Beck, *Risk Society: Towards a New Modernity* (Sage, 1992). It is of course my overall point that non-universality and ever more demarcated political lines are *fundamental* to climate politics. I can hardly give Beck a full analysis here, but his far-fetched arguments like the "both-and" for industry are fully in line with, say, Hayek's dethronement of the political. Elsewhere (Chaudhary "Emancipation, Domination…") I have discussed Critical Theory in relation to climate, but Beck, in my view, should not be understood at a Critical Theorist.

190 See earlier note on psychology; this is the first topic addressed in Chapter 4, as all of these models incline *against* politics and *for* conflict avoidance.

191 Steffen et al., "Planetary boundaries: Guiding human development on a changing planet," *Science* 347, no. 6223 (2015); there are vast literatures on rethinking risk in ecology, mainstream and heterodox economics, sociology, and more; a full discussion of these is beyond the scope of this work.

192 Indeed, it is remarkable how much the Dajian framework implies this kind of logic, well within planetary boundaries, but not the *least* or most stringent as some degrowth scholars suggest. Raworth's application is similar, if pushing towards the upper bounds.

193 While I do not share their enthusiasm for E.O. Wilson, Half-

Earth theory, or veganism, Troy Vettese and Drew Pendergrass resuscitate this concept from William Morris within its proper conditions in their own fascinating climate analysis. As we will return to in the final chapter, they (and, well, me) are hardly alone in looking to this kind of temporal freedom across the world.

194 Adapted from Chaudhary, "It's Already Here"

195 Arsel, "Climate change and class conflict…".

## Chapter 4: The Exhausted of the Earth

1   See Antonovsky, *Unravelling the Mystery of Health: How People Manage Stress and Stay Well* (Jossey-Bass, 1987); Antonovsky, *Health, Stress, and Coping* (Jossey-Bass, 1979); Adrian DuPlessis Van Breda, "Resilience Theory: A Literature Review," *South African Military Health Service, Military Psychological Institute, Social Work Research & Development*, Report no. MPI/R/104/12/1/4 (Oct 2001); and Bengt Lindström and Monica Eriksson, "Contextualizing salutogenesis and Antonovsky in public health development," *Health Promotion International* 21, no. 3 (2006). Before this time, Antonovsky argued for a rather social approach, on vague, quasi-Marxian lines, perhaps better categorized as ethnocratic labor Zionism. However, his personal politics hardly matter as every subsequent reformulation, no matter attempted qualification, slipped even *further* into internal disposition.

2   See Werner, "Vulnerability and Resiliency: A Longitudinal Study of Asian Americans from Birth to Age 30," paper presented at the 9th Biennial Meeting of the International Society for the Study of Behavioural Development, Tokyo (1987), here 12 and 27.

3   C.S. Holling, "Resilience and stability of ecological systems," *Annual Review of Ecology and Systematics* 4 (1973); there does not appear to be direct influence on the parallel development, although both derive from older concepts of mechanical resilience in engineering or physics, or resilience

in mathematics. As I discuss here, in recent years there has been obvious cross fertilization between the concepts, although largely *outside* scholarly discussion. In a critique of resilience policy, geographers Danny MacKinnon and Kate Driscoll Derickson refer to this discourse as the "'grey literature' produced by government agencies, think tanks, consultancies and environmental interest groups" (MacKinnon and Derickson, "From resilience to resourcefulness: A critique of resilience policy and activism," *Progress in Human Geography* 37, no. 2 (2012), here 254). A direct scholarly example is the far better known Canadian psychologist Albert Bandura, who's self-efficacy concept, Antonovsky drew some parallels with, who is described in textbooks alongside all the thinkers describes throughout this section, and who's work is found throughout current IPCC reports.

4    Carl Folke et al., "Social-ecological resilience and biosphere-based sustainability science," *Ecology and Society* 21, no. 3 (2016).

5    See, for example, Nilufar Matin and Richard Taylor, "Emergence of human resilience in coastal ecosystems under environmental change," *Ecology and Society* 22, no. 2 (2015); Siambabala Bernard Manyena, "The concept of resilience revisited" *Disasters* 30, no. 4 (2006); Ana Raquel Nunes, "Exploring the interactions between vulnerability, resilience and adaptation to extreme temperatures," *Natural Hazards* 109 (2021); Susara E. Van der Merwe et al., "A framework for conceptualizing and assessing the resilience of essential services produced by socio-technical systems," *Ecology and Society* 23, no. 2 (2018). Prominent ecological resilience theorists like Folke and Adger have made many calls for such bridging work; of course, such bridging work *is* necessary to both understand climate and ecology across natural and social boundaries.

6    For just a few "grey literature" examples, see: Susan Clayton et al., "Mental Health and Our Changing Climate: Impacts, Implications, and Guidance," *American Psychological Association and ecoAmerica* (2017); Helen Boon et al., "Recovery from

disaster: Resilience, adaptability and perceptions of climate change," *National Climate Change Adaptation Research Facility, Australia* (2012); World Health Organization report, "Promotion of mental well-being 2017" and "Health promotion glossary of terms 2021," as well as other WHO and World Bank initiatives.

7   See UNISDR report "UNISDR Terminology on Disaster Risk Reduction" 2009, 24. Note how this UN definition both reverts to a quasi-homeostatic model and synthesizes socioecological resilience and psychosocial resilience. The most recent IPCC definition, "the capacity of interconnected social, economic and ecological systems to cope with a hazardous event, trend or disturbance, responding or reorganising in ways that maintain their essential function, identity and structure," drops the individual, although that level remains in the text, and removes absorb but "cope" and "coping" throughout the report often function in the same way. The 2012 IPCC "Managing the Risks of Extreme Events and Disasters to Advance Climate Change Adaptation. A Special Report of Working Groups I and II of the Intergovernmental Panel on Climate Change" definition of resilience explicitly cites Werner as one of the three main sources for the concept (see Endnote 2 above).

8   It should be emphasized that the IPCC AR6 report tries to push back against these interpretations — particularly in contrast to earlier reports — and emphasizes that "assets" and "resources" should largely be interpreted materially as opposed to internal states fairly vociferously. Nonetheless, they are omnipresent and are found throughout the report, often through region or topic specific articles, mixing of definition, or employing adjacent concepts (self-efficacy for example from positive psychology). Above all "grey literature" sources dominate from American government, NGO, UN, and WHO among other sources, not least previous IPCC reports and special reports. The special report on disaster risk mentioned in Endnote 7 builds largely on the work of the 2010 *Behavioral Science Perspectives on Resilience* from CARRI (ed. Fran H. Norris) which features Antonovsky, Werner, Norman Garmezy (another influential

early psychological resilience theorist who categorized the maladaptive as "externalizers," citing poorly behaved children and political activists) not to mention more recent and extreme theorists like George Bonnano.

9    Kathleen Tusaie and Janyce Dyer, "Resilience: A historical review of the construct," *Holistic Nursing Practice* 18, no. 1 (2004); also an example of *trying* to apply resilience in good faith.

10   Sarah Bracke, "Bouncing Back: Vulnerability and Resistance in Times of Resilience," in *Vulnerability in Resistance* (Duke University Press, 2016).

11   Tusaie and Dyer, "Resilience…".

12   See Joan Martinez-Alier, "Mapping ecological distribution conflicts: The EJAtlas," *The Extractive Industries and Society* 8, no. 4 (2021), and ibid. *The Environmentalism of the Poor* (Edward Elgar Publishing, 2002).

13   Murat Arsel, Bengi Akbulut, and Fikret Adaman, "Environmentalism of the malcontent: anatomy of an anti-coal power plant struggle in Turkey," *The Journal of Peasant Studies* 42, no. 2 (2015).

14   Arnim Scheidel, "Environmental conflicts and defenders: a global overview," *Global Environmental Change* 63 (2020), 5.

15   Joan Martinez-Alier et al, "Is there a global environmental justice movement?" *Journal of Peasant Studies* 46, no. 3 (2016) 5.

16   Şahan Savaş Karataşli et al., "A New Global Tide of Rising Social Protest? The Early Twenty-first Century in World Historical Perspective," paper presented at the Eastern Sociological Society Annual Meeting, Baltimore (2018).

17   David Clark and Patrick Regan, "Mass Mobilization Protest Data," *Harvard Dataverse* (database), Harvard University (2021).

18   See Karataşli et al., "A New Global Tide…".

19   Organization for Economic Cooperation and Development, "Perspectives on Global Development 2021: From Protest to Progress?" here and above 17.

20   See Allianz press release, "Businesses need to prepare for a rise in social unrest incidents," June 14, 2022.

21   The term is Claus Offe's.

22  See Karataşli, "The twenty-first century revolutions and internationalism: a world historical perspective," *Globalizations* 16, no. 7 (2019), here 992.

23  Fanon, *The Wretched of the Earth*, 86.

24  See Crary, *24/7*, and William E. Scheuerman, *Liberal Democracy and the Social Acceleration of Time* (Johns Hopkins University Press, 2004); see also notes on speed in Chapter 2.

25  For a longer discussion of this argument, see Chaudhary, "In the Court of the Centrist King: Emmanuel Macron and Authoritarian Liberalism," *Political Research Associates*, Nov 28, 2017.

26  Malm, *How to Blow Up a Pipeline*, 161.

27  Considering it is actually *difficult* to read much of Fanon, particularly the opening section of *The Wretched of the Earth* as anything *other* than a mode of radical realism, it is striking how infrequent serious treatments of Fanon as a realist are. This was more common in earlier studies (see Gerald Tucker's 1969 study *The Political Thought of Machiavelli and Fanon* (McGill University Libraries)) and only occasionally appears in contemporary literature. Discussions of Fanon as a realist, when they do occur, often follow a limited understanding of realism as in popular caricatures of Machiavelli. A contemporary reading that is closest to my own understanding of Fanon as radical realist can be found in my colleague Geo Maher's "Decolonial realism: Ethics, politics, and dialectics in Fanon and Dussel" (*Contemporary Political Theory* 13, no. 1 (2014)) which explicitly connects Fanon's political thought to Raymond Geuss's definitions of radical realism in *Philosophy and Real Politics* (Princeton University Press, 2008). Jean-Paul Sartre's famous introduction did not help matters; it is in many ways a radical *misreading* of Fanon — in particular Sartre claims Fanon celebrates irrational violence of the colonized as against reason. This is precisely the opposite of Fanon's argument: irrational violence is impractical and unproductive. It is the difficult task of political organization to bring such outbursts into a rational framework (mirroring his psychological practice as well).

Fanon also argues that this "dignity" argument is temporary and passing — for anyone who reads the book through, Fanon argues violence might be necessary but it is corrosive to all parties involved. Fanon criticized Sartre rather forcefully in parts of *Black Skin, White Masks*, although he was also a great admirer. Nonetheless, it remains something of a mystery why he asked Sartre to write an introduction. Fanon never really read the intro or had the opportunity to provide commentary. As Beauvoir recounts in her memoirs, Sartre took a lackadaisical approach and Fanon would only be shown a copy when he was quite literally on his death bed, largely catatonic. This has been deeply unfortunate as many readings of Fanon (Arendt's perhaps most famously) have taken Sartre's preface as a trustworthy summary, not actually following Fanon's argument through to the end.

28  See examples of actual vs. imagined resource depletion in Chapter 2.

29  Marx, *Capital Vol. I*, 538.

30  Ryan Flavelle, "Help or Harm: Battle Exhaustion and the RCAMC During the Second World War," *Journal of Military and Strategic Studies* 9, no. 4 (2007), 4.

31  See Flavelle, "Help or Harm…," here 5 and 7.

32  Rabinbach, *The Human Motor*, here 153 and 20.

33  Jonathan Malesic, *The End of Burnout* (University of California Press, 2022), 194–196.

34  Rabinbach, *The Human Motor*, 4.

35  See International Labour Organization, "Workplace Stress: A Collective Challenge" (2016); *Gallup Workplace* report, "Perspective on Employee Burnout" (2019); *Gallup Workplace* report, "State of the Global Workplace" (2022).

36  *Gallup*, "Global Emotions" (2022).

37  Ibid.; while *Gallup* surveys of course have all the limitations discussed in the earlier chapters, here again they are fascinating as they puncture so many "common sense" images of the world. Furthermore, *Gallup* has little incentive to juke the stats one way or another. The data is equally valuable to their audiences whether the portrait is grim or optimistic.

38  Helen Herrman et al., "Time for united action on depression: a *Lancet*-World Psychiatric Association Commission," *The Lancet Commissions* 399, no. 10,328 (2022), 968.

39  Francisco Javier Carod-Artal and Carolina Vázquez-Cabrera, "Burnout Syndrome in an International Setting," in *Burnout for Experts* (Springer, 2012), 18.

40  COVID-19 Mental Disorders Collaborators, "Global prevalence and burden of depressive and anxiety disorders in 204 countries and territories in 2020 due to the COVID-19 pandemic," *The Lancet* 398, no. 10312 (2021), 1704; Luke Kemp et al., "Climate Endgame: Exploring catastrophic climate change scenarios," *PNAS* 119, no. 34 (2022), 4.

41  Herrman et al., "Time for united action on depression…," 964.

42  Sighard Neckel, Anna Katharina Schaffner, and Greta Wagner (eds.), *Burnout, Fatigue, Exhaustion: Interdisciplinary Perspectives on a Modern Affliction* (Palgrave Macmillan, 2017).

43  Toscano, "Antiphysis/Antipraxis: Universal Exhaustion and the Tragedy of Materiality," *Mediations* 31, no. 2 (2018), 125.

44  Hal, *The Burnout Society* (Stanford University Press, 2015).

45  See Odette K. Lawler et al., "The COVID-19 pandemic is intricately linked to biodiversity loss and ecosystem health," *The Lancet Planetary Health* 5, no. 11 (2021): "Multiple human-mediated environmental changes and activities have been found to be key drivers of zoonotic disease emergence, promoting the conditions in which zoonoses can emerge. Such drivers include, for example, land-use change, intensive livestock production, wildlife trade, and anthropogenic climate change, all of which have been linked to multiple zoonotic disease outbreaks in humans." Or Rob Wallace, *Dead Epidemiologists: On the Origins of COVID-19* (Monthly Review Press, 2020); Eamon Whalen, "The Unemployed Epidemiologist Who Predicted the Pandemic," *The Nation*, Aug 30, 2021; as discussed in Chapter 3, the vast scientific consensus on the origins of COVID-19 agree on zoonotic spillover. For a recent comprehensive review, see Jonathan E. Pekar et al., "The molecular epidemiology of multiple zoonotic origins of SARS-CoV-2": https://www.ncbi.

nlm.nih.gov/pmc/articles/PMC9348752/, accessed 8/29/23; and a follow up in preprint, Pekar et al., "The recency and geographical origins of the bat viruses ancestral to SARS-CoV and SARS-Cov-2," https://pubmed.ncbi.nlm.nih.gov/37502985/, accessed 8/29/23; the "lab leak" conspiracy theory was rekindled through a bipartisan effort under President Biden which produced evaluations from US national security agencies, like the CIA and the FBI. Even still, most security agencies found no evidence and the most confident were fairly inconclusive. However, news media covered the report as if it contained new evidence (see Office of the Director of National Intelligence report on "Potential Links Between the Wuhan Institute of Virology and the Origin of the COVID-19 Pandemic," June 2023). Meanwhile, the scientific consensus is so strong that most research has moved on into questions of how to better address the more frequent pandemics caused by climate change, questions of science communications, and conspiracy theories.

46  Brennan, *Exhausting Modernity: Grounds for a New Economy* (Routledge, 2000), 12.

47  Mark Fisher, *Ghosts of My Life: Writings on Depression, Hauntology and Lost Futures* (Zer0 Books, 2014), 8.

48  As quoted in Ignacio M. Sánchez Prado, "Reading Benjamin in Mexico: Bolívar Echeverría and the Tasks of Latin American Philosophy," *Discourse* 32, no. 1 (2010), 47; see also Farhana Sultana "The unbearable heaviness of climate coloniality," *Political Geography* 99 (2022): "Overexploitation of human and more-than-human systems is long-standing. The crumbling infrastructure, inadequate social safety nets, lack of access to clean water and sanitation, the racialized classed bodies laboring the land or waiting for water, breathing air pollution and choking, feeling the heat and toiling in it, facing housing and educational disparities, are affective and embodied experiences of communities and individuals. These are complex forms of abjection, precarity, uncertainty, exhaustion, trauma, stress among those deemed disposable."

49  World Economic Forum Global Risk Report 2023, 9; general

map on page 10 and health and environment map on page 32.
While the empirical data is strong, the risk analysis employed by
the WEF is drawn from the same resilience literatures discussed
above.

50 See Catalyst report, "Remote-Work Options Can Boost
Productivity and Curb Burnout": https://www.catalyst.org/
reports/remote-work-burnout-productivity/, accessed 8/29/23.

51 Zambelli, "Geoethics — from an ethics of exhaustion to one of
abundance," *EGU Blogs*, Jun 30, 2020.

52 Ronald Fischer and Diana Boer, "What Is More Important for
National Well-Being: Money or Autonomy? A Meta-Analysis of
Well-Being, Burnout, and Anxiety Across 63 Societies," *Journal
of Personality and Social Psychology* 101, no. 1 (2011): after
study-level effects were controlled, both wealth (Model 2) and
individualism (Model 3) were significantly related to EE when
entered separately.

53 See Greta Wagner, "Exhaustion and Euphoria: Self-Medication
with Amphetamines," in: Sighard Neckel et al (eds.), *Burnout,
Fatigue, Exhaustion* (Palgrave Macmillan, 2017).

54 Vikram Patel et al., "Chronic fatigue in developing countries:
population based survey of women in India," *BMJ* 330 (2005).

55 Ronald Labonté and Ted Schrecker, "Globalization and social
determinants of health: Promoting health equity in global
governance (part 3 of 3)," *Globalization and Health* 3 (2007).

56 Neckel et al, *Burnout, Fatigue, Exhaustion.*

57 James Davies (ed.), *The Sedated Society: The Causes and
Harms of Our Psychiatric Drug Epidemic* (Palgrave Macmillan,
2017); and simultaneously, when controlled by capital, they
are channeled only to some actors and not others. The under-
treatment of African American patients, for example, should be
understood as a kind of social abandonment, a phenomenon we
will return to in examining "surplus populations."

58 While not as pronounced as in the climate sciences, there
is a significant neuroscientific and psychiatric literature
developing — outside of social theoretical and psychoanalytic
analyses — specifically locating both the proliferation of these

health trends and the complex questions of how they should be treated explicitly as a matter of the contemporary social and ecological conditions of capitalism. See, for example, Danae Kokorikou et al., "Testing hypotheses about the harm that capitalism causes to the mind and brain: a theoretical framework for neuroscience research," *Frontiers in Sociology* 8 (2023), or the medical contributions to the collection Davies (ed.), *The Sedated Society*.

59  For example, Brazil is technically part of the contested "treatment gap" but actually is officially prescribed benzodiazepines at one of the highest rates in the world (similar to the US) and is likely the second highest consumer of stimulants like methylphenidate (Ritalin) and amphetamines; see Francisco Ortega and Manuela Rodrigues Müller, "Global Mental Health and Pharmacology: The Case of Attention Deficit and Hyperactivity Disorders in Brazil," *Frontiers in Sociology* 5 (2020). India similarly is often discussed as having an alarmingly low availability of antidepressants (regardless of efficacy), but medical researchers on the ground point out that SSRIs and SNRIs are readily and cheaply procured throughout the country, although without total social care (and official record keeping); see Kalman Applbaum, "Solving global mental health as a delivery problem: toward a critical epistemology of the solution," in *Re-Visioning Psychiatry* (Cambridge University Press, 2017), here 556.

60  Michelle Addison et al., "Exploring pathways into and out of amphetamine type stimulant use at critical turning points: a qualitative interview study," *Health Sociology Review* 30, no. 2 (2021).

61  Ibid.

62  See Applbaum, "Solving global mental health…," 550.

63  See Ruth Brauer et al, "Global psychotropic medicine consumption in 65 countries and regions from 2008 to 2019," *The Lancet Psychiatry* 8, no. 12 (2021) and WHO reports on global mental health.

64  Indeed there are cases — as in the Indian state of Kerala, a case

we will return to — where psychosocial, locally embedded, and pharmacological approaches are all embraced as part of the widely noted Communist Party of India (Marxist) social development program, which emphasizes both top-down resource distribution and free or low-cost access to the largest number of per capita clinics and psychiatrists of any Indian state (mirroring the equally prolific primary care network), as well as horizontal volunteer door-to-door outreach, counseling, and village-based support. This kind of organization is part of the partial success of the famous "Kerala model," which is largely ignored by the Western left and critiqued by both Western and Indian Hindutva elites as aggressively at odds with "traditional" practices. That Keralan Communism — precarious at best within the context of Indian national neofascism — is extraordinarily popular and successful seems uninteresting to such commentators.

65  Simar Singh Bajaj, Lwando Maki, and Fatima Cody Stanford, "Vaccine apartheid: global cooperation and equity," *The Lancet* 399, no. 10334 (2022); See also Giorgo Agamben's paranoid biopolitical rants about public health for a good illustration of the real limits of biopolitical critique. In fact, the ongoing pandemic (and the high probability of greater frequency and intensity of such zoonotic pandemics as climate change worsens) has prompted many to embrace programs of a left, care-based, and social biopolitics. Again, it is remarkable that some critiques of psychiatric regimes return to preaching resilience in the precise terms described above. Meanwhile, the combination of lean production networks and poor regulatory governance has left some easily 20–40 *million* Americans without access to their rather (an)hedonic productivity stimulants (Dylan Scott, "The ongoing, unnecessary Adderall shortage, explained," *Vox*, Apr 10, 2023; the CDC only counts prescriptions, not patients, so the total of 41.4 million is probably the upper limit. But that is also only for Adderall).

66  See, for example, Fanon's chapter "The Black Man and Psychopathology" in *Black Skin, White Masks* in which he

discusses Lacan (and other Lacanian analysts) on the "mirror stage" and gender. At the same time, he applies a Marxist critique to Lacan's theory for lacking any sense of historical and economic environment. (A critique I have always found germane.) He also pillories Jung for the similarly ahistorical and supposedly universal theory of archetypes. He also critiques Sartre for his misunderstanding all racism as similar; Fanon points out (echoing, interestingly, arguments from Adorno) that anti-Semitism and anti-Black racism are not constitutively, functionally, or structurally similar. Fanon's existentialism is far closer to Beauvoir's — a fact that perhaps even Fanon did not realize. Unfortunately, the relatively recent *Alienation and Freedom*, a collection largely of Fanon's psychiatric works as well as other documents, letters, etc., while containing material of extraordinary value, has been heavily distorted by Robert Young to downplay Fanon's politics and Marxism in particular. Young insists Fanon read Sartre's preface, but letters in the volume show Fanon clearly died before having a chance to read it (even worse than Beauvoir's account). The collection includes letters about Fanon's Italian reception and publication which was entirely within left-Marxist reception (read alongside Benjamin and Gramsci in fact). This goes unmentioned in the commentaries. Young writes a chronology that doesn't match the actual documents in the same volume. Co-editor Jean Khalifa compiles a list of Fanon's library but, bizarrely, creates a separate list of Marxist writings on the grounds that she (or Young) doesn't believe Fanon read them. Unsurprisingly, Fanon's vast library is dominated by psychiatric books and journals, but far and away the largest number of texts are by Mao. Marginalia clearly critical of Sartre are carefully explained to be salutary to the reader. All of this fits within the now fashionable academic tendency to read Fanon entirely out of his political and historical context. As I've argued in earlier work, it is essentially impossible to overstate just how much anti-colonial thought was written within an overall milieu which took some form of Marxism for granted. With Fanon, one needn't go past the plain text itself.

67  Indeed, while many analyses of Fanon are either in a mode of demonization (for example, Arendt's critique) or domestication (where Fanon's radicality is subordinated to an idealist moral universe of largely conceptual "decolonization"), he has also had constructive critics. From a Marxist perspective, the Vietnamese thinker Nguyen Nghe highlights Fanon's power in understanding the mobilization of affect even as he finds shortcomings in his existentialism and in places where his discussion of violence and armed struggle seems to become more than strategic and passing. (See further discussion in Hussein Bulhan's *Fanon and the Psychology of Oppression*.) The notes on violence might be overstated, as Fanon himself is far more circumspect than is often acknowledged. Fanon's violence is fundamentally — before any question of recognition — *realist*, as I discuss here.

68  Fanon, *The Wretched of the Earth*, 40.

69  Ibid., 19.

70  Frantz Fanon, *Toward the African Revolution* (Grove Press, 1994), 6.

71  Ranajit Guha (ed.), *Subaltern Studies, Vol. 1* (Oxford University Press, 1982).

72  Fanon, *The Wretched of the Earth*, 5.

73  See Laura D. Kubzansky et al, "Affective States in Health," in: *Social Epidemiology*, 2nd edition (Oxford University Press, 2014).

74  Teresa Brennan, *The Transmission of Affect* (Cornell University Press, 2004), 1.

75  Ibid., 3.

76  Fanon, *The Wretched of the Earth*, 46.

77  Ibid., 40.

78  The legal scholar Nica Siegel has also traced an account of Fanon, his therapeutics, and exhaustion in "Fanon's Clinic: Revolutionary Therapeutics and the Politics of Exhaustion" (*Polity* 55, no. 1 (2023)). I am much in agreement with her arguments about not trying to periodize or isolate Fanon's clinical work from his political work and on the fascinating theme of exhaustion that runs through all of Fanon's work.

However, I think she significantly downplays Fanon's political realism, his heterodox Marxism, and other radical elements.

79 Stuart Hall, *The Hard Road to Renewal: Thatcherism and the Crisis of the Left* (Verso, 1988).

80 Fanon, *The Wretched of the Earth*, 86.

81 Kathi Weeks, *The Problem with Work* (Duke University Press, 2011), 19–20 and 236, Footnote 11.

82 Kathi Weeks, *Constituting Feminist Subjects* (Verso, 2018), 7.

83 Marx, "The Holy Family," in Robert Tucker (ed.), *The Marx-Engels Reader* (Norton, 1978), 134; Ato Sekyi-Otu, *Fanon's Dialectice of Experience* (Harvard University Press, 1997), 98.

84 Karataşli et al., "A New Global Tide…".

85 Fanon, *The Wretched of the Earth*, 159.

86 Ibid., 54–55.

87 Ibid., 21.

88 Robinson, *Black Marxism*, 310.

89 See Davis, "Millenarian Revolutions," in *Late Victorian Holocausts*.

90 Marx, *Capital Vol. I*, 548.

91 Saito, *Karl Marx's Ecosocialism* (Monthly Review Press, 2017), 122.

92 Fanon, *The Wretched of the Earth*, 2; "Summary for Policymakers of IPCC Special Report on Global Warming of 1.5°C"

93 One of the events in non-European politics that caused Marx to reconsider his early stagism and rather silly catch-all "Asiatic Mode of Production".

94 See for example similar discussions of COVID and China in Lin Chun and Riofrancos; Kate Aronoff et al, "A Reparative Politics for the Climate Crisis: A Roundtable," *Dissent*, Spring 2021.

95 See Karataşli, "The twenty-first century revolutions…"; see also Jodi Dean, *Crowds and Party* (Verso, 2016).

96 Gramsci, *Prison Notebooks*, 148–149.

97 See Ruth Wilson Gilmore, *Abolition Geography* (Verso, 2022).

98 Cédric Durand, "Zero-Sum Game," *NLR Sidecar*, Nov 17, 2021.

99 Martin Wolf, "The UK's future depends on improving economic performance," *Financial Times*, Apr 16, 2023.

100 Matthieu Bellon and Emanuele Massett, "Economic Principles for Integrating Adaptation to Climate Change into Fiscal Policy," *IMF Staff Climate Notes* (2022), 3.

101 See Tweet at https://twitter.com/kmac/status/ 1647262746747236352, accessed 8/29/23.

102 This was first discussed to my knowledge by the Open Woods Collective, including Bosworth and my BISR colleague Sophie Lewis (see Open Woods Collective "Après moi le déluge! Fossil fuel abolitionism and the carbon bubble – part 2," libcom.org, Apr 21, 2014; see also Chris Hayes in *The Nation* and Matt Karp for *Jacobin*, among others.

103 See Piketty, *Capital in the 21ˢᵗ Century*; see also Henwood on Piketty:"The Top of the World," *Bookforum*, Apr/May 2014 I have calculated this as well through the Madison Project Database, with some difficulty, as one must reach back into the older reports and data that are found in legacy files and sites. And other obvious cases (like the Russian Revolution) are clearly lower, confirming the intuition in Piketty, and the comparison in these analyses.

104 Gramsci, *Prison Notebooks*, 229.

105 See, for example, Richard McAlexander's extraordinary dissertation "The Politics of Anticolonial Resistance: Violence, Nonviolence, and the Erosion of Empire" (Columbia University) which confirms in decidedly non-radical, statistical analysis what qualitative history has noted for years: in the Indian context and beyond violent strategies and spontaneous violence are far more effective than nonviolence. Even research working *within* the hegemonic peace studies framework records that Gandhian nonviolence was largely a failure when it was working on its own (a few years in the early 30s); see also Rikhil Bhavnani and Saumitra Jha, "Gandhi's Gift: Lessons for Peaceful reform from India's struggle for democracy," *The Economics of Peace and Security* 9, no. 1 (2014). Remarkably, the authors still try to hew to the doxa that even the most straightforward textbook reading of Indian independence (see Metcalf and Metcalf, *A Concise History of Modern India* (Cambridge University Press,

2012)) relates. They use Chenoweth criteria to manage to define the mass violence eruption after the Quit India campaign as a "failure." Almost everyone else admits that this, as much as World War II, was the final nail in the Raj's coffin.

106 Fanon, *The Wretched of the Earth*, 51.

107 Nikhil Pal Singh, "America's crisis-industrial complex," *New Statesman*, Jun 30, 2022.

108 (In the terms of mainstream political science. From a Marxist, radical democratic, or even some egalitarian *liberal* standards, the US is simply not a democracy.)

109 Raymond Geuss, *Changing the Subject: Philosophy from Socrates to Adorno* (Harvard University Press, 2017), 153.

110 Raymond Geuss and Quentin Skinner, "Introduction," *Marx: Later Political Writings* (Cambridge University Press).

111 Singh, "America's Crisis-Industrial Complex," *New Statesman*, Jun 30, 2022.

112 Raymond Geuss, *Reality and Its Dreams* (Harvard University Press, 2016).

113 My translation. Geuss provides the original German, but I have not yet been able to find the whole original poem: "*Wem angesichts der Vergiftung unserer Erde nichts einfällt als die Frage nach dem Bruttosozialprodukt Dem habe ich nichts zu sagen.*"

114 Clifford Krauss, "Evo Morales of Bolivia Accepts Asylum in Mexico," *New York Times,* Nov 11, 2019.

115 Ajl, *A People's Green New Deal*

116 Anria, *When Movements Become Parties* (Cambridge University Press, 2018), 9.

117 Ibid.

118 Carwil Bjork-James, "Mass Protest and State Repression in Bolivian Political Culture: Putting the Gas War and the 2019 Crisis in Perspective," Research Working Paper, Human Rights Program, Harvard Law School (2020), 25.

119 Ibid., 33–35.

120 International Criminal Court, "Situation in the Plurinational State of Bolivia Final Report" (2022).

121 Ibid., 57 and 63.

122 Edwin F. Ackerman, *Origins of the Mass Party: Dispossession and the Party-Form in Mexico and Bolivia in Comparative Perspective* (Oxford University Press, 2021), conclusion.

123 Ibid., 146; see also Anria, *When Movements*, 14.

124 See Battistoni, Riofrancos, Ansel.

125 See Chaudhary, "Toward a Critical 'State Theory' for the Twenty-First Century," in *The Future of the State: Philosophy and Politics*, (Rowman Littlefield, 2020).

126 Mark Fisher, *k-punk: The Collected and Unpublished Writings of Mark Fisher (2004–2016)*, ed. Darren Ambrose (London: Repeater Books, 2018), 475; italics in original.

127 See Gilmore, *Abolition Geography*.

128 See Isaac Ankrah et al., "Is energy transition possible for oil-producing nations? Probing the case of a developing economy," *Cleaner Production Letters* 4, no. 3 (2023); Reza Hafezi et al., "Iran's approach to energy policy towards 2040: a participatory scenario method," *Foresight* 25, no. 5 (2023).

129 See Olúfemi O. Táíwò in Aronoff et al., "A Reparative Politics…".

130 Many Western economic sources now try to refute the carbon leakage or carbon haven hypothesis. (See, for example, Hannah Ritchie's claim that despite the US and interestingly Europe in particular being glaringly obvious net *importers* of emissions, and even upper-middle-income states like China being clear net *exporters*, "China is no longer a large emitter because it produces goods for the rest of the world" (Ritchie, "How do $CO_2$ emission compare when we adjust for trade?" *Our World in Data*, Oct 7, 2019)). While this doesn't even match her own massaged statistics, she calculates this on a global average (where most countries are profoundly poor) and even then distorts China's consumption emissions (28 mwh per capita) as roughly like Western European ones (Germany, 59 mwh; France, 54 mwh) when in fact they are below even relatively efficient if non-wealthy states like Spain (34 mwh) and Portugal (30 mwh) (see Ritchie, "How much energy do countries consume when we take offshoring into account?" *Our World in Data*, Dec 7, 2021). The US of course stands on top with a mind boggling

97 mwh. Furthermore, this sleight of hand, which I've seen throw off even the most politically engaged climate scientists, doesn't even line up with their own data on energy embodied in trade. Ritchie makes the astounding claim that in states like Germany essentially absolute decoupling is occurring. When accounting for energy embodied in trade we see Germany's imported emissions *increasing* as of 2020 to 617 terawatt hours and Chinese exported emissions vastly *higher* than ten years ago, at -2,221 terawatt hours. These glaring discrepancies are unexplained. But other researchers studying energy embodied in trade, have explained this carefully. While lower than the past, 15% of Chinese emissions are still exported. (Lixia Guo and Xiaoming Ma,"Research on global carbon emission flow and unequal environmental exchanges among regions," *IOP Conf Series: Earth and Environmental Science* 781 (2021), 4. Furthermore, when economic value is accounted, Europe in particular, again the illusion of the Environmental Kuznets Curve, is an astounding beneficiary: "In the 6 regions, EU 28 greatly promoted this inequality in international trade. In the process of trading with other 5 regions, EU 28 not only transferred carbon emissions, but also did not make any economic compensation for the pollution outsourcing. Instead, it gained value added inflows from other regions. In contrast, China was the region that suffered the most in global trade. In the trade with the other 5 regions, China was always the region receiving carbon emissions, but hardly obtained the equivalent inflow of value added." Europe grabs double the value from all trade with China and almost *triple* from the rest of the world without China. Meanwhile, a different team notes that energy embodied in trade in Belt and Road Initiative countries has increased almost 50%, even while energy intensity has decreased to varying but converging degrees across all countries. In other words, China is now exporting some of its emissions to BRI countries, along with efficient technologies, largely still supplying the voracious appetites of the overdeveloped wealthy (see Dimitrio Pappas et al., "Energy and carbon intensity: A

study on the cross-country industrial shift from China to India and SE Asia," *Applied Energy* 225 (2018)). Meanwhile, upward consumption emissions in low- and middle-income BRI countries including China is simply a measure of development from poverty to moderate income, or from moderate to still well-below "sufficiency" income (even disregarding the *wealth* accumulated in places like the US and EU). Despite agreeing on many points, one of the reasons I don't self-ascribe to degrowth is that this kind of reality of international political economy and genuine geopolitics is rarely treated with serious care.

131 While the subfield of International Relations is fraught to say the least (having long served as specifically constructed around US security interests), a host of IR analyses, even within these bounds, have observed the Chinese pivot as a response to — surprising apparently even to elite Chinese actors — the American shift from one of cooperation and engagement with the PRC to one in which China is portrayed in terms of an existential national security threat. (See for a range of examples from largely Western perspectives: Catalin Badea "U.S.-China Relations through the Perspective of Social-Constructivism," *Journal Studia Europaea* 51, no. 2 (2006); Ryan Hass "How China is responding to escalating strategic competition with the US," *Brookings*, Mar 1, 2021; and from a more Chinese perspective Zhaohui Wang "Understanding the Belt and Road Initiative from the Relational Perspective," *Chinese Journal of International Review* 3, no. 1 (2021). See also Ernesto Gallo — again from a Eurocentric perspective — on the absurdity of China as a military threat and again demonstrating Chinese policy as largely constructivist or relational (to use Wang's fourth category) to US and European belligerence: "Is China's Belt and Road Initiative Really What Interpreters Make of it?" *OXPOL Blog*, Aug 30, 2019, and Matthew Anzarouth, "China's Belt and Road Initiative: A Bumpy, Promising Path," *Harvard Political Review*, Jul 28, 2022: https://web.archive.org/web/20230410153948/https://harvardpolitics.com/belt-and-road/, accessed 97/23.

132 See Xingwei Li et al., "How does the Belt and Road policy affect the level of green development? A quasi-natural experimental study considering the $CO_2$ emission intensity of construction enterprises," *Humanities and Social Sciences Communications* 9 (2022); Ziyi Ma and Yu Ma, "What's After Coal? Accelerating China's Overseas Investment in Renewables," *World Resources Institute*, Jan 31, 2023; Frontier Services Group News, "BRI a win-win plan, not neocolonial tool," Sep 7, 2021; Amitai Etzioni, "Is China a New Colonial Power?" *The Diplomat*, Nov 9, 2020.

133 See Minqi Li, "China: Imperialism or Semi-Peripher?" *Monthly Review*, Jul 1, 2021. From a Marxist perspective and on China's superior carbon efficiency up to at least 2017, see Melanie Hart et al., "Everything You Think You Know About Coal in China Is Wrong," *Center for American Progress* report, May 15, 2017.

134 See Ajit Singh, "The myth of 'debt-trap diplomacy' and realities of Chinese development finance," *Third World Quarterly* 42, no. 2 (2021); Kevin Acker et al., "Debt Relief with Chinese Characteristics," Working Paper no. 2020/39, China Africa Research Initiative, School of Advanced International Studies, Johns Hopkins University (2020).

135 See again Minqi Li in *Monthly Review*: "if China were to become a core country in the capitalist world system, the existing core countries would have to give up most of the surplus value they are currently extracting from the periphery. It is inconceivable that the core countries would remain economically and politically stable under such a development." Azanrouth writes, in US-centered IR "realist" terms: "Furthermore, in its pursuit of global geopolitical power, China has an incentive to forge strong international relationships and demonstrate a commitment to fair trade and multilateralism [...] to preserve this pathway to global dominance, China has every incentive to avoid being seen as a malicious and predatory outsider by the international community."

136 Michael R. Davidson et al., "Risks of decoupling from China on low-carbon technologies," *Science* 377, no. 6612 (2022); Many Western analysts also simply do not understand the

ways in which China's internal political-economy allows large capital write-offs — like power plants — or simple restriction on marketable goods, in ways inconceivable in purely capitalist countries. If you read many Chinese policy documents this kind of planning is suggested (for example, changing internal consumption patterns through altering what is on offer to be in line with the PRC's climate agenda instead of the nudges of consumption taxes and subsidies).

137 Eyal Weizman, *Forensic Architecture* (Zone Books, 2017), 228–229; As Weizman observes incredulously even some of the most radical environmentalists mistakenly accept the logic of collateral damage which "makes a convenient assumption under which we are all perpetrators of climate change as well as its victims," 254.

138 Pete Volk, "*How to Blow Up a Pipeline*: A perfect blend of radical politics and heist-movie thrills," *Polygon*, Apr 7, 2023; see also Rotten Tomatoes reviews: https://www.rottentomatoes.com/m/how_to_blow_up_a_pipeline/reviews?intcmp=rt-scorecard_tomatometer-reviews, accessed 8/30/23.

139 Jana Winter, "Law-Enforcement Agencies Have Sent 35 Warnings About This Movie," *Rolling Stone*, Apr 21, 2023 (A running gag — both charming and telling in the film — is people reading the book in the background of several shots.)

140 Malm himself seems torn between a logic of universalism and a more radical embrace. He cannot shake the ideal of working class unity — and even cites Huber's arguments that so well fit his model of collective climate pathology. On the one hand, "I can't accept the idea that the working class is part of the enemy — not even coal workers — but on the other hand, I don't really believe in the idea that the organised labour movement will be the main driver of the climate front" (see "Andreas Malm: 'Total, BP, and Shell Will Not Voluntarily Give Up Their Profits. We Must Become Stronger Than Them…,'" *Verso Blog*, Nov 8, 2022); in more recent writings, examined in the last chapter, Malm allusively suggests a "praxis of the Children of Kali" — the fictional terrorist organization in Kim Stanley Robinson's

*The Ministry for the Future* — only to reiterate his tactical
moderation.

141 Sadly, I do not have space here to dive into the voluminous
literature on violence and nonviolence in social movements and
within environmental "politics." But Malm's critique for example
of the "scholarship" of Erica Chenoweth and Maria Stephan
is again if anything too lenient. Their recent "maximalist" set
of conflicts between 1900–2019 (available from the Harvard
Dataverse dataset: https://dataverse.harvard.edu/dataset.
xhtml?persistentId=doi:10.7910/DVN/ON9XND, accessed
8/30/23) is a classic case of nearly every methodological error
possible. Case inclusion and exclusion appears at best random
and, more likely, grounded in the bias of Security Studies. For
example, Indian Independence is excluded in the first data set
and then listed as only partial success in the second. It is also
listed as "nonviolent," when this is historically not true. Almost
no anti-US cases are cited. And no cases on US soil are cited.
The US Civil Rights movement even is noticeably absent as
are labor conflicts across the board. The ANC struggle against
apartheid is categorized as "non-violent"; entirely untrue. Their
wobbly definition of violence includes sabotage and property
destruction but they wish it away when desired. To look to the
Bolivia case, the anti-MAS coup is categorized as *nonviolent*
(an absolute distortion) and successful; the anti-fascist struggle
is categorized again as nonviolent and *unsuccessful*. The IRA
is categorized as only partially successful. Color revolutions
are omnipresent, major ideological and polarized conflicts
are almost categorically excluded except for cases so obvious
(Bolsheviks, the Iranian Revolution, the Algerian Civil War)
as to be impossible to ignore. Even so, all the anti-colonial
movements (except the mischaracterized Indian and South
Africa cases) are successful and violent (this is largely ignored).
There are five entries for Syria but none for the successful (and
violent) establishment of Kurdish autonomous regions. Thus,
the set is for largely "universal" movements with the central
operating principle, again likely inherited from the Security

Studies background, of state stability. A thin quantitative veneer is applied, lending an air of "scientific" rigor, akin to Adorno's notes on how modern astrology will attempt to legitimate itself through adopting the appearance of science. It is not simply that these studies have led a wide swath of analysts astray; their ubiquitous approval is at least partially the function of confirmation bias among movement scholars who already *want* the findings to be true. Chenoweth and Stephan's work counts among the crowning achievements of American academic cooptation and, for lack of a better word, sabotage of radical movements around the world.

142 Maria Elena Martinez-Torres and Peter M. Rosset, "La Vía Campesina: The Evolution of a Transnational Movement," *Global Policy Forum*, Feb 8, 2010.

143 EJAtlas, "Women against 'green deserts' (eucalyptus monoculture), R.G. do Sul, Brazil,": https://ejatlas.org/conflict/women-agaist-the-expansion-of-eucalyptus-monoculture, accessed 8/30/23.

144 See Asbjorn Osland and Joyce S. Osland, "Aracruz Celulose: best practices icon but still at risk," *International Journal of Manpower* 28, no. 5 (2007).

145 See again EJAtlas: https://ejatlas.org/conflict/women-agaist-the-expansion-of-eucalyptus-monoculture

146 EJOLT, "Algeria cancels fracking plans": http://www.ejolt.org/home/a-success-story/, accessed 8/30/23.

147 See EJAtlas, "Anti-Fracking Uprising in Ain Salah, Algeria": https://ejatlas.org/conflict/anti-fracking-uprising-in-ain-salah, accessed 8/30/23; Pierre Longeray, "Violence Flares Over Halliburton's Fracking Tests in Algeria," *Vice*, Mar 5, 2015

148 I would hazard that the asymmetry between the violence of corporate and state forces and that of protestors is part of why the language is vague; alongside the unconscious ideological image of the spiritual, nonviolent, non-Westerner, particularly when the research is taken up in analysis.

149 Francisco Venes et al., "Not victims, but fighters: A global overview on women's leadership in anti-mining struggles,"

*Journal of Political Ecology 30*, no. 1 (2023); the 30% represents 21 out of the 62 qualifying cases, located through the EJAtlas. This is just one among many studies that break out of the prevailing academic prejudice. See, for example, Hamza Hamouchene, "Extractivism and Resistance in North Africa," *Transnational Institute* (2019).

150 See Durand, "Zero-Sum Game."

151 Stefan Gössling and Andreas Humpe, "Millionaire spending incompatible with 1.5 °C ambitions," *Cleaner Production Letters* 4, no. 1 (2022). The calculation of high-end consumption excludes what could be considered the more determinate questions of investment and production, possibly the only consumption level at which behavioral changes actually have serious significant impact. These rates of consumption hold true even accounting for carbon emissions alone, as the authors note, and excluding material footprint or even other GHG emissions.

152 See, for example, Wibecke Brun, "Cognitive components in risk perception: Natural versus manmade risks," *Journal of Behavioral Decision Making* 5, no. 2 (1992); Michael Siegrist and Bernadette Sütterlin, "Human and nature-caused hazards: the affect heuristic causes biased decisions," *Risk Analysis* 34, no. 8 (2014); Gea Hoogendoorn, Bernadette Sütterlin, and Michael Siegrist, "The climate change beliefs fallacy: the influence of climate change beliefs on the perceived consequences of climate change," *Journal of Risk Research* 23, no. 12 (2020), among many.

153 See myriad negative dimension discussions in in Arjen Boin (ed.), *Crisis Management* (Sage, 2008); there is also a growing recognition of this problem in interdisciplinary climate science literatures. See, for example, Emmanuel Raju, Emily Boyd, Friederike Otto, "Stop blaming the climate for disasters," *Communications Earth & Environment* 3 (2022); or Ksenia Chmutina et al., "Language Matters: Dangers of the 'Natural Disaster' Misnomer," Contributing Paper to Global Assessment Report on Disaster Risk Reduction 2019.

154 Bonnie L. Green, "Psychological responses to disasters: Conceptualization and identification of high-risk survivors," *Psychiatry and Clinical Neurosciences* 52, no. S1 (1998).

155 Siegrist and Sütterlin, "Human and nature-caused hazards…".
156 Hoogendoorn et al., "The climate change beliefs fallacy…".
157 Shaul Kimhi et al., "Resilience protective and risk factors as prospective predictors of depression and anxiety symptoms following intensive terror attacks in Israel," *Personality and Individual Difference* 159 (2020).
158 George A. Bonnano et al., "Weighing the Costs of Disaster: Consequences, Risks, and Resilience in Individuals, Families, and Communities," *Psychological Science in the Public Interest* 11, no. 1 (2010).
159 Ibid.
160 Ibid. See also for example Bena Labadee and Eleanor Bennett's chapter "Recognizing Normal Psychological Reactions to Disasters" in the handbook *Mental Health and Psychosocial Support in Disaster Situations in the Caribbean* (Pan American Health Organization, 2012): "If a disaster is caused by human actions, survivors tend to struggle with deliberate human-on-human violence or human error as causal agents. Recovery is hampered by blame and anger, evoked by the perception that the event was preventable and a sense of betrayal by a fellow human(s)."
161 Steven Breakall and Eric Thornton, "How police leaders can set officers on a pathway to resilience," *Police1*, Jan 3, 2023.
162 On "common enemy" theories see, for example, John Drury, "Recent developments in the psychology of crowds and collective behaviour," *Current Opinion in Psychology* 35 (2020), or Donatella della Porta, "Radicalization: A Relational Perspective," *Annual Review of Political Science* 21 (2018) and "Capitalism, Class, Contention," *Global Dialogue 10.1*, Feb 21, 2020; on the problem with abstract "enemies," see examples like "The fusing power of natural disasters: An experimental study": "negative experiences that are not caused by a defined enemy, such as natural disasters, should be less fusing than negative experiences that pit groups against each other."
163 Sebastian Bamberg et al., "Environmental protection through societal change: What psychology knows about collective

*Climate Action* — and what it needs to find out," in *Psychology and Climate Change* (Academic Press, 2018); it should be noted that, despite this error, Bamberg et al. are not focusing on crisis or risk management but instead on genuine social and political action. However, their analysis is severely impeded by this fundamental causal error. The number of studies, handbooks, and reports that argue for collective responsibility (and almost always individual, market-based behavioral action) is simply too vast to cite them here.

164 When Berlant writes of "crisis-shaped subjectivity," they have more support in both the managerial and critical social-psychological literatures than probably imagined. Jodi Dean's connection of Elias Canetti and party formation finds its counterpart in crisis management discourse *fearing* those exact dynamics; "incandescent" passions is from Gramsci.

165 W.E.B. Du Bois, *Black Reconstruction.*

166 See Nick Estes, *Our History is the Future* (Verso, 2019).

167 Illan rua Wall, "'No Justice, No Peace': Black Radicalism and the Atmospheres of the Internal Colony," *Theory, Culture & Society* (2022).

168 Fanon, *The Wretched of the Earth.*

169 Maggie FitzGerald, "Violence and Care: Fanon and the Ethics of Care on Harm, Trauma, and Repair," *Philosophies* 7, no. 3 (2022).

170 My colleagues Anthony Allesandrini and Amrita Ghosh address this issue well in a discussion for *Inverse Journal*, "On Frantz Fanon, Postcolonial and Middle Eastern Studies, and Palestine and Kashmir — Anthony Alessandrini in Conversation with Amrita Ghosh," May 14, 2021.

171 *Beauvoir, Fanon, and the Existential Ethics of Liberation: An Anticolonial Inheritance for New Revolutions*; Dhanvanatri's dissertation is an extraordinary accomplishment of both textual analysis, philosophical rigor, and innovation. However, I think she downplays their shared Marxist frameworks and overplays Beauvoir's existentialism. Beauvoir's and Fanon's existentialism furthermore is grounded in *material* and *structural* constraints as opposed to transhistorical claims about human ontology.

Scholars like José Esteban Muñoz and my colleague Adriana Gariga Lopez have explored similar questions through the lens of the intertwining of care and activism in queer struggles.

172 Mike Davis, *Old Gods New Enigmas.*

173 Dean, "The Question of Organization," 831.

174 https://www.versobooks.com/blogs/news/4773-angela-davis-the-political-prisoner

175 Feng, "Power beyond powerlessness: Miners, activists, and bridging difference in the Appalachian coalfields" (2020).

176 https://grist.org/protest/what-happened-to-the-war-on-coal-in-west-virginia/

177 See Battistoni, *Sustaining Life on this Planet* (2020); on the work of reproduction; interestingly, in the lost encounter between Beauvoir and Fanon this distinction is lost. After all, Fanon's *primary* occupation for the FLN was as a doctor — a caregiver; his political theorizing and representation were secondary. Beauvoir, meanwhile, known for her disdain for the restriction of women to domesticity — echoing Marxist feminists of the early twentieth century — recognized that "there is often, in women, a kind of caring for others that is inculcated in them by education, and which should be eliminated when it takes the form of slavery. But caring about others, the ability to give to others, to give of your time, your intelligence — this is something women should keep, and something that men should learn to acquire." Quoted in "Care Ethics and Paternalism: A Beauvoirian Approach."

178 https://www.ualrpublicradio.org/npr-news/npr-news/2022-10-29/how-big-coal-companies-avoid-cleaning-up-their-messes

179 Jeng.

180 https://www.ft.com/content/7aa77038-62b6-4761-9ce3-1e0647ffb607; see also Wright and Nyberg.

181 Jason Moore has made this point well.

182 Marx, *Capital* vol. I, 381. Emphasis added.

183 In the World Economic Forum's 2023 Risk Report, a direct line is drawn between "hotspots" of "conflict and violence" and

"shifts" in "biodiversity patterns." Again, the *crisis management* literature is often far more attuned to these questions than left political theory has been.

184 Smale et al., "Marine heatwaves threaten global biodiversity and the provision of ecosystem services" (2019); Xu et al., "An increase in marine heatwaves without significant changes in surface ocean temperature variability" (2022); Amaya et al., "Bottom marine heatwaves along the continental shelves of North America" (2023).

185 See refs in Chapter 1.

186 https://www.latimes.com/lifestyle/image/story/2022-07-25/ mike-davis-reflects-on-life-activism-climate-change-bernie-sanders-aoc-los-angeles-politics?s=03

187 https://newleftreview.org/issues/ii126/articles/mike-davis-trench-warfare

188 https://newleftreview.org/sidecar/posts/thanatos-triumphant

189 Laleh Khalili, "Pacifying Urban Insurrections," *Historical Materialism* 25, no. 2 (August 3, 2017).

190 Khalili, "Thinking about Violence," 2014.

191 Fanon, *The Wretched of the Earth*, 183. (Fanon places his clinical writings in Soviet cortico-visceral languages as set against the anti-historical idealism of mainstream psychology.)

192 A logic discussed by my colleagues Suzanne Schneider and Patrick Blanchfield.

193 Neumann, *The Authoritarian and the Democratic State*, 264; in this he is following Marx's analysis in "The Eighteenth Brumaire of Louis Napoleon" that saw the left in power focusing on social experimentation and ignoring the organizing of coercive power.

194 Dean, *Communist Horizon*, 60.

195 Ngai, *Ugly Feelings*, 339.

196 Gramsci, *Prison Notebooks*, 232.

197 Myisha Cherry, *The Case for Rage: Why Anger Is Essential to Anti-Racist Struggle* (Oxford University Press, 2021).

198 I think the synthesis Cherry and Delmas suggest is certainly possible, but they underestimate the pull of idealist moralism à la Rawls into depoliticization, as Geuss argues. For those who *need* the moral case, it can be made.

199 Khalili, 2014.

200 https://salvage.zone/communism-the-manifesto-and-hate/

201 https://www.*Dissent*magazine.org/article/hello-to-my-haters-tucker-carlsons-mob-and-me

202 https://www.moviemaker.com/how-to-blow-up-a-pipeline-eco-vigilantism/

203 https://www.polygon.com/reviews/23672727/how-to-blow-up-a-pipeline-movie-review

204 https://collider.com/how-to-blow-up-a-pipeline-review-daniel-goldhaber/

205 https://www.polygon.com/reviews/23672727/how-to-blow-up-a-pipeline-movie-review

206 Walter Benjamin, "Paralipomena to 'On the Concept of History'" in *Walter Benjamin: Selected Writings, 4: 1938–1940*, ed. Howard Eiland and Michael W. Jennings (New York: Schocken Books, 2007).

207 https://www.sefaria.org/Psalms.139.22

208 Both in Surah 2, the Cow.

## Chapter 5: The Long Now

1 Maggie Astor, "Hottest April Day Ever Was Probably Monday in Pakistan," *New York Times*, May 4, 2018, https://www.nytimes.com/2018/05/04/world/asia/pakistan-heat-record.html.

2 Ethan D. Coffel et al., "Temperature and humidity-based projections of a rapid rise in global heat stress exposure during the 21st century," *Environmental Research Letters* 13, no. 1 (2018), 1.

3 While work produced as recently as the 2010s generally projected such thresholds to be frequently crossed by the end of the twenty-first century, Colin Raymond et al. in *Science Advances* point out that they were already being exceeded with some regularity by 2020 "in South Asia, the coastal Middle East, and coastal southwest North America." See Raymond et al., "The emergence of heat and humidity too severe for human tolerance," *Science Advances* 16, no. 19 (2020), 1.

4    Thresholds for children, the elderly, and anyone engaged in strenuous activity (such as manual labor or sport) are considerably lower; see Raymond et al., 8.

5    The contemporary use of the term "Anthropocene" was coined by biologist Eugene F. Stoermer in the 1980s and popularized by the Nobel Prize-winning atmospheric chemist Paul J. Crutzen in the early 2000s. Writing together in the *Global Climate Change* newsletter, Stoermer and Crutzen pointed to the long history of scientific theory and measurement of increasingly global anthropogenic transformations, concluding that "considering these and many other major and still growing impacts of human activities on earth and atmosphere, and at all, including, global, scales, it seems to us more appropriate to emphasize the central role of mankind in geology and ecology by proposing to use the term 'Anthropocene' for the current geologic epoch." See Stoermer and Crutzen, "The Anthropocene," *Global Change Newsletter* 41 (2000), 17. In 2016, a team of natural scientists led by Colin Waters published evidence that "novel stratigraphic signatures support the formalization of the Anthropocene at the epoch level" that "renders the Anthropocene stratigraphically distinct from the Holocene and earlier epochs"; see Waters et al., "The Anthropocene is functionally and stratigraphically distinct from the Holocene," *Science* 351, no. 6269 (2016), 138. By May 2019, the Anthropocene Working Group of the International Union of Geological Sciences had presented evidence for final bureaucratic approval of the classification, and 88% of members voted in approval of an "official" recognition of the epoch; see "Working Group on the 'Anthropocene,'" *Subcommission on Quaternary Stratigraphy*, accessed December 17, 2021, http://quaternary.stratigraphy.org/working-groups/anthropocene/. The Anthropocene as a concept has been critiqued, much as I argue here, for an overly universal read of "human agency." In his 2014 PhD dissertation work, published as *Fossil Capital: The Rise of Steam Power and the Roots of Global Warming* (New York: Verso, 2016), Andreas Malm first posed "Capitalocene" as a corrective — a term also taken up slightly differently by

Jason Moore in *Capitalism in the Web of Life: Ecology and the Accumulation of Capital* (New York: Verso, 2015) — positing the understanding that it is not "all people," but rather Capital that has produced the stratigraphic variation. Others, like Anna Tsing, have proposed "Plantationocene," thinking of not only capital but its broader historical and constitutive social relations, particularly in terms of colonialism and racial capitalism; see Tsing et al., "Anthropologists Are Talking — About the Anthropocene," *Ethnos* 81, no. 3 (2016), 556 ff. In *A Billion Black Anthropocenes or None* (Minneapolis: University of Minnesota Press, 2018), Kathryn Yusoff underscores how the visibility of geologic change is differently perceived across heterogenous, particularly racialized, populations. I find all these critiques deeply productive, and my affinity for Anthropocene is purely practical. Elsewhere I have written of the "intuitive critical theories" of the climate sciences, and "Anthropocene" is the term with the most traction in current climate science; see Chaudhary, "Emancipation, Domination, and Critical Theory in the Anthropocene," in *Domination and Emancipation: Remaking Critique*, ed. Daniel Benson (Lanham, MD: Rowman Littlefield, 2021). This is largely a comradely gesture toward natural scientific fellow travelers, most of whom already understand the centrality of Capital and, sometimes, related historical questions. Such a comradely approach, as McKenzie Wark argues in *Molecular Red: Theory for the Anthropocene* (New York: Verso, 2015), is vital for bringing historical materialism, analytically and politically, into the current moment. My use of the term Anthropocene also acknowledges the anti-romantic nature of any climate mitigation and adaptation scenario that could be understood as emancipatory, i.e. a left-wing climate realism. Even in the most positive projection, what any such scenario proposes — no matter how (wisely) detached from techno-mysticism and Prometheanism — still involves humans as a primary geological force in our global ecological niche. I will return in this text to the question of Anthropocene temporality also in terms of the past. Dating the origin of the

Anthropocene requires thinking not simply about stratigraphic layers but questions of causality that cannot avoid political characterization. The existence of anthropogenic climate change is not a matter of dispute, and it requires that we think of something like the "metabolism" between society and nature that Marx proposed. This is as much a political as a natural scientific question.

Interestingly, the term Anthropocene originates in the Soviet Union in 1928, with the geologist Aleksei Petrovich Pavlov and with precursor theorization by Vladimir Verdanskii — the former largely unknown and the latter acknowledged in passing in texts like Cruetzen and Stoermer's; see Alec Brookes and Elena Fratto, "Towards a Russian Literature of the Anthropocene," *Russian Literature* 114–115 (2020), 1–22. The lines of reasoning around the Anthropocene in Soviet thought are in many ways parallel with that of the later eco-Marxist James O'Connor, each pointing towards what O'Connor calls a "second contradiction" in capitalism between political economy and the environment. Both the early Soviets and O'Connor, and important Critical Theorists also thinking a mode of eco-Marxism like Alfred Schmidt, saw this "contradiction" at best becoming a question of, as Schmidt put it, "mastery by the whole of society of society's mastery of nature"; see Schmidt, *The Concept of Nature in Marx*, translated by Ben Fowkes (London: Verso, 2014). Such a situation is not necessarily dissolved in a classless society.

6   My "political-time" is adapted in part from Sheldon Wolin's concept of "political time," although without his romanticism and particular commitments. See Wolin, "What Time Is It?" *Theory & Event* 1, no. 1 (1997), 1. The experience of climate change splinters in a "temporal disjunction" based, in part, on non-universal time horizons of mitigation and adaptation. The political divide relates to a fuller sense of capital as a socioecological system. As Wolin observes, "political time is out of synch with the temporalities, rhythms, and pace governing economy and culture." But this does not have to be a problem

per se. Rather it can be part of a foundation for genuine political conflict. Wolin's concern in "What Time Is It?" is about the instability and difficulty of such temporalities for a "deliberate" liberal democracy, and he is hostile overall to any overarching "synoptic" political theory — precisely the kind of political theory which my analysis of political-time and climate change is a part of. However, Wolin is astute in his observation about how out of synch political-time can be and what challenges such temporalities might pose. The temporal disjuncture between theory, practice, and socioecological reality is not limited to liberalism but implicates many radical and critical modes of politics, including many versions of "orthodox" Marxism. In *Liberal Democracy and the Social Acceleration of Time* (Baltimore: Johns Hopkins University Press, 2004), William Scheurmann, while also hampered by a conventional account of liberal democracy as the millennium of political form, is particularly attentive to the "empire of speed" that is today's global economy, a decent characterization of one aspect of Capital's socioecological present, and how this speed is at odds with many conceptions of politics. Global Value Chains (GVCs) and supply chains, for example, are specifically structured to elude "the rule of law"; for more, see Chaudhary, "Toward a Critical 'State Theory' for the Twenty-First Century," in *The Future of the State: Philosophy and Politics*, ed. Artemy Magun (Lanham, MD: Rowman Littlefield, 2020). I have addressed it in more detail in "Subjectivity, Affect, and Exhaustion: The Political Theology of the Anthropocene," *Political Theology Network*, February 25, 2019, accessed December 18, 2021, https:// politicaltheology.com/subjectivity-affect-and-exhaustion/.

7    See Chaudhary, "Sustaining What?...".

8    Will Steffen et al., "Trajectories of the Earth System in the Anthropocene," *PNAS* 115, no. 33 (2018), 8,252–59. Many climate scientists have grown far bolder in explicit engagement with capitalism; see, for example, McPherson et al., "Large-scale shift in the structure of a kelp forest ecosystem co-occurs with an epizootic and marine heatwave," *Communications Biology*

4, no. 298 (2021), or Steinberger and Pirgameir, "Roots, Riots, and Radical Change — A Road Less Travelled for Ecological Economics," *Sustainability* 11, no. 7 (2019), 2001.

9    Benjamin, "On the Concept of History," in *Walter Benjamin: Selected Writings*, 4: 1938–1940, ed. Howard Eiland and Michael W. Jennings (New York: Schocken Books, 2007), 395.

10   Christopher J. Smith et al., "Current fossil fuel infrastructure does not yet commit us to 1.5 °C warming," *Nature Communication* 10, no. 101 (2019); this aspect of climate change, although long known, is what garnered significant journalistic attention in the IPCC AR6 "physical sciences" report. Although much coverage implied that this meant nothing could be done and everything is "irreversible," this is not the case presented in the report nor in climate science literatures more broadly.

11   Such "organizing," to adapt Jason Moore's language from *Capitalism in the Web of Life* (New York: Verso, 2015), is *not* synonymous with quasi-mystical Promethean exhortations for technological domination as positive program. It is rather the acknowledgment of the reality reflected in the impetus to thinking the Anthropocene in the first place: humans are now the dominant geological force on the planet and will continue to be in any number of possible scenarios.

12   Berlant, *Cruel Optimism* (Durham: Duke University Press, 2011), 195.

13   Fisher, *Ghosts of My Life*; Fisher, *Capitalist Realism*, 2.

14   Dayna Tortorici, "In the Maze," *n+1* 30, (2018).

15   Nixon, *Slow Violence and The Environmentalism of the Poor* (Cambridge, MA: Harvard University Press, 2011), 20–21. While all the thinkers mentioned here have differing (if overlapping) conceptions of time in the current conjuncture, I should note that Nixon reverts to more conventional stories of time and climate borrowing from "the future" and "the past." Furthermore, in trying to grapple with his central problem — the many challenges in representing what he terms environmental "slow violence" — Nixon is far too dismissive of the fundamental difference of spectacular and political violence

(descriptively and as part of potential positive programs) and particularly misunderstands Frantz Fanon and how well his political thought can be adapted towards climate politics today.

16 Hall, *Selected Political Writings* (Durham: Duke University Press, 2017), 283.

17 I must thank my colleague, Rebecca Ariel Porte, for translating some of my thinking into these two "poles," which I discuss more thoroughly later, and for pushing me to write down some of my temporal ideas and lessons.

18 Wark, "Communicative Capitalism," *Public Seminar*, March 23, 2015, http://publicseminar.org/2015/03/communicative-capitalism/#.VRCKzPnF-So. Such an embrace is also suggested by Kathryn Yusoff: "This geologic prehistory has everything to do with the Anthropocene as a condition of the present; it is the material history that constitutes the present in all its geotraumas and thus should be embraced, reworked, and reconstituted in terms of agency *for* the present, *for the end of this world* and the possibility of others, because the world is already turning to face the storm, writing its weather for the geology next time." See Yusoff, 101; italics in original.

19 Alberto Toscano argues, "The debate around the Anthropocene event, its date of inception, with which I began, ironically marks this short-circuit between supposedly being able to think a geological time scale and being entirely rudderless when it comes to cognizing historical difference in the present." See Toscano, "The World Is Already without Us," *Social Text* 34, no. 2 (2016), 118. The Long Now helps us to think geologic time with historical difference as part of what I have called elsewhere "left-wing climate realism" or "the politics of exhaustion," wherein exhaustion is an affective matrix in which political solidarity is possible *through* (not above, beyond, or against) difference at the present socioecological conjuncture.

20 This is merely an approximate schematic from Marx's class theory, derived largely from texts like "The Communist Manifesto" and "The Eighteenth Brumaire of Louis Napoleon." Restricted to just texts like *The German Ideology* and the *Manifesto*, one could also graph the development and

transformation of modes of production. However, this would require ignoring Marx's revisions (c.f. his response to Vera Zasulich in the preface to the Russian edition of the *Manifesto*) which continued till his death. Marx's theories of politics remained underdeveloped, especially in comparison with his magisterial work in demonstrating the logic and development of Capital itself. All of this, though, is in part Benjamin's point. Marx's method not only needs but requires reexamination. And Benjamin's critique is a further refinement, however radical and unorthodox, of Historical Materialism.

21  Nor does this disjuncture map easily onto an "early" vs. "late" Marx. In this light, it is the early, philosophical writings that appear all the more mechanistic and the later scientific writings all the more open-ended *pace* "structuralist" and "humanist" interpretations alike.

22  Kautsky himself embraced this terminology in the first edition of *The Road to Power*. And not without precedent. Engels's initial work on a proto-Manifesto also utilized the form of a catechism, which Marx wisely abandoned for the now-famous structure of *The Communist Manifesto*. This was not simply economism. Every step of capitalist development, such as imperialism or even the War itself — which most of the socialist parties in Europe joined, with the notable exceptions of the Bolsheviks and the Italians, precipitating the collapse of the International — was just another advance in the ripening evolutionary progress of socialist victory. Benjamin excoriated this naïve faith in progress: politically, economically, and technologically. Kautsky, writing just about a decade after the historical carnage of World War I, and just about another decade before World War II, was still arguing that "the fabulous expansion of the capitalist mode of production in the past hundred years leads us to expect that its transformation into socialist forms will also occur extremely quickly everywhere where the necessary economic and psychic conditions have been created by the successful employment of modern technology" (https://www.marxists.org/archive/kautsky/1929/12/naturesoc.htm). This was December, 1929; not only were the scars of

the war still fresh, but the Great Depression — which struck Germany harder than almost anywhere else — had already begun. "Orthodox" Marxism had turned Marx on his head and, against all evidence, recreated the temporal ideal of the Hegelian stepladder as an article of blind faith. In earlier work, Benjamin had differentiated Kautsky from this tendency in the SPD, but that was in a different historical light and during a time when there was greater overlap between the two thinkers. Kautsky later adopted precisely the attitudes that Benjamin critiques in the "Theses." See Adam Przeworski, *Capitalism and Social Democracy* (Cambridge, UK: Cambridge University Press, 1985), 48; and Jukka Gronow, "Karl Kautsky (1854–1938)," in *Routledge Handbook of Marxism and Postmarxism*, ed. Alex Callinicos et al. (New York: Routledge, 2020), 162.

23  Walter Benjamin, *One-Way Street*, translated by Edmund Jephcott (Cambridge, MA: Harvard/Belknap, 2016), 95. While Bernstein openly called for socialist colonialism, Kautsky treated the war just as he treated the rise of fascism: they were interruptions. Imperialism was demoted from a Hilferding-derived monopoly/finance theory to a kind of bourgeois policy choice. (In contrast to Lenin, but even more so to the most thorough treatment of the subject at the time, Rosa Luxemburg's *The Accumulation of Capital*.)

24  Benjamin, *One-Way Street*, 66.

25  Ibid., 66.

26  It is too far afield from this essay for an in-depth review, but it is worth noting that, understood as such, these formations and ideas represent the clear majority not only of Marxist theory but of socialist, communist, and Marx-influenced political movements the world several times over, and on a mountain of clear political economic, ecological, and experiential evidence. As Fanon presciently suggested at the end of *The Wretched of the Earth*, "Let us leave this Europe which never stops talking of man yet massacres him at every one of its street corners, at every corner of the world." See Fanon, translated by Richard Philcox (New York: Grove Press, 2005), 235.

27 Gershom Scholem, perhaps ironically, introduced Benjamin to
Luxemburg's work in 1915 (see Esther Leslie, *Walter Benjamin:
Overpowering Conformism* (London: Pluto Press, 2000),
32). Benjamin was likely introduced to Lenin beyond casual
knowledge from current events through early conversations
with Ernst Bloch dating back to the late war period, Benjamin's
brother Georg — a member of the KPD — gifting Benjamin
a copy of the first German collection of Lenin's writings, and
more famously through Benjamin's reading of Lukács's *History
and Class Consciousness*, and his tutelage by the Soviet theater
director and popular educator Asja Lacis. References to Lenin
and Bolshevism first appear in the early 1920s in Benjamin's
writing and proliferate over time. The Bolshevik slogan "*Kein
Ruhm dem Sieger, kein Mitleid den Besiegten*" ["No Fame for the
Victor, No Pity for the Vanquished"] was originally the epigraph
for "Thesis XII," which contains Benjamin's interpretation of
it in light of his *Historical Materialism*: "Marx presents it as
the last enslaved class — the avenger that completes the task
of liberation in the name of generations of the downtrodden."
In a line deeply resonant with politics and time in current
socioecological terms, Benjamin continues that radical politics
(including its necessary "hate") is "nourished by the image
of enslaved ancestors rather than by the ideal of liberated
grandchildren." Out of fear of postal interception or censorship,
Benjamin cut the Bolshevik slogan. See Leslie, 200; Benjamin,
*SW4*, 393.

28 Susan Buck-Morss, *The Origin of Negative Dialectics: Theodor W.
Adorno, Walter Benjamin, and the Frankfurt Institute* (New York:
The Free Press, 1977), 62. Parts of Benjamin's interventions in
terms of time are concerned with how to conceive such a history
of Capital, as is the case with his incomplete *Arcades Project*
from which much of his "Theses on the Philosophy of History"
is derived. See Benjamin, *The Arcades Project*, translated by
Howard Eiland and Kevin McLaughlin (Cambridge, MA:
Belknap Press, 2002). To be fair, Marx understood that he was
laying out the methodology of an open science and was not an

undialectical, evolutionary socialist or technological determinist, as he is too often cast.

29 Benjamin, *SW4*, 392; italics in original. The colonial echo in Césaire just a few years later: "I hear the storm. They talk to me about progress, about 'achievements,' diseases cured, improved standards of living [...] [but] Europe is responsible before the human community for the highest heap of corpses in history." Césaire, *Discourse on Colonialism*, as quoted by Branwen Gruffydd Jones in "Time, History, Politics: Anticolonial Constellations," *Interventions* 21, no. 5 (2019), 605. And the ecological echo today: "Solar radiation management would in no way eliminate nature, only raise the stakes in a society that seeks to overmaster it. And onwards the history of capital goes, from one combination to the next, the perils mounting along the curve and, with Benjamin, the debris growing 'towards the sky.' What we call progress is *this* storm'" (italics in original). Malm, *The Progress of this Storm: Nature and Society in a Warming World*, (New York: Verso, 2018), Chapter 5, iBooks.

30 Benjamin, *Arcades*, 473.

31 Benjamin, *SW4*, 396. Despite frequent citations of his work in radical climate literatures, the Anglophone reception of Benjamin has paid relatively little attention to his engagement with the natural sciences of his day; see Christiane von Buelow, "Troping toward Truth: Recontextualizing the Metaphors of Science and History in Benjamin's Kafka Fragment," *New German Critique*, no. 48 (1989), 109–133, and parts of my own "Religions of Doubt" (PhD diss., Columbia University, 2013) for exceptions, even though it is a significant constant in Benjamin's later works, including the Kafka essays, *The Arcades Project*, and the "Theses." In the latter example, no one seems to have tried to discover who the unnamed "biologist" is in "Thesis 18" — for example, the whole section is simply omitted from Michael Löwy's often illuminating *Fire Alarm: Reading Walter Benjamin's 'On the Concept of History'*, translated by Chris Turner (New York: Verso, 2016). The quote that I cite is from Jean Rostand's *Heredity and Racism* (1938) which Benjamin

wrote an unpublished review of in 1939. Benjamin's review can be found in Volume III of his *Gesammelte Schriften* (Frankfurt: Suhrkamp Verlag, 2019). Rostand was an enthusiastic eugenicist and a soft fascist sympathizer who endorsed the 1933 Nazi edict "Prevention of Offspring with Hereditary Diseases," which mandated the forced sterilization of people with mental or physical disabilities. *Heredity and Racism* marked Rostand's rather late break with Nazi sympathies. What clearly interested Benjamin in his work was this turnaround and how Rostand saw it as scientifically necessary to refute spurious theories of biological progress in order to do so. "Progress" was ideological "second nature" smuggled into pseudo-scientific accounts of "first nature" or, simply, nature. Although he explicitly rejected race theory, Rostand remained in his rather successful and celebrated post-War life and fame (long after Benjamin's death) an enthusiastic eugenicist. His Nazi flirtations were completely buried in both Europe and the US, as a 1971 *New York Times* profile reflects in its title, "A Gentle, Rumpled French Biologist Says There's More to Life Than Pure Chance." See John L. Hess, *New York Times*, May 30, 1971, https://www.nytimes.com/1971/05/30/archives/a-gentle-rumpled-french-biologist-says-theres-more-to-life-than.html.

32 Intergovernmental Panel on Climate Change, *Global Warming of 1.5 °C. An IPCC Special Report on the impacts of global warming of 1.5 °C above pre-industrial levels and related global greenhouse gas emission pathways, in the context of strengthening the global response to the threat of climate change, sustainable development, and efforts to eradicate poverty*, ed. V. Masson-Delmotte et al. (2018), in print, 77.

33 Edelman, *No Future: Queer Theory and the Death Drive* (Durham: Duke University Press, 2004), 2–5.

34 Nancy Fraser and Rahel Jaeggi, *Capitalism: A Conversation in Critical Theory* (Cambridge, UK: Polity, 2018).

35 Helen Hester, *Xenofeminism* (Cambridge, UK: Polity Press, 2018), 51–2.

36 Gilmore, *Golden Gulag: Prisons, Surplus, Crisis, and Opposition*

*in Globalizing California* (Oakland: University of California Press, 2007), 74; see also David Neilson and Thomas Stubbs, "Relative surplus population and uneven development in the neoliberal era: Theory and empirical application," *Capital & Class* 35, no. 3 (2001), 435–453.

37 Smith et al. Emphasis added.
38 As of finalizing this book, only the Working Group I report on physical sciences, and the Working Group II report on impacts, adaptation, and vulnerability have been published.
39 IPCC AR6 WG II, 35.
40 Muñoz, *Cruising Utopia: The Then and There of Queer Futurity* (New York: New York University Press, 2009), 3.
41 Contrary to much speculation otherwise, Benjamin was not particularly learned as a Jew, nor was he committed to mystical, kabbalistic interpretations of Judaism personally or in his work. Benjamin did advance a Judaic political theology but in a thoroughly secularized form — as concepts for materialist thought — drawing on relatively common, well-known, and normative Jewish ideas such as the Messiah. Although he was close with Gershom Scholem, he seems to have picked up relatively little from Scholem's scholarship. Benjamin does not write of kabbalistic themes, and indeed critiques all forms of neo-platonism in *The Origin of German Tragic Drama* (New York: Verso, 2009). His uses, for example, of *halachah* and *aggadah* in his essays on Kafka and Krauss drew not on deep scholarly or yeshiva education but on then contemporary essays by Hayyim Bialik that Scholem sent along at Benjamin's request; for more, see Chaudhary, "Religions of Doubt." In this case, Benjamin is advancing one of two prevalent ideas about "the world to come" that have circulated from Second Temple through Talmudic and contemporary times. One view, closer to that of Christianity, holds that the world to come is a spiritual recompence largely unrelated to the Messianic age. The other, which Benjamin advances and which is far more in accord with his Marxist materialism, identifies the world to come with the Messianic age as a physical transformation or

adjustment of "this world." See *olam ha-zeh v'olam ha-ba: This World and the World to Come in Jewish Belief and Practice*, ed. Leonard Greenspoon (West Lafayette, IN: Purdue University Press, 2017); Jacob Neusner, "Death and Afterlife in the Later Rabbinic Sources: The Two Talmuds and Associated Midrash-Compilations," in *Handbook of Oriental Studies: Section 1, The Near and Middle East*, ed. Maribel Fierro et al., vol. 49 (Leiden: Brill, 2022); or even "The World to Come," from Chabad's "My Jewish Learning," accessed December 21, 2021, https://www.myjewishlearning.com/article/the-world-to-come/. For more on Benjamin's use of these concepts, see Chaudhary, "Religions of Doubt."

42 What developed before is natural history. As Deborah Cook notes — bringing together Benjamin, Adorno, and the unlikely support of John Bellamy Foster — "nature and history co-evolve owing to their metabolic interaction." See Cook, *Adorno on Nature* (New York: Routledge, 2011), 77. Benjamin has quite literally a geologic view of time:

As rocks of the Miocene or Eocene in places bear the imprint of monstrous creatures from those ages, so today arcades dot the metropolitan landscape like caves containing the fossil remains of a vanished monster: the consumer of the pre-imperial era of capitalism, the last dinosaur of Europe. On the walls of these caverns their immemorial flora, the commodity, luxuriates and enters, like cancerous tissue, into the most irregular combinations. A world of secret affinities opens up within: palm tree and feather duster, hairdryer and Venus de Milo, prostheses and letter-writing manuals. The odalisque lies in wait next to the inkwell, and priestesses raise high the vessels into which we drop cigarette butts as incense offerings. These items on display are a rebus: how one ought to read here the birdseed in the fixative-pan, the flower seeds beside tile binoculars, the broken screw atop the musical score, and the revolver above the goldfish bowl — is right on the tip of one's tongue. After all, nothing of the lot appears to be new. The goldfish come perhaps from a pond that dried up long ago, the revolver was a corpus delicti, and these scores could hardly have preserved their previous owner from starvation when her last pupils stayed away. (*Arcades*, 540)

Benjamin's quick note takes the form of metaphor but, both in the "Theses" and in the *Arcades*, Benjamin is explicit in their actual dialectical relation and continuity: "No historical category without its natural substance, no natural category without its historical filtration" (*Arcades*, 864). This is wholly consonant with Marx's argument in the preface to the first edition of *Capital* in "which the development of the economic formation of society is viewed as a process of natural history" (92).

43  Owen Hatherley, *Militant Modernism* (Winchester: Zer0 Books, 2008), 6.

44  Ibid., 12.

45  Scott, *Seeing Like a State: How Certain Schemes to Improve the Human Condition Have Failed* (New Haven: Yale University Press, 1999).

46  "An Ecomodernist Manifesto," April 2015, http://www. ecomodernism.org/. Ironically, I discovered while further researching this chapter that Stuart Brand — of Whole Earth Catalog fame and one of the signatures to the *Ecomodernist Manifesto*, whose slogan "we are as gods, and we have to get good at it," is, as political ecologist Anne Fremaux notes, "probably the acme of this Promethean vision" — has also coined a use for the phrase "The Long Now." See *Rethinking the Environment for the Anthropocene*, ed. Manuel Arias-Maldonado and Zev Trachtenberg (Abingdon, UK: Routledge, 2018), 99. However, unsurprisingly, the phrase means almost exactly the opposite of my usage. Brand's "Long Now Foundation" describes its understanding of The Long Now as follows: "Our work encourages imagination at the timescale of civilization — the next and last 10,000 years — a timespan we call *the long now*." (https://longnow.org/). True to Promethean form (see endnote 62), Brand's version of the long now emphasizes

normative understandings of time with a thin veneer of science
fiction painted over the top. The kinds of inquiries into the
political-time of the current socioecological conjunction and
the intensity of this particular moment are wholly absent.
As Fremaux continues: "It seems obvious that those eminent
eco-constructivists, some of them being members of the
Breakthrough Institute, make a confusion between 'human-
induced planetary change' and 'human planetary control,'
thinking that the extent to which we alter the planet gives us
control of it."

47 Armin Grunwald, "Diverging pathways to overcoming the
environmental crisis: A critique of eco-modernism from
a technology assessment perspective," *Journal of Cleaner
Production* 197 (2018), 1854–62. So-called "ecomodernism"
has a far larger footprint within policy and public discourse
than even in technology-focused natural scientific research.
Grunwald's paper is one of only a handful to try to engage
ecomodernism seriously simply because its premises are
so distant, and preposterous, in relation to almost anything
approaching natural scientific consensus: "It is not clear to what
extent the ecomodernist premise that technological progress
promises to make central contributions to a more ecologically
friendly world can be supported by sound arguments or
whether these are subjective convictions and ideologies." He
concludes "The ecomodernist approach grounds on premises
which are not based on knowledge or experience, but rather
on mere belief in the technological advance." (1861). As
Fremaux writes, "Promethean hyper-modernists seem to have
lost sight of the fact that Anthropocene scientific perceptions
are not only concerned with dreams of mastery but also with
nonlinearity, the existence of critical thresholds, bifurcations and
stochasticity, that is with doubt, uncertainty, irreversibility, and
unpredictability" (Fremaux, 26.)

48 One of the reasons why queer theory is such a fertile ground
for continuing to rethink political-time in socioecological
terms is best articulated in Berlant's *Cruel Optimism*. Although

they absolutely pillory the Cult of the Child *avant la lettre* in works like *The Queen of America Goes to Washington City* (Durham: Duke University Press, 1997), in *Cruel Optimism* Berlant alights on the argument that "what it means to take the measure of the impasse of the present: to see what is halting, stuttering, and aching about being in the middle of detaching from a waning fantasy of the good life." (263). What Berlant describes here is the letting go of normative fantasies and of preconceived trajectories, which they derive from queer theory and which is paramount in conceiving of a flourishing global human ecological niche whose parameters are so radically unexpected. In addition to presenting pure mystification as dogged realism, and dogmatic faith as scientific curiosity, this Promethean techno-mysticism, even in its nominally "left" forms, is profoundly *conservative*. While "ecomodernism" in this mold certainly shares affinities with accelerationism and other radical Promethean projects of the left, it is quite unique in this thorough conservativism. Many ecologically oriented thinkers may have serious reservations about Shulamith Firestone or Srnicek and Williams, but such thinkers certainly do not intend to replicate and preserve the corrosive world of bourgeois capitalist normality.

49 Probably the most well-known academic school arrayed against so-called "ecomodernism" is the broad umbrella of "degrowth." While degrowth positions usually demonstrate a lucid engagement with actual ecology, it is (a) difficult to call it a coherent ideology or movement; (b) insofar as it can be so characterized, subject to serious limitations particularly around a seemingly absolute allergy to modernism in any form in many sources; (c) sometimes impoverished in its understandings of politics and political economy; and (d) *shares*, to a degree, a temporal framework (just in reverse) with the Promethean techno-mystics. As mentioned before, many degrowth theories still do not account beyond a gesture for how development *will* occur in underdeveloped geographies (even by new standards) as well as how overdeveloped but highly unequal geographies

will reconfigure beyond simply measures aimed at reducing growth. As many sympathetic natural scientists and social thinkers have noted, such measures on their own are inefficient in addressing climate mitigation and adaptation, even while they are far more realistic about ecology than "ecomodernists." Still, contemporary work in ecological economics, much of it produced from what can largely now be clustered as degrowth perspectives, is invaluable for climate analyses, and "degrowth," despite its unfortunate nomenclature and shortcomings, has made vital interventions, corrections, and continues to develop sharper analyses. See the work of Jason Hickel, Julia Steinberger, among others, for good examples of such contributions. Many degrowth scholars themselves are critical of degrowth's singular and focus on growth (a symptom and not always easy to define) as opposed to a more comprehensive underlying causal set of social relations, i.e., capitalism-as-we-know-it. Such scholars have helped bring far more radical perspectives — from eco-Marxism to eco-feminism — to mainstream natural scientific consideration. I return to these questions via my examination of the debate between the economists Branko Milanović and Kate Raworth later in this chapter. Part of my turn here to reclaiming a real ecomodernism is as a comradely contribution to this developing school as well as overall a vital part of a climate politics.

50  Binod Khadka, "Rammed earth, as a sustainable and structurally safe green building: a housing solution in the era of global warming and climate change," *Asian Journal of Civil Engineering* 21, no. 1 (2020), 119–36; B.V. Venkatarama Reddy, "Sustainable materials for low carbon buildings," *International Journal of Low-Carbon Technologies* 4, no. 3 (2009), 175–81; R. Sivarethinamohan and S. Sujatha, "Broad-Spectrum of Sustainable Living Management Using Green Building Materials," in *Recent Advancements in Geotechnical Engineering*, ed. B. Soundara et al. (Millersville, PA: Materials Research Forum LLC, 2021), 1–8; Ousmane Zoungrana, et al., "The Paradox around the Social Representations of Compressed

Earth Block Building Material in Burkina Faso: The Material for the Poor or the Luxury Material?" *Open Journal of Social Sciences* 9, no. 1 (2021), 50–65; Mohamed A.B. Omer and Takafumi Noguchi, "A conceptual framework for understanding the contribution of building materials in the achievement of Sustainable Development Goals (SDGs)," *Sustainable Cities and Society* 52 (2020), https://doi.org/10.1016/j.scs.2019.101869;Vaishali Sharma, "Building Sustainable and Livable Asian Cities: Learnings from Compressed Stabilized Earth Blocks (CSEB) Constructions in India," https://www.researchgate.net/publication/350459494_Building_Sustainable_and_Livable_Asian_Cities_Learnings_from. This is not the place for a comprehensive review, but is it worth noting that such materials and techniques have both long vernacular and modern use, contributing to contemporary, new research in engineering and architecture. This is just not the kind of scientific, technological, and design research that so-called "ecomodernists" are looking for. At the same time, many of them are *quite* modern in a way that is rarely embraced in degrowth literatures.

51  Fisher, *k-punk*, 566.
52  Moten, *In the Break: The Aesthetics of the Black Radical Tradition* (Minneapolis: University of Minnesota Press, 2003), 25. In the sense that I am arguing here, Moten's discussion of Delaney, Artaud, and Strayhorn marks jazz as the quintessentially Adornan art form, against Adorno's own ill-informed and short-sighted objections.
53  Jalal Al-e Ahmad, *Occidentosis: A Plague from the West*, translated by R. Campbell (Berkeley: Mizan Press, 1984), 34.
54  Getachew, *Worldmaking After Empire* (Princeton, NJ: Princeton University Press, 2019), 13.
55  Quinn Slobodian, *Globalists: The End of Empire and the Birth of Neoliberalism* (Cambridge, MA: Harvard University Press, 2020).
56  See Irving Wohlfrath cited in Richard Leppert, "Introduction," in Theodor Adorno, *Essays on Music* (Berkeley: University of California Press, 2002), 70.

57 Tim Barker, "Other People's Blood," *n+1* 34, (2019).

58 Fisher, *k-punk*, 763.

59 Roko Rumora, "A Utopia of Yugoslav Architecture at MoMA," *Hyperallergic*, September 3, 2018, https://hyperallergic. com/458084/a-utopia-of-yugoslav-architecture-at-moma/

60 Martin Stierli and Vladimir Kulić, *Toward a Concrete Utopia: Architecture in Yugoslavia, 1948–1980* (New York: Museum of Modern Art, 2018), 157.

61 Ibid., 165.

62 The concept of the "Good Anthropocene" is a mainstay of existing techno-mystical "ecomodernism," particularly as popularized by the industry-friendly thinktank The Breakthrough Institute, originating most likely with one of its only actual natural scientists, Erle Ellis. However, the concept has proved ambiguous if still often dominated by its original techno-optimistic, status quo-preserving premises. Many authors have found utility in the term precisely rejecting some of those premises. See, for example, the marine ecologist Carolyn Lundquist et al., "Visions for nature and nature's contributions to people for the 21st century," which explicitly rejects in its energy mix "nuclear as an unnecessary high-risk option" and sets out to question economic growth orthodoxies; or Timon McPhearson et al., "Radical changes are needed for transformations to a good Anthropocene," which repositions the "Good Anthropocene" as a kind of maximally sustainable, *degrowth* position involving numerous Promethean bête noires like serious recognition of biophysical limits, planetary boundaries, and skepticism towards many techno-fixes.

63 Max Ajl, "Auto-centered development and indigenous technics: Slaheddine el-Amami and Tunisian delinking," *The Journal of Peasant Studies* 46, no. 6 (2018), 1240–63. See Chaudhary, "Sustaining What?…" for more on the concept of a sustainable global human ecological niche.

64 Despite the prevalence through political influence of green growth and technological narratives, the IPCC AR6 Working Group II is quite explicit in rejecting several prominent

"ecomodernist" positions — from its long discussions of the
destructive nature of industrial petro-farming vs. agroecological,
agroforestry, and hybrid indigenous and contemporary
methodologies (see Chapter 5) to naming air-conditioning as
a form of "maladaptation" (see Chapters 6 and 7) in which an
adaptation measure actually negatively impacts overall adaptive
capacity (in this case through overall increasing temperatures
and energy consumption in a negative feedback loop). Gesturing
toward the soon-to-be-released Working Group III report
on mitigation, the report also notes how technologically and
energy-intensive methods can also disrupt mitigation efforts. In
contrast, the report highlights precisely passive cooling systems
(see Chapter 6), as I've been discussing here. See also Parisa
Izadpanahi et al., "Lessons from Sustainable and Vernacular
Passive Cooling Strategies Used in Traditional Iranian Houses"
or Rabani et al., "Numerical simulation of an innovated
building cooling system with combination of solar chimney
and water spraying system," among many others. In addition
to its ecological sustainability, as measured by Rabani et al.,
such methods — which, as I discuss here, can also involve more
contemporary design elements — provide cooling comparable
or *greater* than the common "high tech," capitalist developed
solutions, at between 9–14 °C.

65   In contrast, as Alfred Schmidt argued in the 1970 English
preface to his 1962 *The Concept of Nature in Marx*, claims,
including some of his own previous work, that "Marx was
solely concerned to secure quantitative increase in the existing
forms of mastery over nature," were fundamentally flawed.
"On the contrary," Schmidt noted, "Marx wanted to achieve
something qualitatively new: mastery by the whole of society
of society's mastery over nature" (11). Here, Schmidt openly
restates Benjamin's argument from *One-Way Street*, that an
emancipated technology "is the mastery of not nature but of the
*relation* between man and nature" (*OWS*, 95; emphasis added).
In both cases, the "realm of freedom" is correctly identified not
as quantitative cornucopian and standard or endless economic

growth (accumulation), but rather with the achievement of the possibility, withheld by Capital, that society might reconcile with the very nature of which it is but a unique extension and control its own technological and other powers. In contemporary ecological terms, that is coming to terms with what is often called a "safe operating space" for humanity and flourishing in ways that are far removed from capitalist ideals.

66   In the previous discussions of building materials and passive cooling systems, many of the cited studies and the IPCC chapters note the complementarity of quite contemporary design and planning ideas designed around low material and energy cost and which provide comfort and even ecological restoration.

67   Arman Hashemi et al., "Environmental Impacts and Embodied Energy of Construction Methods and Materials in Low-Income Tropical Housing," *Sustainability* 7, no. 6 (2015), 7,866–7,883.

68   Peter Rosset and Miguel Altieri, *Agroecology: Science and Politics* (Bourton-on-Dunsmore: Practical Action Publishing Ltd, 2017), 79–81.

69   Agnes Lee, "The American Dream Is Alive and Well," *New York Times*, May 18, 2020, https://www.nytimes.com/2020/05/18/opinion/inequality-american-dream.html. It is worth looking at the whole report, "AEI Survey on Community and Society: Social Capital, Civic Health, and Quality of Life in the United States," which can be found on the AEI website. The authors loudly trumpet that Americans still believe in "the American Dream," but try to massage the fact that "the American dream" has moved significantly away from conservative (or liberal) principles.

70   GSS 2020, https://gss.norc.org/get-documentation.

71   Gilles Deleuze and Félix Guattari, *Anti-Oedipus: Capitalism and Schizophrenia* (New York: Penguin Books, 1977), 239–40.

72   Ibid., 239–40; this heighten-the-contradictions argument, often attributed to Lenin, is actually here derived from Nietzsche.

73   See Chaudhary, "Sustaining What?…" and "Climate of Socialism."

74 Višnja Kukoč, "Towards a Low-Carbon Future? Construction of Dwellings and Its Immediate Infrastructure in City of Split" (paper presented at Places and Technologies Conference, Belgrade, Serbia, April 2016).

75 See the moral utopianism of "Climate X" in Geoff Mann and Joel Wainwright's *Climate Leviathan: A Political Theory of Our Planetary Future* (New York: Verso, 2017).

76 Fanon, *The Wretched of the Earth*, 2.

77 Al-e Ahmad, *Occidentosis*, 71.

# Note on Methodology

As this book is intended for a general audience, it does not have a formal, academic methodology section. This presents challenges for the specialist reader. At the broadest level, this work embraces a form of methodological dualism, with methodologies of the natural sciences on the one side and complementary critical social, historical materialist, and dialectical approaches on the other. This, though, is a gross oversimplification. Both inform each other even at extremes, while multiple disciplinary methods are embraced closer to the metabolic nexus between society and nature, or, more accurately, society and ecological niche. And at that point, still greater care is needed. I have written about this at great length in my academic chapter "Emancipation, Domination, and Critical Theory in the Anthropocene." From a more Marxological perspective, please consult Kohei Saito's *Marx in the Anthropocene*. Since a book like this must engage data and critique, including from fields — for example, psychology, sociology, economics, ecology — which themselves are host to many methodological disputes, some fuzziness is unavoidable. Here I would ask specialist readers to consult the endnotes for commentary where applicable and generally to think of these as illustrative within the larger presentation. Furthermore, as I discuss in the aforementioned academic chapter (and here, in some ways, within chapter 3), the search for a single unifying science is almost certainly wrongheaded, even if, as I argue, we should understand these areas of inquiry as different dimensions of what is in truth a unity, albeit differently mediated. At the micro-level there are additional problems that go beyond such overviews. For example, some of the discussions of profitability here refer to the term at the level of the firm, while many common debates utilize the term purely in a macroeconomic way: overall profitability. Sometimes the latter meaning is employed, often the former. This will probably be easier for the general reader than for those looking for specific controversies within existing, especially Marxian debates. Similar issues exist for a host of concepts in the social and natural sciences, as well as some specific to Marxian critique. Here I would ask specialist readers again to consult what endnotes and commentary I have provided and to read such sections with the understanding that this is not the place for fully explicating such questions. I hope to write more formally on such concepts and debates in appropriately specialized fora.

# Note on Translations and Images

Images in chapter 3 are from their respective architecture and design studios, cited in the endnotes for the text for each example. There are two exceptions. First, the image of Kampala is by Dennis Wegewijs/Shuttershock and is available for free use. The image of Stepanova's clothing designs is from the article cited in that section by David Ferrero Peláez. All figures in chapter 5 are based on my own original sketches as redesigned by Mark DeLucas for print here. The only exception here is figure 7 which is from Steffen et al., "Trajectories of the Earth System in the Anthropocene" (citation in the notes).

Unless otherwise noted in the text or here, translations are my own. The opening epigraph is from Bertolt Brecht's "Hollywood Elegies" as translated by Adam Kirsch for the Poetry Foundation (and can be found in full on the Poetry Foundation website, available for fair use.) The epigraph for chapter 2, from Forugh Farrokhzad's "Window," is my own translation from the original Persian but builds on previous translations by Hasan Javadi and Susan Sallee (for the collection, Forugh Farrokhzad, *Another Birth and Other Poems*) and Leila Farjami for the Forugh Farrohkzad open forum website. As similarly discussed in the methodology note, there is not space here for a full discussion of theory and methods for translation. Put simply though, here I have emphasized historical context for informing translation in the more obviously political parts, and closeness of style for others. The excerpt from Goethe's "The

Sorcerer's Apprentice" in chapter 3 is from Edwin Zeydel's 1955 translation. The Brecht translations discussed in the endnotes are my own based either on the original German or, in one case, on the German as provided in Raymond Geuss's *Reality and its Dreams*. All original script from the Hebrew Bible, the Babylonian Talmud, and the Passover (Pesach) Haggadah is from Sefaria.org under fair use. Translations are in all cases either theirs or my own. Transliterations in such cases are my own. Quranic verses in chapter 4 are adapted from Ahmed Ali's *Al-Quran: A Contemporary Translation*. Other unmarked translated epigraphs (such as from Marx in chapter 3) are from the respective works cited in the chapter.

# Index

# Index

# Index

Berlant, Lauren (*see also cruel optimism*) 104, 105, 132, 133, 136, 139, 158, 165, 175, 227, 230, 240, 294, 340, 373, 387, 388

Bernstein, Eduard 243, 380

Bezos, Jeff 6, 86–9, 101, 105, 130, 307, 334

Biden, Joseph 207, 286, 355

billionaires 21, 59, 334

biochemicals 48, 51, 60

biodiversity loss 1, 5, 52, 81, 93, 273, 304, 355

biology 94, 184

BlackRock 30

Blanchfield, Patrick 208

Bloch, Ernst 252, 381

Bolivia (*see also Movimiento al Socialismo (MAS)*) 177, 200–203, 209, 211, 212, 217, 370

Bolsonaro, Jair 34, 151

Bonanno, George 215

Bosworth, Kai 132, 135, 136, 201, 213, 231, 232, 362

Bouazizi, Mohamed 65

Bougainville Mining Workers' Union (BMWU) 123, 126, 127

boycotts 197, 218

BP 120, 331

Bracke, Sarah 165

Brazil (*see also Landless Workers Movement (MST)*) 40, 151, 178, 188, 210, 223, 224, 226, 277, 357

Breakthrough Institute 80, 87, 387, 391

Brennan, Teresa 175, 182

Brown, John 219, 220

Buck-Morss, Susan 150

Bulhan, Hussein 185, 360

burnout 63, 171, 173–5

## C

California 49–51, 65, 110, 182, 250

wildfires in 9–11, 21, 42, 267–8

Canetti, Elias 216, 373

capital (*see also accumulation: capital*) 13, 14, 31, 33, 39, 66, 71, 75, 77, 80, 86, 96, 113, 116, 118, 120, 121, 136, 137, 139, 146, 152, 154, 155, 158, 173, 189, 194, 195, 207, 208, 210, 215, 217, 226, 245, 250, 255, 261, 269, 281, 288, 326, 328, 357, 368, 382, 392

financialized 47

fixed 51

fossil 55, 121, 127, 129, 223, 224

global 56, 58, 63

human 49, 51, 291

natural 54

transnational 56, 125

capitalism 2, 6, 16, 21, 25, 26, 51, 52, 54, 55, 57, 61, 69–71, 75, 81, 92, 95, 107, 109, 111, 113, 125, 139, 158, 177, 183, 187, 234, 239, 240, 243, 244, 258, 267, 272, 292, 299, 300, 341, 385

# Index

# Index

# Index

# Index

# Index

# Index

# Index

# Index

# Index

# Acknowledgments

I often joke that this is my second book. Technically, my first is the dissertation stuffed all the way in the back of the deepest drawer at the bottom of my desk. But the joke is slightly more accurate, even if slightly less technically apt: really my first "book" is BISR. Even if I was, let's say in the spirit of about 50% of the studies in here, the lead author, that ongoing book is always a collaboration. And no matter how lonely writing a real book is and feels, it turns out it, too, is a collaboration. BISR is more 1970s Miles; sure, there's a leader, but the band is making most of the music. This is more 1940s Bird; a lot of hands comping to make the soloist shine. (I'm not as good as either, but the bands are). It would be impossible to name every single person I should, but I will try my best to name those I can, who have been the rhythm section for my first actual book.

Work here has been in many editorial hands over the years. Arrangers, orchestrators. My colleague Kali Handleman edited some of the earliest writing that would become this book as well some of the last that would become its final chapter, originally an essay in *Late Light*, BISR's journal; Mark Krotov, Nikil Saval, and Dayna Tortorici edited my very first piece of public writing on climate for *n+1*; Jonathon Sturgeon, Emily Carroll, and Zach Webb all worked on early versions of essays that became the first two chapters here for *The Baffler*; Emily was also — in an unbelievably clutch moment — able to assist on some of my final edits as well. Peter Hogness, the former editor of the Professional Staff Council union newspaper and one of my wife's friends and mentors, gave crucial commentary on some of the thorniest sections. My colleague Mark DeLucas turned some of my comically simple illustrations and graphs into elegant designs worthy of public consumption. And my colleague Lauren K. Wolfe contributed both commentary on early versions as well as some of the most important final edits on citations and endnotes. Without either, this would be a much-diminished project. Andrew Ascherl helped come up with index topics and subtopics — an art I was previously unfamiliar with, and Thomas Evans and Natasha Gilmore gifted me an *absurd* amount of design and architecture books which helped shape the writings here on the built-world. Lauren Cerand continues to counsel me on all things publicity. Tariq Goddard gave me the first full editorial read of this work with not only a whirlwind of perspicacious questions, responses, and observations but some incredibly sage advice, encouragement, and well-advised caution. And Josh Turner has been the painstaking editor for the final journey of this text, not only exhibiting deep insight and editorial finesse but an almost preternatural patience with the length of time this has taken to finalize.

This book has also had several shepherds, or impresarios if you will, over the years, pushing and pulling on the broad picture, till its one worthy of promotion. I will mention just three: Roisin Davis, one of the earliest people outside of BISR to believe in this project; Barnaby Raine, my dear friend and colleague who gave advice and criticism along that path; and finally, Chris Washington, a friend and BISR student, who helped make the final connections between the work and the good people at Repeater.

That list of good people includes, but is not limited to: Tariq and Josh, already named, Rebecca Wright, Carl Neville, Christiana Spens, and Sneha Alexander. It takes not only a certain dedication to radical thought and a clear creative vision but frankly a respectable level of chutzpah to actually roll the dice on a book like this. And see it through to the end. These are the kind of record producers you want. I have taken pretty unusual paths for much of my life, but rarely have I found one of those paths already carved out so well in advance. I've been reading Repeater books as long as they've been in print and, to me, you are all a blessing on the memory of Mark Fisher.

There are countless giants who have influenced me that you've read about already, but my own life's work has underscored the importance of your direct teachers. I've been lucky to have incredible teachers, from Susan Buck-Morss and Diane Rubenstein, who introduced a very young me to Critical Theory and political analysis, to Marc Nichanian, whose tiny and deeply rigorous seminars were a model of learning and scholarship, and my old doctoral advisor, Hamid Dabashi, who's work and support actually made getting through graduate school well over a decade ago now, possible.

I've also had the good fortune to have champions, promoters to keep my tortured metaphor running. McKenzie Wark was one of the first people to push me to publish this and one of the only academics who really supported BISR at our start when we were little more than a scrappy project handing out flyers to study Benjamin at the local bar. When it was clear that my scholarly life was going to be anything but traditional, Wayne Proudfoot stepped in to support me and support BISR as an institute when other professors scrambled away. Andreas Malm, who I only met through socially distanced online presentations (only one of which descended into a now well-documented shitshow), gave encouragement he may not even realize, when he casually and generously rerouted pointed questions about climate politics to me instead of him. Julia Steinberger was one of the first natural scientists who recommended my work which, again, gave me the hope that I was on the right path in writing critical thought that was scientifically legible and credible. I was also fortunate to present some of this work at an annual meeting of the Radical Critical Theory Circle in Nisyros, Greece. Members not only gave crucial critiques and encouraging engagement, but also invited me onto numerous projects where much of my recent, related, more purely academic writing has occurred.

I could easily name every BISR board member or supporter

but will highlight Elizabeth Castelli, Michelle Miller, Miriam Haier, Maryam Newman, Sara Clugage, Noah McCormack, and Ted Kennedy, among so many others. These are people who make our work possible and in turn make books like this possible.

Similarly, I could easily name every member of BISR's faculty. The big band. The very first version of what would become this book was workshopped at BISR under the guidance of our late, beloved colleague Jeffrey Escoffier who not only encouraged me to see how it was, could, and should be a book but who was a model in his life and at our Institute of what radical thought outside of, and often indeed in spite of the academy, could be. There are many of my colleagues named in these pages. And many more not. I'll list those who have most directly impacted this work and my thinking in related areas: among others Nara Roberta Silva, Michael Stevenson, Christine Smallwood, Alyssa Battistoni, Max Ajl, Tony Alessandrini, Sophie Lewis, Soraya Batmanghelichi, Maeve Adams, Patrick Blanchfield, Adriana Gariga-Lopez, Alirio Karina, Isi Litke, Joseph Earl Thomas, Joseph Osmundson, and Nafis Hasan have all, whether through direct advice, classroom comments, casual conversation, or long friendship helped bring this work to fruition.

Raphaele Chappe, an incomparable economic mind, was an early co-author of mine on other writing and helped cement in me an absolute conviction that Critical Theory and empirical analysis not only can but must be connected; Rebecca Ariel Porte has read drafts, lent whole concepts, and been a constant interlocuter on nearly every subject from passacaglia to particle physics for years. In addition to what I've already mentioned, Mark DeLucas and Lauren K. Wolfe particularly helped make the final year of this book possible by taking on even more of BISR's gargantuan administrative challenges.

Abby Kluchin, one of BISR's co-founders, and one of the most brilliant teachers I know, was the first person to introduce me to affect theory, without which this work is unthinkable, and did so in such a way that the concepts and connections to psychological and other modes of thought would be legible to anyone, anywhere, even a grumpy old materialist like me.

And, of course, Suzanne Schneider. Her work and mine have been in dialogue for years, she has been a constant source of learning and insight, and she has been a steady hand, weathering the storm as BISR's deputy director as we've grown. Suzy has read countless versions of parts of this book, not to mention outlines and lectures and all the related ephemera of publishing. Her dedication to intellectual rigor, whether the most detailed historical nuance, the most abstract and obscure concept, or the most engaged commentary is a testament to what the life of the mind should be.

If I could fill pages with every member of BISR's faculty spread across some five continents now, I would. It's more orchestra than band now. And somewhat differently, I don't think I want to besmirch the good names of the many friends who have given me feedback, humored my endless ramblings, or have been patient as I vanished to finish this work. If anyone wants to be mentioned, let me know; I'll add you to a website or, *Baruch Hashem*, a second printing. I will highlight in particular though Graziella Matty, who has been my friend and practically a sister since we first met decades ago, and Robin Varghese who has been a constant companion, towering intellectual influence, and ever faithful and generous friend and support, personally and professionally. I will always be grateful that both were willing to sign on the dotted line as the first BISR board members and have always been there for me and my work all these years. Similarly, my brother Arun Chaudhary and my sister Amanda Chaudhary first introduced me to everything from science fiction to technology, radical politics of many kinds, history, music, and more. And their own idiosyncratic work, from new media and music to professional propaganda and politics, has always been a guide to mine.

The last members of the band to mention are my students. Long before I wrote this material, I studied and taught it.

Teaching working adult students from nearly every walk of life has contributed so much to this project; student questions guide new directions in analysis; their concerns give you a different window on the world. This is true of my direct classes on climate but also on other topics, where my thinking — and many of my students — are haunted by social and ecological devastation. I've had the honor of teaching about climate in its socioecological dimensions in massive events, as we did a few years back with participants from Sunrise and DSA's Ecosocialist and Finance and Debt working groups, or tiny rooms of activist or artists both here and abroad. Without my students, this book would not exist. And it is seeing students contribute the most unique reflection, taking the smallest action, and starting or joining movements that gives me hope that we really can alter the course of our burning world. Editors, impresarios, producers, promoters, teachers, faculty, students, friends: they are the accents; the well-voiced harmony, the quiet contrapuntal line. Every missed note of the book is mine and mine alone. They are the steady groove.

Finally, Dania Rajendra, whose love, patience, care, insight, taste, and brass-tacks, real-world experience has not only made this work possible, she has made my life possible. In reading so many pieces of writing, in listening to so much kvetching, in inevitably absorbing the horrors I was learning about but also in imparting her well-earned wisdom as a labor organizer, a poet, an activist, and a leader; wisdom that runs throughout these pages. I must concede some of the good notes, some of the best, to her. I almost died of COVID-19 in the very early days of the pandemic. If the hospitals in our neighborhood hadn't already stopped intake, I would have been intubated. Her love and care literally kept me alive. I write a lot about hate in this book but very little about love. Well, here it is. My love for Dania reverberates across every measure, if you'll allow this metaphor a little while longer. She helped me see so many years ago that I didn't want the academic job on offer but was already doing the work I wanted to, now including this book. And I, in turn,

Acknowledgments

believe in hers and hope that my notes add to her songs. That's a not very funny punch line to the joke but it is a great cadence to finish the composition. Love is not, I think, a political emotion. It is *better* than that. Politics is the paltry, sad, necessary prelude to a better life, beautifully imperfect, most of us will never see. I count myself among the blessed to have caught a glimpse of that because of her.

בְּכָל־דּוֹר וָדוֹר חַיָּב אָדָם לִרְאוֹת אֶת־עַצְמוֹ כְּאִלּוּ הוּא יָצָא מִמִּצְרָיִם.

# Repeater Books

is dedicated to the creation of a new reality. The landscape of twenty-first-century arts and letters is faded and inert, riven by fashionable cynicism, egotistical self-reference and a nostalgia for the recent past. Repeater intends to add its voice to those movements that wish to enter history and assert control over its currents, gathering together scattered and isolated voices with those who have already called for an escape from Capitalist Realism. Our desire is to publish in every sphere and genre, combining vigorous dissent and a pragmatic willingness to succeed where messianic abstraction and quiescent co-option have stalled: abstention is not an option: we are alive and we don't agree.